Topics in

DIFFERENTIAL GEOMETRY AND CALCULUS OF VARIATIONS

AF792428

Topics in

DIFFERENTIAL GEOMETRY AND CALCULUS OF VARIATIONS

By

PARMANAND GUPTA

M.Sc. M.Phil(Maths)

M.A. (Economics)

Head, Deptt. of Mathematics

Indira Gandhi National College, Ladwa

KURUKSHETRA UNIVERSITY

Haryana

FIREWALL MEDIA

(An Imprint of Laxmi Publications Pvt.Ltd.)

An ISO 9001:2015 Company

BENGALURU • CHENNAI • GUWAHATI • HYDERABAD

KOCHI • KOLKATA • LUCKNOW • RANCHI

NEW DELHI

Topics in Differential Geometry and Calculus of Variations

Copyright © by Laxmi Publications (P) Ltd.

All rights reserved including those of translation into other languages. In accordance with the Copyright (Amendment) Act, 2012, no part of this publication may be reproduced, stored in a retrieval system, translated into any other language or transmitted in any form or by any means, electronic, mechanical, photocopying, recording or otherwise. Any such act or scanning, uploading, and or electronic sharing of any part of this book without the permission of the publisher constitutes unlawful piracy and theft of the copyright holder's intellectual property. If you would like to use material from the book (other than for review purposes), prior written permission must be obtained from the publishers.

Printed and bound in India

Typeset at : Goswami Associates, Delhi

First Edition : 2012, *Reprint :* 2014, 2018, 2023; *Edition* : 2026

FDI-3098-359- TOPICS IN DIFF GEO CAL-GUP

Price: ₹ 359.00

Limits of Liability/Disclaimer of Warranty: The publisher and the author make no representation or warranties with respect to the accuracy or completeness of the contents of this work and specifically disclaim all warranties. The advice, strategies, and activities contained herein may not be suitable for every situation. In performing activities adult supervision must be sought. Likewise, common sense and care are essential to the conduct of any and all activities, whether described in this book or otherwise. Neither the publisher nor the author shall be liable or assumes any responsibility for any injuries or damages arising from here. The fact that an organization or Website if referred to in this work as a citation and/or a potential source of further information does not mean that the author or the publisher endorses the information the organization or Website may provide or recommendations it may make. Further, readers must be aware that the Internet Websites listed in this work may have changed or disappeared between when this work was written and when it is read.

All trademarks, logos or any other mark such as USP, Amanda, Golden Bells, Firewall Media, A-Cube, Triunity, Mathstick, Engpot, Laxmi appearing in this work are trademarks and intellectual property owned by or licensed to Laxmi Publications, its subsidiaries or affiliates. Notwithstanding this disclaimer, all other names and marks mentioned in this work are the trade names, trademarks or service marks of their respective owners.

Branches

Bengaluru	080-26 75 69 30
Chennai	044-24 34 47 26
Guwahati	0361-254 36 69
Hyderabad	040-27 55 53 83
Kochi	0484-405 13 03
Kolkata	033-40 04 77 79
Lucknow	0522-430 36 13
Ranchi	0651-224 24 64

PUBLISHED IN INDIA BY

FIREWALL MEDIA
(An Imprint of Laxmi Publications Pvt.Ltd.)

An ISO 9001:2015 Company
113, GOLDEN HOUSE, GURUDWARA ROAD, DARYAGANJ,
NEW DELHI - 110002, INDIA

Corporate Office : B-33, Sector-5, Noida,
Gautambuddha Nagar (U.P.) - 201 301, INDIA
Telephone : +91-120-44 46 900
www.laxmipublications.com
info@laxmipublications.com

C—R/026/08
Printed at: Giriraj Offset Printer, Delhi.

Contents

Chapters *Pages*

SECTION A: DIFFERENTIAL GEOMETRY

SECTION B: CALCULUS OF VARIATIONS

Preface

This book on "Differential Geometry and Calculus of Variations" has been written according to new guidelines and syllabus issued by UGC for various universities of India and all over the world.

In writing this book, the author has placed great emphasis on the fundamental concepts and methods of proofs of articles. I have provided sufficient number of illustrative examples on each concept. To reinforce the concepts developed in the definitions and articles, several exercises have been given at proper places.

Salient features of the present book are:

- Well explained concepts.
- Sufficient number of solved examples.
- Well graded exercises.
- Hints for selected questions.

Suggestions for the improvement of the book from dear students and teachers colleagues are most welcome.

—AUTHOR

Noticeable Concepts

- **Use of Greek Letters**

α	alpha	ϕ	phi	κ	kappa	τ	tau
β	beta	ψ	psi	μ	mu	χ	chi
γ	gamma	ξ	xi	ν	nu	ω	omega
δ	delta	η	eta	π	pi	Δ	cap. delta
ε	epsilon	ζ	zeta	ρ	rho		
i	iota	λ	lambda	σ	sigma		
θ	theta	Γ	cap. gamma	Σ	cap. sigma		

- **Some Important Notations**

$\in$	belongs to	$\Leftrightarrow$	implies and implied by
$\notin$	does not belong to	iff	if and only if
$\Rightarrow$	implies	$\cup$	union
		$\cap$	intersection

- **Useful Data**

$\sqrt{2} = 1.4142$	$\frac{1}{\pi} = 0.3183$	1 rad. = 57° 17′ 45″	$\log_{10}e = 0.4343$
$\sqrt{3} = 1.732$	$e = 2.7183$	1° = 0.0174 rad.	$\log_e 2 = 0.6931$
$\pi = 3.1416$	$\frac{1}{e} = 0.3679$	$\log_e 10 = 2.3026$	$\log_e 3 = 1.0986$

- **Some Mathematical Symbols**

=	equals	~	is of the order of magnitude of
≈	equals approximately	∝	is proportional to
≠	is not equal to	lim	the limit of
≡	is identical to, is defined as	Σ	the sum of
>	is greater than	∫	the integral of
>>	is much greater than	Δx	the change or difference in x
≥	is greater than or equal to	$\lvert x \rvert$	the absolute value or magnitude of x
<	is less than	$\bar{x}$	the average value of x
<<	is much less than	df/dx	the derivative of f with respect to x
≤	is less than or equal to	$\partial f/\partial x$	the partial derivative of f with respect to x

SECTION A

DIFFERENTIAL GEOMETRY

1

Curves in Space

1. INTRODUCTION

Differential geometry is that branch of geometry which is treated using the methods of calculus. In particular, we investigate curves and surfaces in space in differential geometry. Differential geometry plays an important role in engineering designs, geodesy, geograph and space travel. Formulae regarding vector algebra and vector calculus are frequently used in the study of differential geometry.

2. BRANCHES OF DIFFERENTIAL GEOMETRY

There are two branches of differential geometry.

(*i*) **Local Differential Geometry.** In this branch of differential geometry, we study the properties of curves and surfaces in space which depend only upon points close to a particular point of the geometric figure under consideration.

(*ii*) **Global Differential Geometry.** In this branch of differential geometry, we study the properties of curves and surfaces in space which involve the entire geometric figure under consideration.

In the present course, we shall study some of the fundamentals of local differential geometry.

3. FUNCTIONS OF CLASS C^m

A scalar valued (or vector valued) function defined on an interval I belongs to class C^m on the interval I if the mth order derivative of the function exists and is continuous on I.

The class of continuous functions is denoted by C^0. The class of functions having derivatives of all orders is denoted by C^∞.

If a function belongs to the class C^m, then that function is called a **C^m function.**

We know that a vector function is continuous or has a derivative if and only if all components of the functions are continuous or have derivatives.

$\therefore$ A vector function $\mathbf{f}(t) = f_1(t)\mathbf{i} + f_2(t)\mathbf{j} + f_3(t)\mathbf{k}$ belongs to C^m on I if and only if its components $f_1(t)$, $f_2(t)$ and $f_3(t)$ belong to C^m on I.

Remark 1. We know that a differential function is always continuous.

$\therefore$ If a function belongs to C^m, then it belongs to C^k for all $k \le m$.

Remark 2. In printing work, the vector quantity $\mathbf{f}$ is depicted by using bold letter. In writing work, the vector $\mathbf{f}$ is written as $\vec{f}$ or $\underline{f}$.

Example 1. *Show that the vector function* $\mathbf{f}(t) = (\cos t)\mathbf{i} + t^3\,\mathbf{j} + t^{5/3}\,\mathbf{k}$, $-\infty < t < \infty$ *belongs to* C^1 *on* $-\infty < t < \infty$ *and not* C^2 *on* $-\infty < t < \infty$.

Sol. We have $\mathbf{f}(t) = (\cos t)\mathbf{i} + t^3\mathbf{j} + t^{5/3}\mathbf{k}$

$\therefore$ $$\dot{\mathbf{f}}(t) = (-\sin t)\mathbf{i} + 3t^2\mathbf{j} + \frac{5}{3}\,t^{2/3}\mathbf{k} \qquad \left(\dot{\mathbf{f}} = \frac{d\mathbf{f}}{dt}\right)$$

$-\sin t$, $3t^2$, $\frac{5}{3}t^{2/3}$ are continuous functions of t, where $-\infty < t < \infty$.

$\therefore$ $\dot{\mathbf{f}}(t)$ is continuous on $-\infty < t < \infty$.

$\therefore$ $\mathbf{f}(t)$ belongs to C^1 on $-\infty < t < \infty$.

Also, $$\ddot{\mathbf{f}}(t) = (-\cos t)\mathbf{i} + 6t\mathbf{j} + \frac{10}{9t^{1/3}}\mathbf{k}$$

The function $\frac{10}{9t^{1/3}}$ is not continuous at $t = 0$.

$\therefore$ The scalar function $t^{5/3}$ does not belong to C^2 on $-\infty < t < \infty$.

$\therefore$ $\mathbf{f}(t)$ does not belong to C^2 on $-\infty < t < \infty$.

Remark. $\mathbf{f}(t)$ belongs to C^m for all $m \geq 0$ on any interval not containing '0'.

4. CURVE IN SPACE

A **curve in space** is defined as the locus of a point whose position vector relative to a fixed origin may be expressed as a function of a single parameter.

Thus, a curve C in space may be represented by a vector function

$\mathbf{r}(t) = x(t)\mathbf{i} + y(t)\mathbf{j} + z(t)\mathbf{k}$, where t is a parameter. Here $\mathbf{r}(t)$ is the position vector of the point P on the curve C and $x(t), y(t), z(t)$ are the cartesian coordinates of the point P. To each value t' of t there correspond a unique point of the curve C with position vector $\mathbf{r}(t')$ and cartesian coordinates $(x(t'), y(t'), z(t'))$.

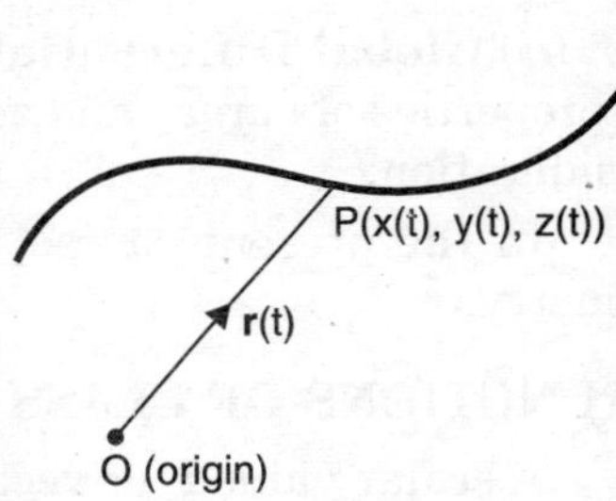

As t increases, the direction of travelling along the curve C is called the **positive sense** on the curve C. Also as t decreases, the direction of travelling along the curve C is called the **negative sense** on the curve C.

If a curve in space lies wholly in a plane, then it is called a **plane curve.**

If a curve in space does not lie wholly in a plane then it is called a **skew curve** or a **tortous curve** or a **twisted curve.**

Example 2. *Show that the curve in space* $\mathbf{r}(t) = a \cos t\,\mathbf{i} + b \sin t\,\mathbf{j} + 0\mathbf{k}$ *is a plane curve.*

Sol. We have $\mathbf{r}(t) = a \cos t\mathbf{i} + b \sin t\mathbf{j} + 0\mathbf{k}$.

$\therefore$ Let (x, y, z) be the coordinates of the point with position vector $\mathbf{r}(t)$.

$\therefore$ $$x = a \cos t, \quad y = b \sin t, \quad z = 0$$

$\therefore$ $$\frac{x^2}{a^2} + \frac{y^2}{b^2} = 1,\ z = 0 \qquad (\because\ \cos^2 t + \sin^2 t = 1)$$

This represents an ellipse in the xy-plane.

$\therefore$ The given curve is a plane curve.

5. REGULAR CURVE

A curve $\mathbf{r} = \mathbf{r}(t)$, $a \le t \le b$ is called a **regular curve** if

(*i*) $\dot{\mathbf{r}}(t)$ exists and is continuous on $a \le t \le b$ *i.e.,* $\mathbf{r}(t)$ is of class C^1 on $a \le t \le b$.

(*ii*) $\dot{\mathbf{r}}(t) \neq \mathbf{0}$ for all t in $a \le t \le b$.

For example, consider the curve

$$\mathbf{r} = \mathbf{r}(t) = 3t\mathbf{i} + t^4\mathbf{j} + 2\mathbf{k}, \; -\infty < t < \infty.$$

Here $\dot{\mathbf{r}}(t) = 3\mathbf{i} + 4t^3\mathbf{j} + 0\mathbf{k}$

$\dot{\mathbf{r}}(t)$ is continuous on $-\infty < t < \infty$ and also non-zero.

$\therefore$ The given curve is a regular curve.

Remark. If $\mathbf{r} = \mathbf{r}(t) = x(t)\mathbf{i} + y(t)\mathbf{j} + z(t)\mathbf{k}$ is a regular curve then, $\frac{dx}{dt}, \frac{dy}{dt}, \frac{dz}{dt}$ are never zero simultaneously.

6. SIMPLE CURVE

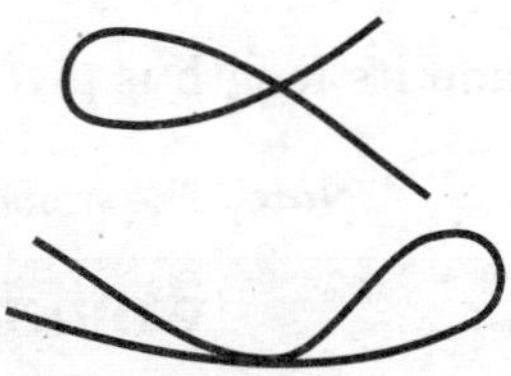

Curves with multiple points

A curve $\mathbf{r} = \mathbf{r}(t)$, $a \le t \le b$ is called a **simple curve** if

(*i*) $\mathbf{r} = \mathbf{r}(t)$, $a \le t \le b$ is a regular curve.

(*ii*) $t_1 \neq t_2 \Rightarrow \mathbf{r}(t_1) \neq \mathbf{r}(t_2)$ *i.e.,* the curve is without points at which the curve intersects or touches itself.

Remark. A point where a curve intersects or touches itself is called a **multiple point.**

EXERCISE 1.1

1. Show that the function $f(t) = t^2 + t^{5/2}$, belongs to:
   (*i*) C^2 on $(-\infty, \infty)$ (*ii*) C^3 on $(1, 4)$.
2. Show that the function $\mathbf{f}(t) = 3t^4\mathbf{i} + 6t^9\mathbf{j} + \mathbf{k}$ belongs to C^∞ on $(-\infty, \infty)$.
3. If the vector functions $\mathbf{f}$ and $\mathbf{g}$ belong to C^m on I, then show that the vector functions $\mathbf{f} + \mathbf{g}$, $\mathbf{f} \cdot \mathbf{g}$, $\mathbf{f} \times \mathbf{g}$ also belong to C^m on I.
4. If $\mathbf{a}$ and $\mathbf{b}$ are constant vectors, then show that the curve in space $\mathbf{r}(t) = \mathbf{a} + t\mathbf{b}$ is a plane curve.
5. Show that the curve in space $\mathbf{r}(t) = 4 \sin t\mathbf{i} + 0\mathbf{j} + 3 \cos t\mathbf{k}$, $-\infty < t < \infty$ is a plane curve.
6. Show that the curve in space $\mathbf{r}(t) = 2t^2\mathbf{i} + (1 + t^3)\mathbf{j} + 7t\mathbf{k}$, $-\infty < t < \infty$ is a regular curve.

---

7. ARC OF A CURVE

An **arc** of a curve is the portion of the curve between any two points of the curve. For simplicity, we shall say 'curve' for curves as well as for arcs.

8. LENGTH OF A CURVE

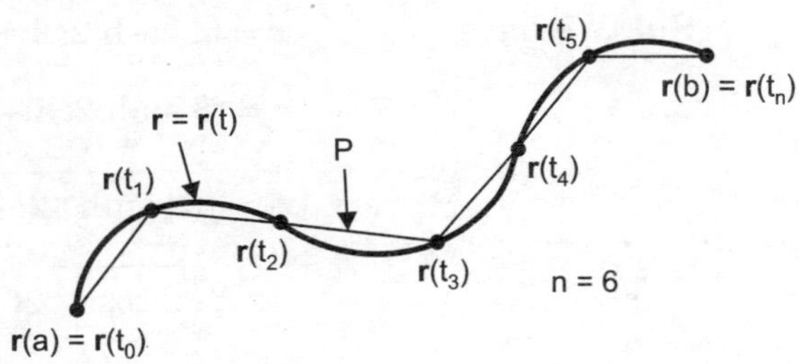

The length of a curve is defined in terms of the lengths of approximating polygonal arcs. Let $\mathbf{r} = \mathbf{r}(t)$, $a \le t \le b$ be the given curve.

Let $a = t_0 < t_1 < ..... < t_n = b$

be a subdivision of the interval $a \le t \le b$. This subdivision determines a sequence of points

$$\mathbf{r}_0 = \mathbf{r}(t_0),\ \mathbf{r}_1 = \mathbf{r}(t_1),\ \ldots\ldots,\ \mathbf{r}_n = \mathbf{r}(t_n).$$

These points are joined in sequence to form an approximating polygonal arc P as shown in the figure.

$$\therefore \quad \text{Length of} \quad P = \sum_{i=1}^{n} |\mathbf{r}_i - \mathbf{r}_{i-1}| = \sum_{i=1}^{n} |\mathbf{r}(t_i) - \mathbf{r}(t_{i-1})|$$

We make subdivisions of the interval arbitrarily small so that the greatest $|t_i - t_{i-1}|$ approaches 0 as $n \to \infty$.

If $\lim_{n\to\infty} \sum_{i=1}^{n} |\mathbf{r}(t_i) - \mathbf{r}(t_{i-1})|$ exists finitely, then the given curve is said to a **rectifiable curve** and the value of this limit is called the **length** of the given curve.

Theorem. If $r = r(t)$, $a \le t \le b$ be a regular curve then prove that this curve is rectifiable and its length is given by the integral $\int_a^b |\dot{r}(t)|\, dt$.

Note. The proof of this theorem is beyond the scope of this book.

WORKING RULES FOR FINDING LENGTH OF THE CURVE $\mathbf{r} = \mathbf{r}(t)$ BETWEEN $a \le t \le b$

Step I. *Find $\dot{\mathbf{r}}$ and then $|\dot{\mathbf{r}}|$.*

Step II. *Evaluate $\int_a^b |\dot{\mathbf{r}}|\, dt$. This gives the required length of the curve.*

Example 1. *Find the length of the helix $\mathbf{r} = (a \cos t)\mathbf{i} + (a \sin t)\mathbf{j} + bt\mathbf{k}$, $0 \le t \le 2\pi$.*

Sol. We have $\mathbf{r} = (a \cos t)\mathbf{i} + (a \sin t)\mathbf{j} + bt\mathbf{k}$, $0 \le t \le 2\pi$.

$$\therefore \quad \dot{\mathbf{r}} = (-a \sin t)\mathbf{i} + (a \cos t)\mathbf{j} + b\mathbf{k}$$

$$\therefore \quad |\dot{\mathbf{r}}| = \sqrt{a^2 \sin^2 t + a^2 \cos^2 t + b^2} = \sqrt{a^2 + b^2}$$

$$\therefore \quad \text{Length of the helix} = \int_0^{2\pi} |\dot{\mathbf{r}}|\, dt$$

$$= \int_0^{2\pi} \sqrt{a^2 + b^2}\, dt = \sqrt{a^2 + b^2}\, t \Big|_0^{2\pi} = 2\pi\sqrt{a^2 + b^2}.$$

Example 2. *Find the length of the curve $\mathbf{r} = (4 \cosh 2t)\mathbf{i} + (4 \sinh 2t)\mathbf{j} + 8t\mathbf{k}$, $0 \le t \le \pi$.*

Sol. We have $\mathbf{r} = (4 \cosh 2t)\mathbf{i} + (4 \sinh 2t)\mathbf{j} + 8t\mathbf{k}$, $0 \le t \le \pi$.

$$\therefore \quad \dot{\mathbf{r}} = (8 \sinh 2t)\mathbf{i} + (8 \cosh 2t)\mathbf{j} + 8\mathbf{k}$$

$$\therefore \quad |\dot{\mathbf{r}}| = \sqrt{64 \sinh^2 2t + 64 \cosh^2 2t + 64}$$

$$= 8\sqrt{2 \cosh^2 2t} = 8\sqrt{2}\, \cosh 2t$$

$\therefore$ Length of the curve $= \int_0^{\pi} |\dot{\mathbf{r}}|\, dt$

$$= \int_0^{\pi} 8\sqrt{2} \cosh 2t\, dt$$

$$= 4\sqrt{2} \sinh 2t \Big|_0^{\pi} = 4\sqrt{2} \sinh 2\pi.$$

Example 3. *Find the length of the semicubical parabola* $\mathbf{r} = t\mathbf{i} + t^{3/2}\mathbf{j}$ *from (0, 0, 0) to (4, 8, 0).*

Sol. We have $\mathbf{r} = t\mathbf{i} + t^{3/2}\mathbf{j}$.

The coordinates of the point with position vector $\mathbf{r}$ are $(t, t^{3/2}, 0)$.

$$t = 0, t^{3/2} = 0 \Rightarrow t = 0 \quad \text{and} \quad t = 4, t^{3/2} = 8 \Rightarrow t = 4.$$

$\therefore$ The given points correspond to the values 0 and 4 of t.

$$\dot{\mathbf{r}} = \mathbf{i} + \frac{3}{2}t^{1/2}\mathbf{j}$$

$$\therefore \quad |\dot{\mathbf{r}}| = \sqrt{1 + \frac{9}{4}t} = \frac{1}{2}\sqrt{4 + 9t}$$

$\therefore$ Length of the given curve

$$= \int_0^4 |\dot{\mathbf{r}}|\, dt = \int_0^4 \frac{1}{2}\sqrt{4+9t}\, dt$$

$$= \frac{1}{2} \cdot \frac{(4+9t)^{3/2}}{(3/2)9}\Bigg|_0^4 = \frac{1}{27}[(40)^{3/2} - (4)^{3/2}]$$

$$= \frac{1}{27}[80\sqrt{10} - 8] = \frac{8}{27}[\sqrt{1000} - 1] = 9.073.$$

Example 4. *Show that the length of the curve* $x = 2a(\sin^{-1} t + t\sqrt{1-t^2})$, $y = 2at^2$, $z = 4at$ *between the points* $t = t_1$ *and* $t = t_2$ *is* $4\sqrt{2}\, a(t_2 - t_1)$.

Sol. We have $x = 2a(\sin^{-1} t + t\sqrt{1-t^2})$, $y = 2at^2$, $z = 4at$.

Let $\mathbf{r}$ be the position vector of the point (x, y, z) on the given curve.

$$\therefore \quad \mathbf{r} = x\mathbf{i} + y\mathbf{j} + z\mathbf{k} = 2a(\sin^{-1} t + t\sqrt{1-t^2})\mathbf{i} + 2at^2\mathbf{j} + 4at\mathbf{k}$$

$$\therefore \quad \dot{\mathbf{r}} = 2a\left(\frac{1}{\sqrt{1-t^2}} + t \cdot \frac{1}{2}(1-t^2)^{-1/2}(-2t) + \sqrt{1-t^2} \cdot 1\right)\mathbf{i} + 4at\mathbf{j} + 4a\mathbf{k}$$

$$= 2a\left(\frac{1}{\sqrt{1-t^2}} - \frac{t^2}{\sqrt{1-t^2}} + \sqrt{1-t^2}\right)\mathbf{i} + 4at\mathbf{j} + 4a\mathbf{k}$$

$$= 2a\left(\sqrt{1-t^2} + \sqrt{1-t^2}\right)\mathbf{i} + 4at\mathbf{j} + 4a\mathbf{k}$$

$$= 4a\sqrt{1-t^2}\,\mathbf{i} + 4at\mathbf{j} + 4a\mathbf{k}$$

$$\therefore \quad |\dot{\mathbf{r}}| = \sqrt{16a^2(1-t^2)+16a^2t^2+16a^2}$$

$$= 4a\sqrt{(1-t^2+t^2+1)} = 4\sqrt{2}\,a$$

$\therefore$ Length of the given curve

$$= \int_{t_1}^{t_2} |\dot{\mathbf{r}}|\,dt = \int_{t_1}^{t_2} 4\sqrt{2}\,a\,dt$$

$$= 4\sqrt{2}at\Big|_{t_1}^{t_2} = 4\sqrt{2}\,a(t_2 - t_1).$$

Example 5. *Find the arc length as a function of* θ *along the epicycloid:*

$$x = (a+b)\cos\theta - b\cos\left(\frac{a+b}{b}\theta\right),\ y = (a+b)\sin\theta - b\sin\left(\frac{a+b}{b}\theta\right),\ z = 0.$$

Sol. The given epicycloid is

$$x = (a+b)\cos\theta - b\cos\left(\frac{a+b}{b}\theta\right),\ y = (a+b)\sin\theta - b\sin\left(\frac{a+b}{b}\theta\right),\ z = 0.$$

$$\therefore \quad \dot{x} = -(a+b)\sin\theta + (a+b)\sin\left(\frac{a+b}{b}\theta\right),$$

$$\dot{y} = (a+b)\cos\theta - (a+b)\cos\left(\frac{a+b}{b}\theta\right) \quad \text{and} \quad \dot{z} = 0$$

Let $\mathbf{r}$ be the position vector of the point (x, y, z) on the epicycloid.

$$\therefore \quad \mathbf{r} = x\mathbf{i} + y\mathbf{j} + z\mathbf{k} \quad \therefore \quad \dot{\mathbf{r}} = \dot{x}\,\mathbf{i} + \dot{y}\,\mathbf{j} + \dot{z}\,\mathbf{k}$$

$$\therefore \quad |\dot{\mathbf{r}}|^2 = \dot{x}^2 + \dot{y}^2 + \dot{z}^2 = (a+b)^2\left(-\sin\theta + \sin\left(\frac{a+b}{b}\theta\right)\right)^2$$

$$+ (a+b)^2\left(\cos\theta - \cos\left(\frac{a+b}{b}\theta\right)\right)^2 + 0^2$$

$$= (a+b)^2\left[2 - 2\sin\theta\sin\frac{a+b}{b}\theta - 2\cos\theta\cos\frac{a+b}{b}\theta\right]$$

$$= (a+b)^2\left[2 - 2\cos\left(\frac{a+b}{b}\theta - \theta\right)\right] = (a+b)^2\left[2 - 2\cos\left(\frac{a}{b}\theta\right)\right]$$

$$= 4(a+b)^2\sin^2\left(\frac{a}{2b}\theta\right)$$

$$\therefore \quad |\dot{\mathbf{r}}| = 2(a+b)\sin\left(\frac{a}{2b}\theta\right)$$

$$\therefore \quad s = \int_0^\theta |\dot{\mathbf{r}}|\,d\theta = \int_0^\theta 2(a+b)\sin\left(\frac{a}{2b}\theta\right)d\theta = -\left.\frac{2(a+b)\cos\left(\dfrac{a}{2b}\theta\right)}{a/2b}\right|_0^\theta$$

$$= -\frac{4(a+b)\,b}{a}\left[\cos\left(\frac{a}{2b}\theta\right) - 1\right] = \frac{4(a+b)b}{a}\left[1 - \cos\left(\frac{a}{2b}\theta\right)\right].$$

Example 6. *Find the length of the curve given by the intersection of the surfaces* $\frac{x^2}{a^2} - \frac{y^2}{b^2} = 1$, $x = a \cosh \frac{z}{a}$ *from the point (a, 0, 0) to the point (x, y, z).*

Sol. We have $$\frac{x^2}{a^2} - \frac{y^2}{b^2} = 1 \quad ...(1)$$

and $$x = a \cosh \frac{z}{a} \quad ...(2)$$

Let $$x = a \cosh t, \quad y = b \sinh t, \quad z = at$$

$\therefore$ (1) and (2) are satisfied.

Let **r** be the position vector of the point (x, y, z) on the given curve.

$$\therefore \quad \mathbf{r} = (a \cosh t)\mathbf{i} + (b \sinh t)\mathbf{j} + at\mathbf{k}$$

$$a \cosh t = a,\ b \sinh t = 0,\ at = 0 \quad \Rightarrow \quad t = 0$$

$\therefore$ The initial point corresponds to $t = 0$.

$$\dot{\mathbf{r}} = (a \sinh t)\mathbf{i} + (b \cosh t)\mathbf{j} + a\mathbf{k}$$

$$\therefore \quad |\dot{\mathbf{r}}| = \sqrt{a^2 \sinh^2 t + b^2 \cosh^2 t + a^2}$$

$$= \sqrt{a^2 (\sinh^2 t + 1) + b^2 \cosh^2 t}$$

$$= \sqrt{a^2 \cosh^2 t + b^2 \cosh^2 t} = \sqrt{(a^2 + b^2)} \cosh t$$

$\therefore$ Length of the given curve

$$= \int_0^t |\dot{\mathbf{r}}|\, dt = \int_0^t \sqrt{a^2 + b^2} \cosh t\, dt$$

$$= \sqrt{a^2 + b^2} \sinh t \Big|_0^t = \sqrt{a^2 + b^2} \sinh t$$

$$= \frac{\sqrt{a^2 + b^2}}{b} y.$$

EXERCISE 1.2

1. Find the length of the helix $\mathbf{r} = (3 \cos t)\mathbf{i} + (3 \sin t)\mathbf{j} + 4t\mathbf{k}$, $0 \le t \le 2\pi$.
2. (*i*) Find the length of one complete turn of the helix :

   $\mathbf{r} = (a \cos t, a \sin t, bt)$, $-\infty < t < \infty$, $a > 0$, $b > 0$.

   (*ii*) Find the length of the helix $\mathbf{r} = a \cos u\mathbf{i} + a \sin u\mathbf{j} + cu\mathbf{k}$, $-\infty < u < \infty$ from the point $(a, 0, 0)$ to the point $(a, 0, 2\pi c)$.
3. Find the length of the curve $\mathbf{r} = (3 \cosh 2t)\mathbf{i} + (3 \sinh 2t)\mathbf{j} + 6t\mathbf{k}$, $0 \le t \le \pi$.
4. Find the length of the catenary $\mathbf{r} = t\mathbf{i} + \cosh t\mathbf{j}$ from $t = 0$ to $t = 1$.
5. Find the length of the curve $\mathbf{r} = (1 + 2t)\mathbf{i} + (2 + t)\mathbf{j} - \mathbf{k}$, $3 \le t \le 7$.
6. Find the length of the curve $\mathbf{r} = (2 + 9t)\mathbf{i} + (1 - 3t)\mathbf{j} + t\mathbf{k}$, $8 \le t \le 15$. Also, verify the result by using the distance formula to find the distance between two given points.

7. Find the length of the semicubical parabola $\mathbf{r} = t\mathbf{i} + t^{3/2}\mathbf{k}$ from (0, 0, 0) to (9, 0, 27).
8. Find the length of the curve $\mathbf{r} = (\sin^{-1} t + t\sqrt{1-t^2})\mathbf{i} + t^2\mathbf{j} + 2t\mathbf{k}$ from $t = 1$ to $t = 3$.
9. Find the length of the curve given by the intersection of the surfaces $x^2 - y^2 = 1$, $x = \cosh z$ from the point (1, 0, 0) to the point (x, y, z).
10. Find the length of the curve $\mathbf{r} = e^t \cos t\mathbf{i} + e^t \sin t\mathbf{j} + e^t\mathbf{k}$, $0 \le t \le \pi$.

Answers

1. 10π **2.** (*i*) $2\pi\sqrt{a^2+b^2}$ (*ii*) $2\pi\sqrt{a^2+c^2}$ **3.** $3\sqrt{2}\sinh 2\pi$
4. sinh 1 **5.** $4\sqrt{5}$ **6.** $7\sqrt{91}$ **7.** 28.73
8. $4\sqrt{2}$ **9.** $\sqrt{2}\,y$ **10.** $\sqrt{3}(e^{\pi}-1)$.

Hint

2. (*i*) The limits in the definite integral are to be t_0 and $t_0 + 2\pi$, where t_0 is any number.

9. ARC LENGTH AS PARAMETER IN REPRESENTATIONS OF CURVES

Let $\mathbf{r} = \mathbf{r}(t)$ be any regular curve. Let $A(t = a)$ be any arbitrary but fixed point on the curve. We define a function s of t as

$$s = s(t) = \int_a^t |\dot{\mathbf{r}}|\, dt \qquad \text{...(1)}$$

$s(t)$ is called the **arc length function** of the curve $\mathbf{r} = \mathbf{r}(t)$. If $t_0 > a$, then $s(t_0)$ is the length of curve between the points with parametric values a and t_0. If $t_0 < a$, then $s(t_0)$ is negative of the length of the curve between the points with parametric values a and t_0. Thus $s(a) = 0$ and for points on one side of A the value of s will be positive ; for points on other side, negative. The choice of the fixed point $A(t = a)$ is arbitrary. Changing point A shall mean changing s by a constant quantity.

For simplicity, the arc length function s is written **arc length**. The use of arc length s as parameter in space curves would help us a lot in studying their curvature and torsion.

By the fundamental theorem of calculus, (1) implies

$$\frac{ds}{dt} = |\dot{\mathbf{r}}| = \left|\frac{d\mathbf{r}}{dt}\right| \qquad \text{...(2)}$$

$$\therefore \quad \left|\frac{d\mathbf{r}}{ds}\right| = \left|\frac{d\mathbf{r}}{dt}\cdot\frac{dt}{ds}\right| = \left|\frac{d\mathbf{r}}{dt}\right|\left|\frac{dt}{ds}\right|$$

$$= \left|\frac{d\mathbf{r}}{dt}\right| \Big/ \left|\frac{ds}{dt}\right| = \left|\frac{d\mathbf{r}}{dt}\right| \Big/ \left|\frac{d\mathbf{r}}{dt}\right| = 1 \qquad \text{(Using (2))}$$

$$\therefore \quad \left|\frac{d\mathbf{r}}{ds}\right| = 1.$$

If the equation of a curve is given in terms of arc length, then we say that the equation of the curve is a **natural representation** of the curve.

If the parameter in the equation of a curve is other than, 'arc length s', then the equation of the curve is called an **arbitrary representation** of the curve.

In general, the geometric quantities along a curve are defined in terms of a natural representation of the curve. By using the chain rule and the relation $\frac{ds}{dt} = \left|\frac{d\mathbf{r}}{dt}\right|$, these quantities can also be derived in terms of any arbitrary parameter.

WORKING RULES FOR WRITING $\mathbf{r} = \mathbf{r}(t)$ IN TERMS OF s

Step I. *Find* $|\dot{\mathbf{r}}|$.

Step II. *Solve* $s = \int_0^t |\dot{\mathbf{r}}|\, dt$ *to find s in terms of t.*

Step III. *Using the relation found in* **Step II**, *find t in terms of s.*

Step IV. *Substitute the value of t in* $\mathbf{r} = \mathbf{r}(t)$ *to get the required natural representation of the given curve.*

Example 1. *Find the equation of the helix* $\mathbf{r} = a \cos t\mathbf{i} + a \sin t\mathbf{j} + ct\mathbf{k}, -\infty < t < \infty$ *in terms of arc length s as parameter.*

Sol. We have

$$\mathbf{r} = a \cos t\mathbf{i} + a \sin t\mathbf{j} + ct\mathbf{k}, \; -\infty < t < \infty \quad ...(1)$$

$$\therefore \quad \dot{\mathbf{r}} = -a \sin t\mathbf{i} + a \cos t\mathbf{j} + c\mathbf{k}$$

$$\therefore \quad |\dot{\mathbf{r}}| = \sqrt{a^2 \sin^2 t + a^2 \cos^2 t + c^2} = \sqrt{a^2 + c^2}$$

Let the point with $t = 0$ be the fixed point for the arc length parameter s.

$$\therefore \quad s = s(t) = \int_0^t |\dot{\mathbf{r}}|\, dt$$

$$= \int_0^t \sqrt{a^2 + c^2}\, dt = \sqrt{a^2 + c^2}\, t$$

$$\therefore \quad t = \frac{s}{\sqrt{a^2 + c^2}}$$

Substituting the value of t in (1), the equation of the given helix in terms of arc length s as parameter is

$$\mathbf{r}(s) = a \cos \frac{s}{\sqrt{a^2 + c^2}}\,\mathbf{i} + a \sin \frac{s}{\sqrt{a^2 + c^2}}\,\mathbf{j} + \frac{cs}{\sqrt{a^2 + c^2}}\,\mathbf{k}.$$

Example 2. *Express the curve* $\mathbf{r} = e^t \cos t\mathbf{i} + e^t \sin t\mathbf{j} + e^t\mathbf{k}, -\infty < t < \infty$ *in terms of arc length s as parameter.*

Sol. We have

$$\mathbf{r} = e^t \cos t\,\mathbf{i} + e^t \sin t\mathbf{j} + e^t\mathbf{k}. \quad ...(1)$$

$$\therefore \quad \dot{\mathbf{r}} = (e^t \cos t - e^t \sin t)\mathbf{i} + (e^t \sin t + e^t \cos t)\mathbf{j} + e^t\mathbf{k}$$

$$= e^t(\cos t - \sin t)\mathbf{i} + e^t (\sin t + \cos t)\mathbf{j} + e^t\mathbf{k}$$

$$\therefore \quad |\dot{\mathbf{r}}| = \sqrt{e^{2t}(\cos t - \sin t)^2 + e^{2t}(\sin t + \cos t)^2 + e^{2t}}$$

$$= e^t \sqrt{\cos^2 t + \sin^2 t - 2 \cos t \sin t + \sin^2 t + \cos^2 t + 2 \sin t \cos t + 1}$$

$$= e^t \sqrt{3}$$

Let the point with $t = 0$ be the fixed point for the arc length parameter s.

$$\therefore \quad s = s(t) = \int_0^t |\dot{\mathbf{r}}|\,dt$$

$$= \int_0^t e^t\sqrt{3}\,dt = \sqrt{3}e^t\Big|_0^t = \sqrt{3}\,(e^t - 1)$$

$$\Rightarrow \quad e^t - 1 = \frac{s}{\sqrt{3}} \quad \Rightarrow \quad e^t = \frac{s}{\sqrt{3}} + 1 \quad \Rightarrow \quad t = \log\left(\frac{s}{\sqrt{3}} + 1\right)$$

Substituting the value of t in (1), the equation of the given curve in terms of arc length s as parameter is

$$\mathbf{r}(s) = \left(\frac{s}{\sqrt{3}} + 1\right)\cos\left(\log\left(\frac{s}{\sqrt{3}} + 1\right)\right)\mathbf{i} + \left(\frac{s}{\sqrt{3}} + 1\right)\sin\left(\log\left(\frac{s}{\sqrt{3}} + 1\right)\right)\mathbf{j} + \left(\frac{s}{\sqrt{3}} + 1\right)\mathbf{k}.$$

EXERCISE 1.3

1. Find the equation of the helix $\mathbf{r} = \cos t\mathbf{i} + \sin t\mathbf{j} + t\mathbf{k}$, $-\infty < t < \infty$ in terms of arc length s as parameter.
2. Find the equation of the curve $\mathbf{r} = e^{2t}\cos t\mathbf{i} + e^{2t}\sin t\mathbf{j} + e^{2t}\mathbf{k}$, $-\infty < t < \infty$ in terms of arc length s as parameter.
3. For the helix $x = a\cos t$, $y = a\sin t$, $z = at\tan\alpha$, show that the length of the curve measured from the point $t = 0$ is $at\sec\alpha$. Also show that $\dfrac{ds}{dt} = a\sec\alpha$.
4. Show that $\mathbf{r} = \dfrac{1}{2}(s + \sqrt{s^2+1})\,\mathbf{i} + \dfrac{1}{2(s + \sqrt{s^2+1})}\mathbf{j} + \dfrac{1}{\sqrt{2}}\log(s + \sqrt{s^2+1})\,\mathbf{k}$ is a natural representation of a curve.

Answers

1. $\mathbf{r} = \cos\dfrac{s}{\sqrt{2}}\mathbf{i} + \sin\dfrac{s}{\sqrt{2}}\mathbf{j} + \dfrac{s}{\sqrt{2}}\mathbf{k}$

2. $\mathbf{r} = \left(\dfrac{2s}{3} + 1\right)\left[\cos\left(\dfrac{1}{2}\log\left(\dfrac{2s}{3} + 1\right)\right)\mathbf{i} + \sin\left(\dfrac{1}{2}\log\left(\dfrac{2s}{3} + 1\right)\right)\mathbf{j} + \mathbf{k}\right].$

Hint

4. $u = s + \sqrt{s^2+1}$ implies $\mathbf{r} = \dfrac{1}{2}u\mathbf{i} + \dfrac{1}{2u}\mathbf{j} + \dfrac{1}{\sqrt{2}}(\log u)\mathbf{k}$

$$\therefore \quad \frac{d\mathbf{r}}{ds} = \frac{d\mathbf{r}}{du}\cdot\frac{du}{ds} = \left(\frac{1}{2}\mathbf{i} - \frac{1}{2u^2}\mathbf{j} + \frac{1}{\sqrt{2}u}\mathbf{k}\right)\left(1 + \frac{s}{\sqrt{s^2+1}}\right)$$

$$\therefore \quad \left|\frac{d\mathbf{r}}{ds}\right| = \sqrt{\frac{1}{4}+\frac{1}{4u^4}+\frac{1}{2u^2}} \cdot \left(\frac{\sqrt{s^2+1}+s}{\sqrt{s^2+1}}\right)$$

$$= \left(\frac{1}{2}+\frac{1}{2u^2}\right)\frac{u}{\sqrt{s^2+1}} = \frac{u^2+1}{2u\sqrt{s^2+1}} = 1.$$

10. TANGENT TO A CURVE

Let C be a curve and P be any point on C. The **tangent** at P to the curve C is the limiting position of a straight line L through P and a point Q of C as Q approaches P along C.

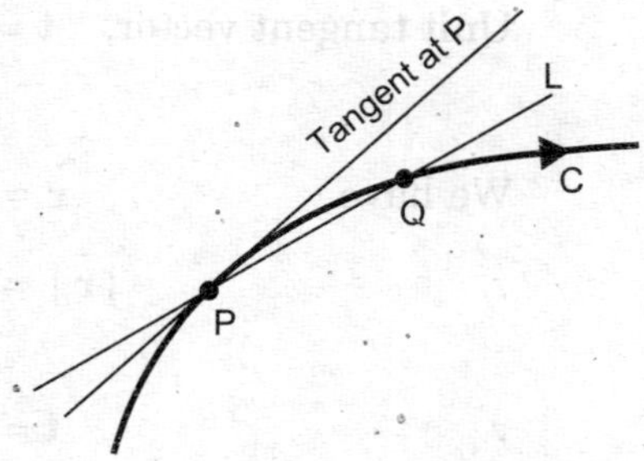

11. UNIT TANGENT VECTOR

Let $\mathbf{r} = \mathbf{r}(t)$ be the equation of a regular curve C in terms of an arbitrary parameter t. Let P and Q be the points on the curve whose position vectors are $\mathbf{r}$ and $\mathbf{r} + \delta\mathbf{r}$ corresponding to the values t and $t + \delta t$ of the parameter respectively.

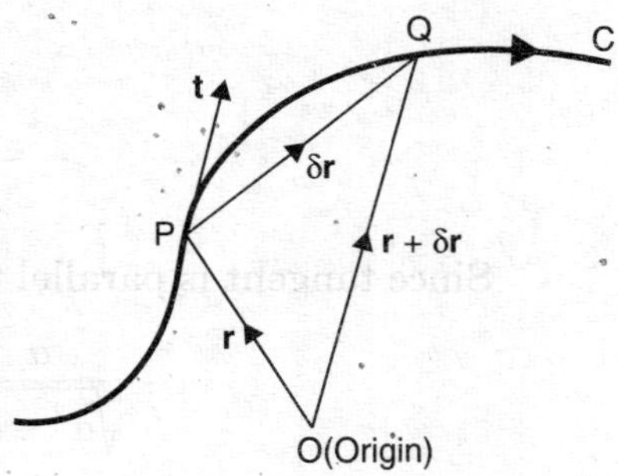

$$\therefore \quad \overrightarrow{PQ} = \overrightarrow{OQ} - \overrightarrow{OP} = (\mathbf{r} + \delta\mathbf{r}) - \mathbf{r} = \delta\mathbf{r}$$

$\therefore$ The quotient $\frac{\delta\mathbf{r}}{\delta t}$ is a vector parallel to the line PQ. Since the given curve is regular, $\mathbf{r}(t)$ has continuous non-zero derivative.

$\therefore$ $\lim_{\delta t \to 0} \frac{\delta\mathbf{r}}{\delta t} = \frac{d\mathbf{r}}{dt}$ exists and is non-zero.

By the definition of a tangent to a curve at a point, the vector $\frac{d\mathbf{r}}{dt}$ *i.e.*, $\dot{\mathbf{r}}(t)$ is parallel to the tangent at the point P.

The vector $\dot{\mathbf{r}}(t)$ is called the **tangent vector** of C at the point P.

The corresponding unit vector $\frac{1}{|\dot{\mathbf{r}}|}\dot{\mathbf{r}}$ is called the **unit tangent vector** of C at the point P and it is denoted by $\mathbf{t}$. The vectors $\dot{\mathbf{r}}$ and $\mathbf{t}$ point in the direction of increasing t. Thus the directions of $\dot{\mathbf{r}}$ and $\mathbf{t}$ are same and depend upon the orientation of the curve C.

In particular, if the equation of the curve C is given in terms of the arc length s, then the tangent vector of C at P is $\frac{d\mathbf{r}}{ds}$.

The vector $\frac{d\mathbf{r}}{ds}$ is denoted by $\mathbf{r}'$. We know that $\frac{d\mathbf{r}}{ds}$ *i.e.*, $\mathbf{r}'$ is a unit vector.

$\therefore$ Unit tangent vector '$\mathbf{t}$' at P = $\mathbf{r}'$.

Example 1. *Find the unit tangent vector* **t** *and direction cosines of the tangent at a point on the circular helix* $x = a \cos t, y = a \sin t, z = bt, -\infty < t < \infty$.

Sol. The given helix is

$$x = a \cos t, y = a \sin t, z = bt, -\infty < t < \infty.$$

Let **r** be the position vector of the point (x, y, z) on the helix.

$$\therefore \quad \mathbf{r} = a \cos t\mathbf{i} + a \sin t\mathbf{j} + bt\mathbf{k}$$

Unit tangent vector, $\mathbf{t} = \dfrac{1}{|\dot{\mathbf{r}}|}\dot{\mathbf{r}}$

We have $\dot{\mathbf{r}} = -a \sin t\mathbf{i} + a \cos t\mathbf{j} + b\mathbf{k}$

$$\therefore \quad |\dot{\mathbf{r}}| = \sqrt{a^2 \sin^2 t + a^2 \cos^2 t + b^2} = \sqrt{a^2 + b^2}$$

$$\therefore \quad \mathbf{t} = \frac{1}{|\dot{\mathbf{r}}|}\dot{\mathbf{r}} = \frac{1}{\sqrt{a^2+b^2}}(-a \sin t\mathbf{i} + a \cos t\mathbf{j} + b\mathbf{k})$$

$$= -\frac{a}{\sqrt{a^2+b^2}} \sin t\mathbf{i} + \frac{a}{\sqrt{a^2+b^2}} \cos t\mathbf{j} + \frac{b}{\sqrt{a^2+b^2}}\mathbf{k}$$

Since tangent is parallel to **t** and **t** is a unit vector, the d.c.'s of the tangent are

$$-\frac{a}{\sqrt{a^2+b^2}} \sin t, \frac{a}{\sqrt{a^2+b^2}} \cos t, \frac{b}{\sqrt{a^2+b^2}}.$$

Example 2. *Show that the tangent vectors along the curve* $\mathbf{r} = at\mathbf{i} + bt^2\mathbf{j} + t^3\mathbf{k}$, *where* $2b^2 = 3a$ *make a constant angle with the vector* $\mathbf{i} + \mathbf{k}$.

Sol. Given curve is $\mathbf{r} = at\mathbf{i} + bt^2\mathbf{j} + t^3\mathbf{k}$.

$$\therefore \quad \dot{\mathbf{r}} = a\mathbf{i} + 2bt\mathbf{j} + 3t^2\mathbf{k}$$

The tangent vector at point 't' is $\dot{\mathbf{r}}$ *i.e.*, $a\mathbf{i} + 2bt\mathbf{j} + 3t^2\mathbf{k}$.

Given vector is $\mathbf{i} + \mathbf{k}$, *i.e.*, $1.\mathbf{i} + 0.\mathbf{j} + 1.\mathbf{k}$

Let θ be the angle between the tangent vector $\dot{\mathbf{r}}$ and the vector $\mathbf{i} + \mathbf{k}$.

$$\therefore \quad \cos\theta = \frac{a(1) + 2bt(0) + 3t^2(1)}{\sqrt{a^2 + 4b^2t^2 + 9t^4}\ \sqrt{1+0+1}}$$

$$= \frac{a + 3t^2}{\sqrt{a^2 + 2(3a)t^2 + 9t^4}\ \sqrt{2}} = \frac{a + 3t^2}{(a + 3t^2)\sqrt{2}} = \frac{1}{\sqrt{2}}$$

$$\therefore \quad \theta = \pi/4, \text{ which is a constant angle.}$$

$\therefore$ The result holds.

12. EQUATION OF THE TANGENT AT A POINT ON A CURVE

Let $\mathbf{r} = \mathbf{r}(t)$ be the equation of a regular curve C in terms of an arbitrary parameter t. Let P($\mathbf{r}$) be any point on the curve. We know that $\dot{\mathbf{r}}$ is the tangent vector at the point P and the tangent at P is parallel to this vector. Let Q be a general point on the tangent at P. Let $\mathbf{R}$ be the position vector of the point Q.

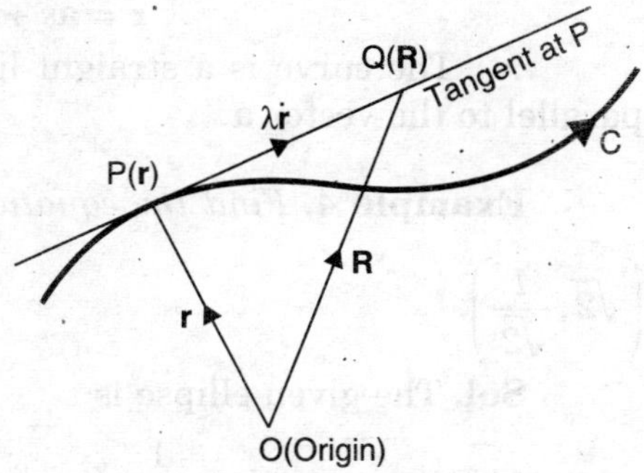

∴ The equation of the tangent at the point P($\mathbf{r}$) is $\mathbf{R} = \mathbf{r} + \lambda\dot{\mathbf{r}}$, where λ is a scalar parameter.

Let the coordinates of P and Q be (x, y, z) and (X, Y, Z) respectively.

∴ $$\mathbf{r} = x\mathbf{i} + y\mathbf{j} + z\mathbf{k} \quad \text{and} \quad \mathbf{R} = \mathrm{X}\mathbf{i} + \mathrm{Y}\mathbf{j} + \mathrm{Z}\mathbf{k}$$

Also $$\dot{\mathbf{r}} = \dot{x}\mathbf{i} + \dot{y}\mathbf{j} + \dot{z}\mathbf{k}$$

∴ The equation of the tangent at P is

$$\mathrm{X}\mathbf{i} + \mathrm{Y}\mathbf{j} + \mathrm{Z}\mathbf{k} = x\mathbf{i} + y\mathbf{j} + z\mathbf{k} + \lambda(\dot{x}\mathbf{i} + \dot{y}\mathbf{j} + \dot{z}\mathbf{k})$$

⇒ $$\mathrm{X} = x + \lambda\dot{x},\ \mathrm{Y} = y + \lambda\dot{y},\ \mathrm{Z} = z + \lambda\dot{z}$$

∴ $$\frac{\mathrm{X} - x}{\dot{x}} = \frac{\mathrm{Y} - y}{\dot{y}} = \frac{\mathrm{Z} - z}{\dot{z}} \quad (= \lambda).$$

These are the cartesian equations of the tangent at the point P(x, y, z). Here $\dot{x}, \dot{y}, \dot{z}$ are direction ratios of the tangent at the point P.

In particular, if the equation of the curve C is given in terms of the arc length s then, $\mathbf{r}'$ is the unit tangent vector of C at P.

∴ The tangent at P is parallel to the vector $\mathbf{r}'$.

∴ The equation of the tangent at P($\mathbf{r}$) is $\mathbf{R} = \mathbf{r} + \lambda\mathbf{r}'$, where $\mathbf{R}$ is the position vector of the general point Q($\mathbf{R}$) on the tangent at P.

The cartesian form of the equations of tangent at P are

$$\frac{\mathrm{X} - x}{x'} = \frac{\mathrm{Y} - y}{y'} = \frac{\mathrm{Z} - z}{z'} \quad (= \lambda)$$

Since $\mathbf{r}' = x'\mathbf{i} + y'\mathbf{j} + z'\mathbf{k}$ is a unit vector, x', y', z' are the direction cosines of the tangent at P.

Example 3. *Show that a curve is a straight line if all tangent lines are parallel.*

Sol. Let $\mathbf{r} = \mathbf{r}(s)$ be the given curve.

∴ Tangent vector = $\mathbf{r}'$

Since tangent line at a point is parallel to the tangent vector at that point and all tangent lines are parallel, the direction of $\mathbf{r}'$ is fixed. Also $\mathbf{r}'$ is a unit vector.

∴ $\mathbf{r}'$ is a non-zero constant vector, say $\mathbf{a}$.

∴ $$\frac{d\mathbf{r}}{ds} = \mathbf{a}$$

Integrating, we get

$$\mathbf{r} = \mathbf{a}s + \mathbf{b}, \text{ where } \mathbf{b} \text{ is a constant vector.}$$

∴ The curve is a straight line passing through the point with position vector **b** and parallel to the vector **a**.

Example 4. *Find the equation of the tangent to the ellipse* $\frac{1}{4}x^2 + y^2 = 1$ *at the point* $\left(\sqrt{2}, \frac{1}{\sqrt{2}}\right)$.

Sol. The given ellipse is

$$\frac{1}{4}x^2 + y^2 = 1.$$

The parametric equations of this ellipse are

$$x = 2 \cos t,\ y = \sin t,\ z = 0$$

Let **r** be the position vector of the point (x, y, z).

$$\therefore \quad \mathbf{r} = x\mathbf{i} + y\mathbf{j} + z\mathbf{k} = 2 \cos t\mathbf{i} + \sin t\mathbf{j} + 0\mathbf{k}$$

$$2 \cos t = \sqrt{2},\ \sin t = \frac{1}{\sqrt{2}} \Rightarrow t = \frac{\pi}{4}$$

∴ The point $\left(\sqrt{2}, \frac{1}{\sqrt{2}}\right)$ in xy-plane corresponds to $t = \frac{\pi}{4}$.

$$\dot{\mathbf{r}} = -2 \sin t\mathbf{i} + \cos t\mathbf{j} + 0\mathbf{k}$$

At $t = \frac{\pi}{4}$,
$$\dot{\mathbf{r}} = -2 \sin \frac{\pi}{4}\mathbf{i} + \cos \frac{\pi}{4}\mathbf{j} + 0\mathbf{k}$$

$$= -\sqrt{2}\,\mathbf{i} + \frac{1}{\sqrt{2}}\mathbf{j} + 0\mathbf{k}$$

∴ The tangent at $\left(\sqrt{2}, \frac{1}{\sqrt{2}}\right)$ passes through $\left(\sqrt{2}, \frac{1}{\sqrt{2}}, 0\right)$ and has d.r.'s $-\sqrt{2}, \frac{1}{\sqrt{2}}, 0$.

∴ The equations of the tangent at $\left(\sqrt{2}, \frac{1}{\sqrt{2}}, 0\right)$ are

$$\frac{x - \sqrt{2}}{-\sqrt{2}} = \frac{y - \frac{1}{\sqrt{2}}}{\frac{1}{\sqrt{2}}} = \frac{z - 0}{0} \quad \text{or} \quad \frac{x - \sqrt{2}}{-\sqrt{2}} = \frac{\sqrt{2}y - 1}{1} = \frac{z}{0}.$$

Example 5. *Find the equation of the tangent to the curve* $x = 1 + t,\ y = -t^2,\ z = 1 + t^2$, $-\infty < t < \infty$ *at the point for which* $t = 2$.

Sol. The given curve is

$$x = 1 + t,\ y = -t^2,\ z = 1 + t^2,\ -\infty < t < \infty.$$

$$t = 2 \Rightarrow x = 1 + 2 = 3,\quad y = -(2)^2 = -4,\quad z = 1 + (2)^2 = 5$$

∴ $t = 2$ corresponds to the point $(3, -4, 5)$ on the curve.

Also, $\dot{x} = 1,\quad \dot{y} = -2t,\quad \dot{z} = 2t$

$$t = 2 \Rightarrow \dot{x} = 1, \dot{y} = -2(2) = -4, \dot{z} = 2(2) = 4$$

∴ The tangent at (3, – 4, 5) passes through (3, – 4, 5) and has d.r.'s 1, – 4, 4.

∴ The equations of the tangent at (3, – 4, 5) are

$$\frac{x-3}{1} = \frac{y-(-4)}{-4} = \frac{z-5}{4} \quad \text{or} \quad x-3 = \frac{y+4}{-4} = \frac{z-5}{4}.$$

Example 6. *Show that the equation of the tangent at any point on the curve whose equation referred to rectangular axes are $x = 3t$, $y = 3t^2$, $z = 2t^3$ makes a constant angle with the line $y = z - x = 0$.*

Sol. Given curve is $x = 3t$, $y = 3t^2$, $z = 2t^3$.

Let **r** be the position vector of the point (x, y, z) on the given curve.

$$\therefore \quad \mathbf{r} = 3t\mathbf{i} + 3t^2\mathbf{j} + 2t^3\mathbf{k}$$

$$\therefore \quad \dot{\mathbf{r}} = 3\mathbf{i} + 6t\mathbf{j} + 6t^2\mathbf{k}$$

The tangent line at the point with parametric value t is parallel to the vector $\dot{\mathbf{r}}$.

∴ D.r.'s of the tangent are 3, $6t$, $6t^2$.

Given line is $y = z - x = 0$

$$\Rightarrow \quad \frac{x}{1} = \frac{y}{0} = \frac{z}{1}$$

∴ D.r.'s of the given line are 1, 0, 1.

Let θ be the angle between the tangent and the given line.

$$\therefore \quad \cos\theta = \frac{3(1) + 6t(0) + 6t^2(1)}{\sqrt{(3)^2 + (6t)^2 + (6t^2)^2}\,\sqrt{1^2 + 0^2 + 1^2}}$$

$$= \frac{3 + 6t^2}{\sqrt{(3 + 6t^2)^2}\,\sqrt{2}} = \frac{1}{\sqrt{2}}$$

$$\therefore \quad \theta = \pi/4, \text{ which is a constant angle.}$$

∴ The result holds.

13. DIRECTION RATIOS OF THE TANGENT AT A POINT ON THE CURVE OF INTERSECTION OF TWO SURFACES

Let the given curve be the intersection of the surfaces

$$F(x, y, z) = 0 \quad \text{...(1)}$$

and

$$G(x, y, z) = 0 \quad \text{...(2)}$$

Eliminating x, y, z from (1) and (2), let the equation of the given curve be

$$\mathbf{r} = \mathbf{r}(t) = x(t)\mathbf{i} + y(t)\mathbf{j} + z(t)\mathbf{k},$$

where t is an arbitrary parameter.

∴ D.r.'s of the tangent to the curve at the point 't' are $\dot{x}, \dot{y}, \dot{z}$.

Differentiating (1) and (2) w.r.t. t, we get

$$\frac{\partial F}{\partial x}\cdot\frac{dx}{dt} + \frac{\partial F}{\partial y}\cdot\frac{dy}{dt} + \frac{\partial F}{\partial z}\cdot\frac{dz}{dt} = 0$$

and

$$\frac{\partial G}{\partial x}\cdot\frac{dx}{dt} + \frac{\partial G}{\partial y}\cdot\frac{dy}{dt} + \frac{\partial G}{\partial z}\cdot\frac{dz}{dt} = 0$$

$$\Rightarrow \qquad (F_x)\dot{x} + (F_y)\dot{y} + (F_z)\dot{z} = 0$$

and
$$(G_x)\dot{x} + (G_y)\dot{y} + (G_z)\dot{z} = 0$$

Solving these equations for $\dot{x}, \dot{y}$ and $\dot{z}$, we get

$$\frac{\dot{x}}{F_yG_z - F_zG_y} = \frac{\dot{y}}{F_zG_x - F_xG_z} = \frac{\dot{z}}{F_xG_y - F_yG_x}$$

$$\therefore \qquad F_yG_z - F_zG_y,\ F_zG_x - F_xG_z,\ F_xG_y - F_yG_x$$

are also d.r.'s of the tangent to the curve given by the equations $F(x, y, z) = 0$, $G(x, y, z) = 0$ at the point t.

Example 7. *Show that the equation of the tangent to the curve of intersection of the ellipsoid* $\frac{x^2}{a^2} + \frac{y^2}{b^2} + \frac{z^2}{c^2} = 1$ *and the confocal* $\frac{x^2}{a^2 - \lambda} + \frac{y^2}{b^2 - \lambda} + \frac{z^2}{c^2 - \lambda} = 1$ *is*

$$\frac{x(X - x)}{a^2(b^2 - c^2)(a^2 - \lambda)} = \frac{y(Y - y)}{b^2(c^2 - a^2)(b^2 - \lambda)} = \frac{z(Z - z)}{c^2(a^2 - b^2)(c^2 - \lambda)},$$

where (x, y, z) is an arbitrary point on the curve.

Sol. The given surfaces are

$$\frac{x^2}{a^2} + \frac{y^2}{b^2} + \frac{z^2}{c^2} - 1 = 0 \qquad \text{...(1)}$$

and
$$\frac{x^2}{a^2 - \lambda} + \frac{y^2}{b^2 - \lambda} + \frac{z^2}{c^2 - \lambda} - 1 = 0 \qquad \text{...(2)}$$

Let the equation of the curve of intersection of given surfaces be $\mathbf{r} = \mathbf{r}(t)$, where t is an arbitrary parameter.

Differentiating the equations (1) and (2) w.r.t. t, we get

$$\frac{2x}{a^2}\dot{x} + \frac{2y}{b^2}\dot{y} + \frac{2z}{c^2}\dot{z} = 0$$

and
$$\frac{2x}{a^2 - \lambda}\dot{x} + \frac{2y}{b^2 - \lambda}\dot{y} + \frac{2z}{c^2 - \lambda}\dot{z} = 0$$

$$\Rightarrow \qquad \frac{x\dot{x}}{a^2} + \frac{y\dot{y}}{b^2} + \frac{z\dot{z}}{c^2} = 0$$

and
$$\frac{x\dot{x}}{a^2 - \lambda} + \frac{y\dot{y}}{b^2 - \lambda} + \frac{z\dot{z}}{c^2 - \lambda} = 0$$

Solving these equations for $\dot{x}, \dot{y}$ and $\dot{z}$, we get

$$\frac{\dot{x}}{\frac{yz}{b^2(c^2 - \lambda)} - \frac{yz}{c^2(b^2 - \lambda)}} = \frac{\dot{y}}{\frac{zx}{c^2(a^2 - \lambda)} - \frac{zx}{a^2(c^2 - \lambda)}} = \frac{\dot{z}}{\frac{xy}{a^2(b^2 - \lambda)} - \frac{xy}{b^2(a^2 - \lambda)}}$$

$$\Rightarrow \quad \frac{\dot{x}}{\dfrac{\lambda yz(b^2-c^2)}{b^2c^2(c^2-\lambda)(b^2-\lambda)}} = \frac{\dot{y}}{\dfrac{\lambda zx(c^2-a^2)}{c^2a^2(a^2-\lambda)(c^2-\lambda)}} = \frac{\dot{z}}{\dfrac{\lambda xy(a^2-b^2)}{a^2b^2(b^2-\lambda)(a^2-\lambda)}}$$

$$\Rightarrow \quad \frac{\dot{x}}{\dfrac{a^2(b^2-c^2)(a^2-\lambda)}{x}} = \frac{\dot{y}}{\dfrac{b^2(c^2-a^2)(b^2-\lambda)}{y}} = \frac{\dot{z}}{\dfrac{c^2(a^2-b^2)(c^2-\lambda)}{z}}$$

$\therefore$ D.r.'s of the tangent at the point (x, y, z) on the curve are

$$\frac{a^2(b^2-c^2)(a^2-\lambda)}{x}, \frac{b^2(c^2-a^2)(b^2-\lambda)}{y}, \frac{c^2(a^2-b^2)(c^2-\lambda)}{z}.$$

$\therefore$ The equations of the tangent at the point (x, y, z) on the curve are

$$\frac{X-x}{\dfrac{a^2(b^2-c^2)(a^2-\lambda)}{x}} = \frac{Y-y}{\dfrac{b^2(c^2-a^2)(b^2-\lambda)}{y}} = \frac{Z-z}{\dfrac{c^2(a^2-b^2)(c^2-\lambda)}{z}}$$

or
$$\frac{x(X-x)}{a^2(b^2-c^2)(a^2-\lambda)} = \frac{y(Y-y)}{b^2(c^2-a^2)(b^2-\lambda)} = \frac{z(Z-z)}{c^2(a^2-b^2)(c^2-\lambda)}.$$

14. NORMAL PLANE TO A CURVE

Let C be a curve and P be any point on C. The **normal plane** at P to the curve C is the plane passing through P and perpendicular to the tangent at P.

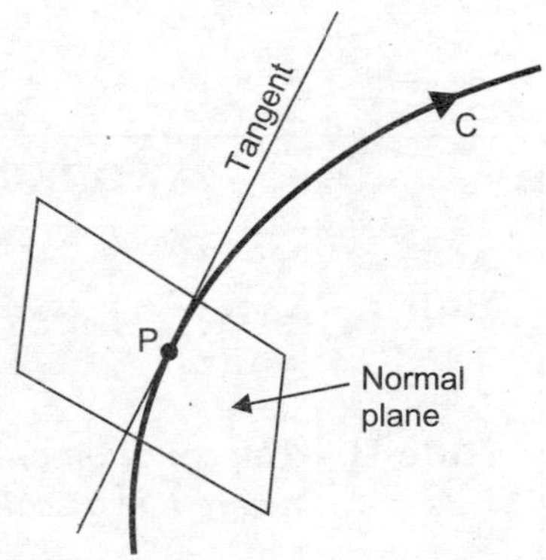

15. EQUATION OF THE NORMAL PLANE AT A POINT ON A CURVE

Let $\mathbf{r} = \mathbf{r}(t)$ be the equation of a regular curve C, where t is an arbitrary parameter. Let P($\mathbf{r}$) be any point on the curve. We know that the tangent at P is parallel to the tangent vector $\dot{\mathbf{r}}$.

Let Q be a general point on the normal plane at P. Let **R** be the position vector of the point Q.

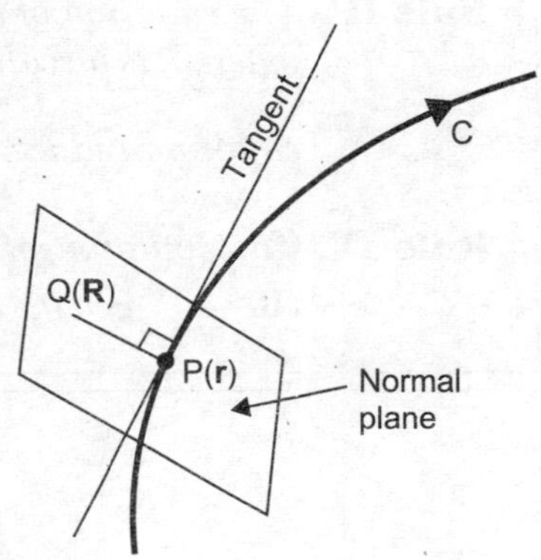

$\therefore$ $\overrightarrow{\mathbf{PQ}}$ and $\dot{\mathbf{r}}$ are perpendicular.

$$\Rightarrow \quad \overrightarrow{\mathbf{PQ}} \cdot \dot{\mathbf{r}} = 0 \Rightarrow (\mathbf{R}-\mathbf{r}) \cdot \dot{\mathbf{r}} = 0 \quad \text{...(1)}$$

This represents the equation of the normal plane at the point P($\mathbf{r}$).

(1) $\Rightarrow$ $(\mathbf{R}-\mathbf{r}) \cdot \dfrac{\dot{\mathbf{r}}}{|\dot{\mathbf{r}}|} = 0. \Rightarrow (\mathbf{R}-\mathbf{r}) \cdot \mathbf{t} = 0$

$\therefore$ The equation of the normal plane at the point P(**r**) can also be written as

$$(\mathbf{R}-\mathbf{r}) \cdot \mathbf{t} = 0.$$

Example 8. *Find the equation of the normal plane to the curve* $\mathbf{r} = t\mathbf{i} + t^2\mathbf{j} + t^3\mathbf{k}$ *at* $t = 1$.

Sol. The given curve is

$$\mathbf{r} = t\mathbf{i} + t^2\mathbf{j} + t^3\mathbf{k}.$$

$\therefore$ $\mathbf{r}(1) = 1.\mathbf{i} + (1)^2\,\mathbf{j} + (1)^3\,\mathbf{k} = \mathbf{i} + \mathbf{j} + \mathbf{k}$

$\therefore$ $t = 1$ corresponds to the point (1, 1, 1) on the curve.

Also $\dot{\mathbf{r}} = \mathbf{i} + 2t\mathbf{j} + 3t^2\mathbf{k}$

$\therefore$ $\dot{\mathbf{r}}(1) = \mathbf{i} + 2\mathbf{j} + 3\mathbf{k}$

$\therefore$ The equation of the normal plane at (1, 1, 1) is $(\mathbf{r} - \mathbf{r}(1)) \cdot \dot{\mathbf{r}}(1) = 0$.

$\Rightarrow$ $(\mathbf{r} - (\mathbf{i} + \mathbf{j} + \mathbf{k})) \cdot (\mathbf{i} + 2\mathbf{j} + 3\mathbf{k}) = 0$

$\Rightarrow$ $\mathbf{r} \cdot (\mathbf{i} + 2\mathbf{j} + 3\mathbf{k}) - (1(1) + 1(2) + 1(3)) = 0$

$\Rightarrow$ $\mathbf{r} \cdot (\mathbf{i} + 2\mathbf{j} + 3\mathbf{k}) = 6.$

WORKING RULES FOR SOLVING PROBLEMS

Rule I. $\mathbf{t} = \dfrac{\dot{\mathbf{r}}}{|\dot{\mathbf{r}}|} = \mathbf{r}'$ *is the unit tangent vector.*

Rule II. *The equation of the tangent to the curve* $\mathbf{r} = \mathbf{r}(t)$ *at the point P(**r**) is* $\mathbf{R} = \mathbf{r} + \lambda\dot{\mathbf{r}}$, *where λ is a scalar parameter. If* $\mathbf{R} = X\mathbf{i} + Y\mathbf{j} + Z\mathbf{k}$ *and* $\mathbf{r} = x\mathbf{i} + y\mathbf{j} + z\mathbf{k}$, *then this equation reduces to* $\dfrac{X-x}{\dot{x}} = \dfrac{Y-y}{\dot{y}} = \dfrac{Z-z}{\dot{z}}\ (= \lambda)$.

Rule III. *The equation of the tangent to the curve* $\mathbf{r} = \mathbf{r}(s)$ *at the point P(**r**) is* $\mathbf{R} = \mathbf{r} + \lambda\mathbf{r}'$, *where λ is a scalar parameter. If* $\mathbf{R} = X\mathbf{i} + Y\mathbf{j} + Z\mathbf{k}$ *and* $\mathbf{r} = x\mathbf{i} + y\mathbf{j} + z\mathbf{k}$, *then this equation reduces to* $\dfrac{X-x}{x'} = \dfrac{Y-y}{y'} = \dfrac{Z-z}{z'}\ (= \lambda)$.

Rule IV. *The equation of the normal plane to the curve* $\mathbf{r} = \mathbf{r}(t)$ *at the point P(**r**) is* $(\mathbf{R}-\mathbf{r}) \cdot \dot{\mathbf{r}} = 0$ *or equivalently* $(\mathbf{R}-\mathbf{r}) \cdot \mathbf{t} = 0$.

EXERCISE 1.4

1. Find the unit tangent vector **t** and the direction cosines of the tangent to the helix $x = a\cos t, y = a\sin t, z = bt, -\infty < t < \infty$ at the point, where $t = \pi/4$.
2. Find the unit tangent vector **t** and the direction cosines of the tangent to the helix $x = a\cos t$, $y = a\sin t, z = at, -\infty < t < \infty$ at the point, where $t = \pi/3$.
3. Find the unit tangent vector **t** to the curve $\mathbf{r} = t\mathbf{i} + t^3\mathbf{j}$ at the point (1, 1, 0).
4. Find the unit tangent vector **t** to the curve $\mathbf{r} = \cos t\mathbf{i} + 2\sin t\mathbf{j}$ at the point $(1/2, \sqrt{3}, 0)$.
5. Find the unit tangent vector **t** to the curve $\mathbf{r} = \cosh t\mathbf{i} + \sinh t\mathbf{j}$ at the point (5/3, 4/3, 0).
6. Find the unit tangent vector **t** to the curve $\mathbf{r} = \log\cos t\mathbf{i} + \log\sin t\mathbf{j} + \sqrt{2}\,t\mathbf{k}$ at the point 't'.
7. Find the equation of the tangent to the curve $x = 1 + t, y = -t^2, z = 1 + t^2, -\infty < t < \infty$ at the point for which (*i*) $t = 1$ (*ii*) $t = 5$.
8. Find the equation of the tangent to the helix $\mathbf{r} = (a\cos t, a\sin t, bt), -\infty < t < \infty$ at the point 't'.
9. Find the equation of the tangent to the curve $\mathbf{r} = t\mathbf{i} + t^3\mathbf{j}$ at the point (1, 1, 0).
10. Find the equation of the tangent to the curve $\mathbf{r} = \cos t\mathbf{i} + 2\sin t\mathbf{j}$ at the point $\left(\frac{1}{2}, \sqrt{3}, 0\right)$.
11. Find the equation of the tangent to the curve $\mathbf{r} = \cosh t\mathbf{i} + \sinh t\mathbf{j}$ at the point $\left(\frac{5}{3}, \frac{4}{3}, 0\right)$.
12. Find the point of intersection of the xy-plane and the tangent line to the curve $\mathbf{r} = (1 + t)\,\mathbf{i} - t^2\mathbf{j} + (1 + t^3)\,\mathbf{k}$ at $t = 1$.
13. Show that the tangent at any point on the curve $\mathbf{r} = at\mathbf{i} + bt^2\mathbf{j} + t^3\mathbf{k}$, $2b^2 = 3a$ makes a constant angle with the line $x - z = 0, y = 0$.
14. Find the equation of the normal plane to the curve $\mathbf{r} = (1 + t)\mathbf{i} - t^2\mathbf{j} + (1 + t^3)\mathbf{k}$ at $t = 1$.
15. Find the point of intersection of the xy-plane and the normal plane to the curve $\mathbf{r} = (\cos t)\mathbf{i} + (\sin t)\mathbf{j} + t\mathbf{k}$ at the point $t = \frac{\pi}{2}$.

Answers

1. $-\frac{a}{\sqrt{2(a^2+b^2)}}\mathbf{i} + \frac{a}{\sqrt{2(a^2+b^2)}}\mathbf{j} + \frac{b}{\sqrt{a^2+b^2}}\mathbf{k}\,;\; -\frac{a}{\sqrt{2(a^2+b^2)}}, \frac{a}{\sqrt{2(a^2+b^2)}}, \frac{b}{\sqrt{a^2+b^2}}$

2. $-\frac{\sqrt{6}}{4}\mathbf{i} + \frac{\sqrt{2}}{4}\mathbf{j} + \frac{\sqrt{2}}{2}\mathbf{k}\,;\; -\frac{\sqrt{6}}{4}, \frac{\sqrt{2}}{4}, \frac{\sqrt{2}}{2}$

3. $\frac{1}{\sqrt{10}}\mathbf{i} + \frac{3}{\sqrt{10}}\mathbf{j} + 0\mathbf{k}$

4. $-\frac{\sqrt{3}}{\sqrt{7}}\mathbf{i} + \frac{2}{\sqrt{7}}\mathbf{j} + 0\mathbf{k}$

5. $\frac{4}{\sqrt{41}}\mathbf{i} + \frac{5}{\sqrt{41}}\mathbf{j} + 0\mathbf{k}$

6. $-\sin^2 t\mathbf{i} + \cos^2 t\mathbf{j} + \sqrt{2}\,\sin t\cos t\mathbf{k}$

7. (*i*) $\frac{x-2}{1} = \frac{y+1}{-2} = \frac{z-2}{2}$ (*ii*) $\frac{x-6}{1} = \frac{y+25}{-10} = \frac{z-26}{10}$

8. $\frac{x - a\cos t}{-a\sin t} = \frac{y - a\sin t}{a\cos t} = \frac{z - bt}{b}$

9. $\mathbf{r} = \mathbf{i} + \mathbf{j} + \lambda(\mathbf{i} + 3\mathbf{j})$

10. $\mathbf{r} = \frac{1}{2}\mathbf{i} + \sqrt{3}\mathbf{j} + \lambda\left(-\frac{\sqrt{3}}{2}\mathbf{i} + \mathbf{j}\right)$

11. $\mathbf{r} = \frac{5}{3}\mathbf{i} + \frac{4}{3}\mathbf{j} + \lambda\left(\frac{4}{3}\mathbf{i} + \frac{5}{3}\mathbf{j}\right)$

12. $\left(\frac{4}{3}, \frac{1}{3}, 0\right)$

14. $\mathbf{r}\cdot(\mathbf{i} - 2\mathbf{j} + 3\mathbf{k}) - 10 = 0$

15. $(-\pi/2, k, 0), -\infty < k < \infty$.

16. MOVING TRIHEDRON OF A CURVE

Let $\mathbf{r} = \mathbf{r}(s)$ be the equation of a regular curve C with arc length s as parameter. We assume that $\mathbf{r}''(s)$ exists and $|\mathbf{r}''(s)| \neq 0$. We know that $\mathbf{r}'(s)$ equals the unit tangent vector $\mathbf{t}(s)$ at the point $\mathbf{r}(s)$ on the curve C.

$$\therefore \quad \mathbf{t} = \mathbf{r}'$$

We define $$\mathbf{n} = \frac{\mathbf{t}'}{|\mathbf{t}'|}.$$

$\mathbf{n}$ is meaningful, because $|\mathbf{t}'| = |\mathbf{r}''| \neq 0$.

Also, $\mathbf{t}$ is a unit vector.

$$\Rightarrow \quad |\mathbf{t}| = 1 \quad \Rightarrow \quad |\mathbf{t}|^2 = 1 \quad \Rightarrow \quad \mathbf{t} \cdot \mathbf{t} = 1$$

$$\Rightarrow \quad \mathbf{t} \cdot \mathbf{t}' + \mathbf{t}' \cdot \mathbf{t} = 0 \quad \Rightarrow \quad 2\mathbf{t} \cdot \mathbf{t}' = 0 \quad \Rightarrow \quad \mathbf{t} \cdot \mathbf{t}' = 0$$

$\Rightarrow$ $\mathbf{t}'$ is perpendicular to $\mathbf{t}$ $\Rightarrow$ $\mathbf{n}$ is perpendicular to $\mathbf{t}$.

Also, $$|\mathbf{n}| = \left|\frac{\mathbf{t}'}{|\mathbf{t}'|}\right| = \frac{1}{|\mathbf{t}'|}|\mathbf{t}'| = 1$$

$\therefore$ $\mathbf{n}$ is a unit vector and is perpendicular to the unit tangent vector $\mathbf{t}$.

$\therefore$ $\mathbf{n}$ lies in the normal plane at the point under consideration. The vector $\mathbf{n}$ is called the **unit principal normal vector** to the curve C at the point $\mathbf{r}(s)$. In terms of $\mathbf{r}$, we have

$$\mathbf{n} = \frac{\mathbf{r}''}{|\mathbf{r}''|}. \qquad (\because \quad \mathbf{t} = \mathbf{r}')$$

We define $$\mathbf{b} = \mathbf{t} \times \mathbf{n}.$$

$$\therefore \quad |\mathbf{b}| = |\mathbf{t} \times \mathbf{n}| = |\mathbf{t}||\mathbf{n}| \sin\frac{\pi}{2} = 1 \times 1 \times 1 = 1$$

Also by the definition of vector cross product, $\mathbf{b}$ is perpendicular to vectors $\mathbf{t}$ and $\mathbf{n}$ both and the vectors $\mathbf{t}$, $\mathbf{n}$ and $\mathbf{b}$ form a right handed triad.

The vector $\mathbf{b}$ is called the **unit binormal vector** to the curve C at the point $\mathbf{r}(s)$.

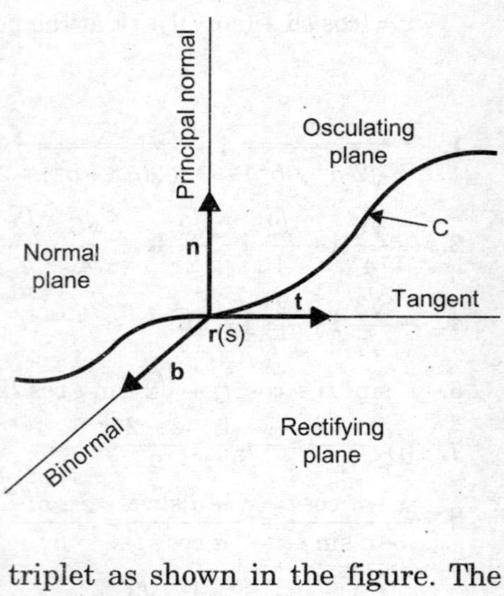

$\therefore$ We have

unit tangent vector, $\mathbf{t} = \mathbf{r}'$

unit principal normal vector $\mathbf{n} = \dfrac{\mathbf{t}'}{|\mathbf{t}'|} = \dfrac{\mathbf{r}''}{|\mathbf{r}''|}$

$$(\because \quad |\mathbf{t}'| = |\mathbf{r}''| \neq 0)$$

unit binormal vector, $\mathbf{b} = \mathbf{t} \times \mathbf{n}$

$$= \mathbf{r}' \times \frac{\mathbf{r}''}{|\mathbf{r}''|} = \frac{\mathbf{r}' \times \mathbf{r}''}{|\mathbf{r}''|}.$$

Thus ($\mathbf{t}$, $\mathbf{n}$, $\mathbf{b}$) forms a right-handed orthonormal triplet as shown in the figure. The triplet ($\mathbf{t}$, $\mathbf{n}$, $\mathbf{b}$) is called the **moving trihedron** of the given curve $\mathbf{r} = \mathbf{r}(s)$.

Remark. Since the unit vectors $\mathbf{t}$, $\mathbf{n}$, $\mathbf{b}$ form a right handed triad, we have

(*i*) $\mathbf{t} \cdot \mathbf{n} = 0 \quad \mathbf{n} \cdot \mathbf{b} = 0 \quad \mathbf{b} \cdot \mathbf{t} = 0$

(*ii*) $\mathbf{t} \times \mathbf{n} = \mathbf{b} \quad \mathbf{n} \times \mathbf{b} = \mathbf{t} \quad \mathbf{b} \times \mathbf{t} = \mathbf{n}$.

The straight lines in the directions of **t**, **n** and **b** are respectively called the **tangent,** the **principal normal** and the **binormal** of the curve C at the point **r**(s). The equations of the tangent at the point **r**(s) is $\mathbf{R} = \mathbf{r} + \lambda\mathbf{t}$, where **R** is a general point on the tangent and λ is a scalar. The equation of the principal normal at the point **r**(s) is $\mathbf{R} = \mathbf{r} + \lambda\mathbf{n}$, where **R** is a general point on the principal normal and λ is a scalar. The equation of the binormal at the point **r**(s) is $\mathbf{R} = \mathbf{r} + \lambda\mathbf{b}$, where **R** is a general point on the binormal and λ is a scalar.

We know that the unit vectors **n** and **b** are both perpendicular to the unit vector **t**.

∴ The normal plane of C at **r** is parallel to the vectors **n** and **b** both at the point **r** and its equation is $(\mathbf{R} - \mathbf{r}) \cdot \mathbf{t} = 0$, where **R** is a general point on the normal plane.

If **R** is the position vector of a general point on the normal plane at the point **r**, then the vectors $\mathbf{R} - \mathbf{r}$, **n** and **b** lie in the normal plane and are thus coplanar vectors.

$$\therefore \qquad [\mathbf{R} - \mathbf{r} \quad \mathbf{n} \quad \mathbf{b}] = 0.$$

This also gives the equation of the normal plane at the point **r.**

The plane through the point **r** and parallel to the vectors **t** and **b** at the point **r** is called the **rectifying plane** of C at the point **r.** The rectifying plane is perpendicular to the vector **n.**

∴ The equation of the rectifying plane at the point **r** is $(\mathbf{R} - \mathbf{r}) \cdot \mathbf{n} = 0$, where **R** is a general point on the rectifying plane.

If **R** is the position vector of a general point on the rectifying plane at the point **r**, then the vectors $\mathbf{R} - \mathbf{r}$, **t** and **b** lie in the rectifying plane and are thus coplanar vectors.

$$\therefore \qquad [\mathbf{R} - \mathbf{r} \quad \mathbf{t} \quad \mathbf{b}] = 0$$

This also gives the equation of the rectifying plane at the point **r.**

The plane through the point **r** and parallel to the vectors **t** and **n** at the point **r** is called the **osculating plane** of C at the point **r**. The osculating plane is perpendicular to the vector **b.**

∴ The equation of the osculating plane at the point **r** is $(\mathbf{R} - \mathbf{r}) \cdot \mathbf{b} = 0$, where **R** is a general point on the osculating plane.

If **R** is the position vector of a general point on the osculating plane at the point **r**, then the vectors $\mathbf{R} - \mathbf{r}$, **t** and **n** lie in the osculating plane and are thus coplanar vectors.

$$\therefore \qquad [\mathbf{R} - \mathbf{r} \quad \mathbf{t} \quad \mathbf{n}] = 0$$

This also gives the equation of the osculating plane at the point **r.**

Thus at each point **r** on the curve C we have the following three characteristic lines and three characteristic planes :

Tangent	$\mathbf{R} = \mathbf{r} + \lambda\mathbf{t}$		
Principal normal	$\mathbf{R} = \mathbf{r} + \lambda\mathbf{n}$		
Binormal	$\mathbf{R} = \mathbf{r} + \lambda\mathbf{b}$		
Normal plane	$(\mathbf{R} - \mathbf{r}) \cdot \mathbf{t} = 0$	or	$[\mathbf{R} - \mathbf{r} \quad \mathbf{n} \quad \mathbf{b}] = 0$
Rectifying plane	$(\mathbf{R} - \mathbf{r}) \cdot \mathbf{n} = 0$	or	$[\mathbf{R} - \mathbf{r} \quad \mathbf{t} \quad \mathbf{b}] = 0$
Osculating plane	$(\mathbf{R} - \mathbf{r}) \cdot \mathbf{b} = 0$	or	$[\mathbf{R} - \mathbf{r} \quad \mathbf{t} \quad \mathbf{n}] = 0.$

17. CARTESIAN EQUATIONS OF CHARACTERISTIC LINES AND PLANES

Let $\mathbf{r} = \mathbf{r}(s)$ be the equation of a regular curve with arc length s as parameter. We assume that $\mathbf{r}''(s)$ exists and $|\mathbf{r}''(s)| \neq 0$. Let $\mathbf{r}(s) = x(s)\mathbf{i} + y(s)\mathbf{j} + z(s)\mathbf{k}$.

$$\therefore \qquad \mathbf{t} = \mathbf{r}' = x'\mathbf{i} + y'\mathbf{j} + z'\mathbf{k}$$

and

$$\mathbf{t}' = x''\mathbf{i} + y''\mathbf{j} + z''\mathbf{k}$$

$$\therefore \quad \mathbf{n} = \frac{\mathbf{t}'}{|\mathbf{t}'|} = \frac{1}{\sqrt{x''^2 + y''^2 + z''^2}}(x''\mathbf{i} + y''\mathbf{j} + z''\mathbf{k})$$

$$= \frac{x''}{\sqrt{x''^2 + y''^2 + z''^2}}\mathbf{i} + \frac{y''}{\sqrt{x''^2 + y''^2 + z''^2}}\mathbf{j} + \frac{z''}{\sqrt{x''^2 + y''^2 + z''^2}}\mathbf{k}$$

$$\mathbf{b} = \mathbf{t} \times \mathbf{n}$$

$$= \begin{vmatrix} \mathbf{i} & \mathbf{j} & \mathbf{k} \\ x' & y' & z' \\ \dfrac{x''}{\sqrt{x''^2 + y''^2 + z''^2}} & \dfrac{y''}{\sqrt{x''^2 + y''^2 + z''^2}} & \dfrac{z''}{\sqrt{x''^2 + y''^2 + z''^2}} \end{vmatrix}$$

$$= \frac{1}{\sqrt{x''^2 + y''^2 + z''^2}} \begin{vmatrix} \mathbf{i} & \mathbf{j} & \mathbf{k} \\ x' & y' & z' \\ x'' & y'' & z'' \end{vmatrix}$$

$$= \frac{1}{\sqrt{x''^2 + y''^2 + z''^2}} ((y'z'' - y''z')\mathbf{i} + (z'x'' - z''x')\mathbf{j} + (x'y'' - x''y')\mathbf{k})$$

$\therefore$ The equation of the tangent $\mathbf{R} = \mathbf{r} + \lambda\mathbf{t}$ at the point $\mathbf{r}$ reduces to

$$\frac{X - x}{x'} = \frac{Y - y}{y'} = \frac{Z - z}{z'}.$$

The equation of the principal normal $\mathbf{R} = \mathbf{r} + \lambda\mathbf{n}$ at the point $\mathbf{r}$ reduces to

$$\frac{X - x}{x''} = \frac{Y - y}{y''} = \frac{Z - z}{z''},$$

because x'', y'', z'' are d.r.'s of the principal normal at the point (x, y, z).

The equation of the binormal $\mathbf{R} = \mathbf{r} + \lambda\mathbf{b}$ at the point $\mathbf{r}$ reduces to

$$\frac{X - x}{y'z'' - y''z'} = \frac{Y - y}{z'x'' - z''x'} = \frac{Z - z}{x'y'' - x''y'},$$

because $y'z'' - y''z'$, $z'x'' - z''x'$, $x'y'' - x''y'$ are d.r.'s of the binormal at the point (x, y, z).

The equation of the normal plane $(\mathbf{R} - \mathbf{r}) \cdot \mathbf{t} = 0$ at the point $\mathbf{r}$ reduces to

$$(X - x)x' + (Y - y)y' + (Z - z)z' = 0.$$

The equation of the rectifying plane $(\mathbf{R} - \mathbf{r}) \cdot \mathbf{n} = 0$ at the point $\mathbf{r}$ reduces to

$$(X - x)x'' + (Y - y)y'' + (Z - z)z'' = 0.$$

The equation of the osculating plane $(\mathbf{R} - \mathbf{r}) \cdot \mathbf{b} = 0$ at the point $\mathbf{r}$ reduces to

$$(X - x)(y'z'' - y''z') + (Y - y)(z'x'' - z''x') + (Z - z)(x'y'' - x''y') = 0.$$

In the above equations, the point $\mathbf{R}$ with coordinates (X, Y, Z) is a general point on the corresponding line (or plane).

18. VALUES OF UNIT VECTORS t, n AND b ALONG A CURVE GIVEN IN TERMS OF AN ARBITRARY PARAMETER

Let $\mathbf{r} = \mathbf{r}(t)$ be the equation of a regular curve C, where t is an arbitrary parameter. We assume that $\ddot{\mathbf{r}}(t)$ exists and $|\ddot{\mathbf{r}}(t)| \neq 0$.

In terms of arc length s, we have

$$\mathbf{t} = \mathbf{r}', \quad \mathbf{n} = \frac{\mathbf{t}'}{|\mathbf{t}'|} \quad \text{and} \quad \mathbf{b} = \mathbf{t} \times \mathbf{n}.$$

(*i*) $$\mathbf{t} = \mathbf{r}' = \frac{d\mathbf{r}}{ds} = \frac{d\mathbf{r}}{dt} \cdot \frac{dt}{ds} = \frac{\dot{\mathbf{r}}}{\frac{ds}{dt}} = \frac{\dot{\mathbf{r}}}{|\dot{\mathbf{r}}|} \qquad \left(s = \int_a^t |\dot{\mathbf{r}}|\, dt \;\Rightarrow\; \frac{ds}{dt} = |\dot{\mathbf{r}}|\right)$$

(*ii*) $$\mathbf{t}' = \frac{d\mathbf{t}}{ds} = \frac{d\mathbf{t}}{dt} \cdot \frac{dt}{ds} = \frac{\dot{\mathbf{t}}}{\frac{ds}{dt}} = \frac{\dot{\mathbf{t}}}{|\dot{\mathbf{r}}|}$$

and $$|\mathbf{t}'| = \left|\frac{\dot{\mathbf{t}}}{|\dot{\mathbf{r}}|}\right| = \frac{|\dot{\mathbf{t}}|}{|\dot{\mathbf{r}}|}$$

$$\therefore \quad \mathbf{n} = \frac{\mathbf{t}'}{|\mathbf{t}'|} = \frac{\dot{\mathbf{t}}/|\dot{\mathbf{r}}|}{|\dot{\mathbf{t}}|/|\dot{\mathbf{r}}|} = \frac{\dot{\mathbf{t}}}{|\dot{\mathbf{t}}|}$$

(*iii*) $$\mathbf{b} = \mathbf{t} \times \mathbf{n} = \frac{\dot{\mathbf{r}}}{|\dot{\mathbf{r}}|} \times \frac{\dot{\mathbf{t}}}{|\dot{\mathbf{t}}|} = \frac{\dot{\mathbf{r}} \times \dot{\mathbf{t}}}{|\dot{\mathbf{r}}||\dot{\mathbf{t}}|}$$

$\therefore$ In terms of arbitrary parameter t, we have

$$\mathbf{t} = \frac{\dot{\mathbf{r}}}{|\dot{\mathbf{r}}|}, \quad \mathbf{n} = \frac{\dot{\mathbf{t}}}{|\dot{\mathbf{t}}|} \quad \text{and} \quad \mathbf{b} = \frac{\dot{\mathbf{r}} \times \dot{\mathbf{t}}}{|\dot{\mathbf{r}}||\dot{\mathbf{t}}|}.$$

WORKING RULES FOR SOLVING PROBLEMS

Let $\mathbf{r} = \mathbf{r}(s)$ be the equation of a regular curve with arc length s as parameter. Let $\mathbf{r}''(s)$ exists and $|\mathbf{r}''(s)| \neq 0$.

Rule I. (*i*) $\mathbf{t} = \mathbf{r}' = x'\mathbf{i} + y'\mathbf{j} + z'\mathbf{k}$

(*ii*) $$\mathbf{n} = \frac{\mathbf{t}'}{|\mathbf{t}'|} = \frac{1}{\sqrt{x''^2 + y''^2 + z''^2}}(x''\,\mathbf{i} + y''\,\mathbf{j} + z''\,\mathbf{k})$$

(*iii*) $$\mathbf{b} = \mathbf{t} \times \mathbf{n} = \frac{1}{\sqrt{x''^2 + y''^2 + z''^2}}((y'z'' - y''z')\mathbf{i} + (z'x'' - z''x')\mathbf{j} + (x'y'' - x''y')\mathbf{k})$$

Rule II. *Equation of tangent:*

(*i*) $\mathbf{R} = \mathbf{r} + \lambda\mathbf{t}$ (*ii*) $$\frac{X - x}{x'} = \frac{Y - y}{y'} = \frac{Z - z}{z'}$$

Rule III. *Equation of principal normal:*

(*i*) $\mathbf{R} = \mathbf{r} + \lambda\mathbf{n}$ (*ii*) $$\frac{X - x}{x''} = \frac{Y - y}{y''} = \frac{Z - z}{z''}$$

Rule IV. *Equation of binormal:*

(*i*) $\mathbf{R} = \mathbf{r} + \lambda\mathbf{b}$ (*ii*) $$\frac{X - x}{y'z'' - y''z'} = \frac{Y - y}{z'x'' - z''x'} = \frac{Z - z}{x'y'' - x''y'}$$

Rule V. *Equation of normal plane:*

(*i*) $(\mathbf{R} - \mathbf{r}) \cdot \mathbf{t} = 0$ (*ii*) $[\mathbf{R} - \mathbf{r} \quad \mathbf{n} \quad \mathbf{b}] = 0$

(*iii*) $(X - x)x' + (Y - y)y' + (Z - z)z' = 0$

Rule VI. *Equation of rectifying plane:*

(*i*) $(\mathbf{R} - \mathbf{r}) \cdot \mathbf{n} = 0$ (*ii*) $[\mathbf{R} - \mathbf{r} \quad \mathbf{t} \quad \mathbf{b}] = 0$

(*iii*) $(X - x)x'' + (Y - y)y'' + (Z - z)z'' = 0$

Rule VII. *Equation of osculating plane:*

(*i*) $(\mathbf{R} - \mathbf{r}) \cdot \mathbf{b} = 0$ (*ii*) $[\mathbf{R} - \mathbf{r} \quad \mathbf{t} \quad \mathbf{n}] = 0$

(*iii*) $(X - x)(y'z'' - y''z') + (Y - y)(z'x'' - z''x') + (Z - z)(x'y'' - x''y') = 0.$

Theorem 1. *Let $r = r(s)$ be the equation of a regular curve with arc length s as parameter. If r'' exists and $|r''| \neq 0$ at a point r, prove that the equation of the osculating plane at the point r is*

$$[R - r \quad r' \quad r''] = 0.$$

Proof. We know that the plane through the point **r** and parallel to the vectors **t** and **n** at the point **r** is the osculating plane at the point **r**.

We have $$\mathbf{r}' = \frac{d\mathbf{r}}{ds} = \mathbf{t}$$

∴ $\mathbf{r}'$ is parallel to the osculating plane.

Also $$\mathbf{r}'' = \frac{d}{ds}(\mathbf{t}) = \mathbf{t}' = |\mathbf{t}'|\,\mathbf{n} \qquad \left(\because \mathbf{n} = \frac{\mathbf{t}'}{|\mathbf{t}'|}\right)$$

∴ $\mathbf{r}''$ is parallel to the osculating plane.

Let **R** be the position vector of a general point on the osculating plane at the point **r**.

∴ The vector $\mathbf{R} - \mathbf{r}$, $\mathbf{r}'$ and $\mathbf{r}''$ lie in the osculating plane and are thus coplanar vectors.

∴ $[\mathbf{R} - \mathbf{r} \quad \mathbf{r}' \quad \mathbf{r}''] = 0.$

This is the equation of the required osculating plane.

Corollary. If $\mathbf{R} = X\mathbf{i} + Y\mathbf{j} + Z\mathbf{k}$ and $\mathbf{r} = x\mathbf{i} + y\mathbf{j} + z\mathbf{k}$, then $\mathbf{r}' = x'\mathbf{i} + y'\mathbf{j} + z'\mathbf{k}$ and $\mathbf{r}'' = x''\mathbf{i} + y''\mathbf{j} + z''\mathbf{k}$.

∴ The equation of the osculating plane is $$\begin{vmatrix} X - x & Y - y & Z - z \\ x' & y' & z' \\ x'' & y'' & z'' \end{vmatrix} = 0.$$

Theorem 2. *Let $\mathbf{r} = \mathbf{r}(t)$ be the equation of a regular curve, where t is an arbitrary parameter. If $\ddot{\mathbf{r}}$ exists and $|\ddot{\mathbf{r}}| \neq 0$ at a point **r**, prove that the equation of the osculating plane at the point **r** is*

$$[\mathbf{R} - \mathbf{r} \quad \dot{\mathbf{r}} \quad \ddot{\mathbf{r}}] = 0.$$

Proof. We know that the plane through the point **r** and parallel to the vectors **t** and **n** at the point **r** is the osculating plane at the point **r**.

We have $$\dot{\mathbf{r}} = \frac{d\mathbf{r}}{dt} = \frac{d\mathbf{r}}{ds} \cdot \frac{ds}{dt} = \mathbf{t}\dot{s} = \dot{s}\mathbf{t}$$

∴ $\dot{\mathbf{r}}$ is parallel to the osculating plane.

Also, $$\ddot{\mathbf{r}} = \frac{d}{dt}(\dot{s}\,\mathbf{t}) = \dot{s}\frac{d\mathbf{t}}{dt} + \frac{d\dot{s}}{dt}\mathbf{t}$$

$$= \ddot{s}\mathbf{t} + \dot{s}\dot{\mathbf{t}} = \ddot{s}\mathbf{t} + \dot{s}|\dot{\mathbf{t}}|\mathbf{n} \qquad \left(\because\ \mathbf{n} = \frac{\dot{\mathbf{t}}}{|\dot{\mathbf{t}}|}\right)$$

$\therefore$ $\ddot{\mathbf{r}}$ lies in the plane of **t** and **n**.

$\therefore$ $\ddot{\mathbf{r}}$ is parallel to the osculating plane.

Let **R** be the position vector of a general point on the osculating plane at the point **r**.

$\therefore$ The vectors $\mathbf{R} - \mathbf{r}$, $\dot{\mathbf{r}}$ and $\ddot{\mathbf{r}}$ lie in the osculating plane and are thus coplanar vectors.

$$\therefore \qquad [\mathbf{R} - \mathbf{r} \quad \dot{\mathbf{r}} \quad \ddot{\mathbf{r}}] = 0.$$

This is the equation of the required osculating plane.

Corollary. If $\mathbf{R} = X\mathbf{i} + Y\mathbf{j} + Z\mathbf{k}$ and $\mathbf{r} = x\mathbf{i} + y\mathbf{j} + z\mathbf{k}$, then $\dot{\mathbf{r}} = \dot{x}\,\mathbf{i} + \dot{y}\,\mathbf{j} + \dot{z}\,\mathbf{k}$

and $$\ddot{\mathbf{r}} = \ddot{x}\,\mathbf{i} + \ddot{y}\,\mathbf{j} + \ddot{z}\,\mathbf{k}.$$

$\therefore$ The equation of the osculating plane is $\begin{vmatrix} X - x & Y - y & Z - z \\ \dot{x} & \dot{y} & \dot{z} \\ \ddot{x} & \ddot{y} & \ddot{z} \end{vmatrix} = 0.$

Example 1. *For the curve $x = 3t$, $y = 3t^2$, $z = 2t^3$, (i) show that any plane meets it in three points and (ii) find the equation of the osculating plane at the point t_1.*

Sol. The given curve is

$$x = 3t,\ y = 3t^2,\ z = 2t^3.$$

(*i*) Let $ax + by + cz + d = 0$ be any plane in space.

Putting $x = 3t$, $y = 3t^2$, $z = 2t^3$, we get $3at + 3bt^2 + 2ct^3 + d = 0$

$$\Rightarrow \qquad 2ct^3 + 3bt^2 + 3at + d = 0$$

This is a cubic equation in t and gives three values of t.

$\therefore$ The plane $ax + by + cz + d = 0$ meets the given curve in three points.

(*ii*) We have $\qquad x = 3t,\ y = 3t^2,\ z = 2t^3$

$$\therefore \qquad \dot{x} = 3,\ \dot{y} = 6t,\ \dot{z} = 6t^2 \ \text{ and } \ \ddot{x} = 0,\ \ddot{y} = 6,\ \ddot{z} = 12t$$

Let (x, y, z) be a general point on the osculating plane at the point t_1.

$\therefore$ The equation of the osculating plane is

$$\begin{vmatrix} x - 3t_1 & y - 3t_1^2 & z - 2t_1^3 \\ 3 & 6t_1 & 6t_1^2 \\ 0 & 6 & 12t_1 \end{vmatrix} = 0$$

$$\begin{vmatrix} x - 3t_1 & y - 3t_1^2 & z - 2t_1^3 \\ 1 & 2t_1 & 2t_1^2 \\ 0 & 1 & 2t_1 \end{vmatrix} = 0$$

$$(x - 3t_1)(2t_1^2) - (y - 3t_1^2)(2t_1) + (z - 2t_1^3)(1) = 0$$

$$\Rightarrow \qquad 2t_1^2 x - 2t_1 y + z = 2t_1^3.$$

Example 2. *Find the equation of the osculating plane to the curve $x = 2 \log t$, $y = 4t$, $z = 2t^2 + 1$ at the point t.*

Sol. The given curve is

$$x = 2 \log t,\ y = 4t,\ z = 2t^2 + 1.$$

$$\therefore \quad \dot{x} = \frac{2}{t},\ \dot{y} = 4,\ \dot{z} = 4t \quad \text{and} \quad \ddot{x} = -\frac{2}{t^2},\ \ddot{y} = 0,\ \ddot{z} = 4$$

Let (X, Y, Z) be a general point on the osculating plane at the point t.

$\therefore$ The equation of the osculating plane is

$$\begin{vmatrix} X - x & Y - y & Z - z \\ \dot{x} & \dot{y} & \dot{z} \\ \ddot{x} & \ddot{y} & \ddot{z} \end{vmatrix} = 0.$$

$$\Rightarrow \quad \begin{vmatrix} X - 2\log t & Y - 4t & Z - (2t^2 + 1) \\ \frac{2}{t} & 4 & 4t \\ -\frac{2}{t^2} & 0 & 4 \end{vmatrix} = 0$$

$$\Rightarrow \quad \begin{vmatrix} X - 2\log t & Y - 4t & Z - (2t^2 + 1) \\ 1 & 2t & 2t^2 \\ 1 & 0 & -2t^2 \end{vmatrix} = 0$$

$$\Rightarrow \quad (X - 2\log t)(-4t^3) - (Y - 4t)(-4t^2) + (Z - 2t^2 - 1)(-2t) = 0$$

$$\Rightarrow \quad 2t^2(X - 2\log t) - 2t(Y - 4t) + Z - 2t^2 - 1 = 0$$

$$\Rightarrow \quad 2t^2 X - 2tY + Z = 4t^2 \log t - 6t^2 + 1.$$

Example 3. *Let* $\mathbf{r} = \mathbf{r}(t)$ *be the equation of a regular curve. By using the equation* $[\mathbf{R} - \mathbf{r}\ \mathbf{r}'\ \mathbf{r}''] = 0$, *show that the equation of the osculating plane at the point* $\mathbf{r}$ *is* $[\mathbf{R} - \mathbf{r}\ \dot{\mathbf{r}}\ \ddot{\mathbf{r}}] = 0$.

Sol. Given equation of the osculating plane at point $\mathbf{r}$ is

$$[\mathbf{R} - \mathbf{r}\ \ \mathbf{r}'\ \ \mathbf{r}''] = 0 \qquad \text{...(1)}$$

$$\mathbf{r}' = \frac{d\mathbf{r}}{ds} = \frac{d\mathbf{r}}{dt} \cdot \frac{dt}{ds} = \frac{\dot{\mathbf{r}}}{\dot{s}}$$

$$\mathbf{r}'' = \frac{d\mathbf{r}'}{ds} = \frac{d}{ds}\left(\frac{\dot{\mathbf{r}}}{\dot{s}}\right) = \frac{d}{dt}\left(\frac{\dot{\mathbf{r}}}{\dot{s}}\right) \cdot \frac{dt}{ds} = \frac{\dot{s}\,\ddot{\mathbf{r}} - \ddot{s}\,\dot{\mathbf{r}}}{\dot{s}^2} \cdot \frac{1}{\dot{s}} = \frac{1}{\dot{s}^2}\ddot{\mathbf{r}} - \frac{\ddot{s}}{\dot{s}^3}\dot{\mathbf{r}}$$

$$\therefore \quad (1) \Rightarrow \quad \left[\mathbf{R} - \mathbf{r} \quad \frac{1}{\dot{s}}\dot{\mathbf{r}} \quad \left(\frac{1}{\dot{s}^2}\ddot{\mathbf{r}} - \frac{\ddot{s}}{\dot{s}^3}\dot{\mathbf{r}}\right)\right] = 0$$

$$\Rightarrow \quad \left[\mathbf{R} - \mathbf{r}\ \frac{1}{\dot{s}}\dot{\mathbf{r}}\ \frac{1}{\dot{s}^2}\ddot{\mathbf{r}}\right] - \left[\mathbf{R} - \mathbf{r}\ \frac{1}{\dot{s}}\dot{\mathbf{r}}\ \frac{\ddot{s}}{\dot{s}^3}\dot{\mathbf{r}}\right] = 0$$

$$\Rightarrow \quad \frac{1}{\dot{s}^3}[\mathbf{R} - \mathbf{r}\ \dot{\mathbf{r}}\ \ddot{\mathbf{r}}] - \left(-\frac{\ddot{s}}{\dot{s}^4}\right)[\mathbf{R} - \mathbf{r}\ \dot{\mathbf{r}}\ \dot{\mathbf{r}}] = 0$$

($\because$ Determinant with two equal rows is zero)

$$\Rightarrow \qquad \frac{1}{\dot{s}^3}[\mathbf{R}-\mathbf{r}\ \ \dot{\mathbf{r}}\ \ \ddot{\mathbf{r}}] + \frac{\ddot{s}}{\dot{s}^4}.0 = 0$$

$$\Rightarrow \qquad \frac{1}{\dot{s}^3}[\mathbf{R}-\mathbf{r}\ \ \dot{\mathbf{r}}\ \ \ddot{\mathbf{r}}] = 0$$

$$\Rightarrow \qquad [\mathbf{R}-\mathbf{r}\ \ \dot{\mathbf{r}}\ \ \ddot{\mathbf{r}}] = 0.$$

∴ The result holds.

Example 4. *For the curve $x = 4a\cos^3 t$, $y = 4a\sin^3 t$, $z = 3c\cos 2t$, find*

(i) the equation of the principal normal at the point t.

(ii) the equation of the osculating plane at the point t.

Sol. (*i*) Let **r** be the position vector of the point (x, y, z) on the given curve.

$$\therefore \quad \mathbf{r} = x\mathbf{i} + y\mathbf{j} + z\mathbf{k} = 4a\cos^3 t\,\mathbf{i} + 4a\sin^3 t\,\mathbf{j} + 3c\cos 2t\,\mathbf{k}$$

$$\therefore \quad \mathbf{r}' = \frac{d\mathbf{r}}{dt}\frac{dt}{ds} = (-12a\cos^2 t\sin t\,\mathbf{i} + 12a\sin^2 t\cos t\,\mathbf{j} - 6c\sin 2t\,\mathbf{k})\frac{dt}{ds}$$

$$= 12\sin t\cos t(-a\cos t\,\mathbf{i} + a\sin t\,\mathbf{j} - c\mathbf{k})\frac{dt}{ds} \qquad \text{(Using } \sin 2t = 2\sin t\cos t)$$

$$\therefore \quad |\mathbf{r}'| = 12\sin t\cos t\sqrt{a^2\cos^2 t + a^2\sin^2 t + c^2}\,\frac{dt}{ds}$$

$$\Rightarrow \quad 1 = 12\sin t\cos t\sqrt{a^2+c^2}\,\frac{dt}{ds} \qquad (\because\ |\mathbf{t}| = |\mathbf{r}'| = 1)$$

$$\therefore \quad \frac{ds}{dt} = 12\sin t\cos t\sqrt{a^2+c^2}$$

$$\therefore \quad \mathbf{r}' = 12\sin t\cos t(-a\cos t\,\mathbf{i} + a\sin t\,\mathbf{j} - c\mathbf{k}).\frac{1}{12\sin t\cos t\sqrt{a^2+c^2}}$$

$$= \frac{1}{\sqrt{a^2+c^2}}(-a\cos t\,\mathbf{i} + a\sin t\,\mathbf{j} - c\mathbf{k})$$

$$\therefore \quad \mathbf{r}'' = \frac{1}{\sqrt{a^2+c^2}}(a\sin t\,\mathbf{i} + a\cos t\,\mathbf{j})\frac{dt}{ds}$$

$$= \frac{a}{\sqrt{a^2+c^2}}(\sin t\,\mathbf{i} + \cos t\,\mathbf{j})\frac{1}{12\sin t\cos t\sqrt{a^2+c^2}}$$

$$= \frac{a}{12(a^2+c^2)}(\sec t\,\mathbf{i} + \operatorname{cosec} t\,\mathbf{j})$$

$$\therefore \quad \mathbf{n} = \frac{\mathbf{t}'}{|\mathbf{t}'|} = \frac{\mathbf{r}''}{|\mathbf{r}''|} = \frac{a}{12(a^2+c^2)|\mathbf{r}''|}(\sec t\,\mathbf{i} + \operatorname{cosec} t\,\mathbf{j})$$

∴ D.r.'s of principal normal are $\sec t$, $\operatorname{cosec} t$, 0.

∴ The equations of the principal normal at the point t are

$$\frac{x - 4a\cos^3 t}{\sec t} = \frac{y - 4a\sin^3 t}{\operatorname{cosec} t} = \frac{z - 3c\cos 2t}{0}.$$

(*ii*) The equation of the osculating plane at the point t is

$$\begin{vmatrix} x - 4a\cos^3 t & y - 4a\sin^3 t & z - 3c\cos 2t \\ -\dfrac{a}{\sqrt{a^2+c^2}}\cos t & \dfrac{a}{\sqrt{a^2+c^2}}\sin t & \dfrac{-c}{\sqrt{a^2+c^2}} \\ \dfrac{a}{12(a^2+c^2)}\sec t & \dfrac{a}{12(a^2+c^2)}\operatorname{cosec} t & 0 \end{vmatrix} = 0$$

$$\Rightarrow \quad \begin{vmatrix} x - 4a\cos^3 t & y - 4a\sin^3 t & z - 3c\cos 2t \\ -a\cos t & a\sin t & -c \\ \sec t & \operatorname{cosec} t & 0 \end{vmatrix} = 0$$

$$(x - 4a\cos^3 t)(c\operatorname{cosec} t) - (y\ 4a\sin^3 t)(c\sec t)$$
$$+ (z - 3c\cos 2t)(-a\sin t\cos t - a\sin t\cos t) = 0$$

$$\Rightarrow \quad c\operatorname{cosec} t \cdot x - c\sec t \cdot y - 2\sin t\cos t\, z$$
$$= 4a\sin t\cos^3 t - 4a\sin^3 t\cos t - 6ac\sin t\cos t\cos 2t$$

$$\Rightarrow \quad c\operatorname{cosec} t \cdot x - c\sec t \cdot y - 2\sin t\cos t\, z$$
$$= 2a\sin 2t\cos^2 t - 2a\sin 2t\sin^2 t - 3ac\sin 2t\cos 2t.$$

Example 5. *Find the vectors* **t**, **n** *and* **b** *along the curve* $\mathbf{r} = (3t - t^3)\mathbf{i} + 3t^2\mathbf{j} + (3t + t^3)\mathbf{k}$.

Sol. We have $\mathbf{r} = (3t - t^3)\mathbf{i} + 3t^2\mathbf{j} + (3t + t^3)\mathbf{k}$.

$$\therefore \quad \dot{\mathbf{r}} = (3 - 3t^2)\mathbf{i} + 6t\mathbf{j} + (3 + 3t^2)\mathbf{k}$$

or $$\dot{\mathbf{r}} = 3[(1 - t^2)\mathbf{i} + 2t\mathbf{j} + (1 + t^2)\mathbf{k}]$$

$$|\dot{\mathbf{r}}| = 3\sqrt{(1-t^2)^2 + 4t^2 + (1+t^2)^2}$$
$$= 3\sqrt{2 + 2t^4 + 4t^2} = 3\sqrt{2}\,(1 + t^2)$$

$$\therefore \quad \mathbf{t} = \frac{\dot{\mathbf{r}}}{|\dot{\mathbf{r}}|} = \frac{1}{3\sqrt{2}(1+t^2)}\,.\,3[(1 - t^2)\mathbf{i} + 2t\mathbf{j} + (1 + t^2)\mathbf{k}]$$

$$= \frac{1-t^2}{\sqrt{2}(1+t^2)}\mathbf{i} + \frac{\sqrt{2}\,t}{1+t^2}\mathbf{j} + \frac{1}{\sqrt{2}}\mathbf{k}$$

$$\dot{\mathbf{t}} = \frac{(1+t^2)(-2t) - (1-t^2)2t}{\sqrt{2}(1+t^2)^2}\mathbf{i} + \frac{\sqrt{2}((1+t^2)\cdot 1 - t\,.\,2t)}{(1+t^2)^2}\mathbf{j} + 0\mathbf{k}$$

$$= \frac{\sqrt{2}}{(1+t^2)^2}\,(-2t\,\mathbf{i} + (1 - t^2)\,\mathbf{j})$$

$$\therefore \quad |\dot{\mathbf{t}}| = \frac{\sqrt{2}}{(1+t^2)^2}\,[4t^2 + (1 - t^2)^2]^{1/2} = \frac{\sqrt{2}}{1+t^2}$$

$$\therefore \quad \mathbf{n} = \frac{\dot{\mathbf{t}}}{|\dot{\mathbf{t}}|} = \frac{\sqrt{2}}{(1+t^2)^2}\,(-2t\mathbf{i} + (1 - t^2)\mathbf{j})\,.\,\frac{1+t^2}{\sqrt{2}} = -\frac{2t}{1+t^2}\mathbf{i} + \frac{1-t^2}{1+t^2}\mathbf{j}$$

$$\mathbf{b} = \mathbf{t} \times \mathbf{n} = \begin{vmatrix} \mathbf{i} & \mathbf{j} & \mathbf{k} \\ \frac{1-t^2}{\sqrt{2}(1+t^2)} & \frac{\sqrt{2}\,t}{1+t^2} & \frac{1}{\sqrt{2}} \\ -\frac{2t}{1+t^2} & \frac{1-t^2}{1+t^2} & 0 \end{vmatrix}$$

$$= \frac{1}{\sqrt{2}(1+t^2)(1+t^2)} \begin{vmatrix} \mathbf{i} & \mathbf{j} & \mathbf{k} \\ 1-t^2 & 2t & 1+t^2 \\ -2t & 1-t^2 & 0 \end{vmatrix}$$

$$= \frac{1}{\sqrt{2}(1+t^2)^2} [-(1-t^4)\mathbf{i} - 2t(1+t^2)\mathbf{j} + (1+t^2)^2\mathbf{k}]$$

$$= \frac{1}{\sqrt{2}(1+t^2)} [(t^2-1)\mathbf{i} - 2t\mathbf{j} + (1+t^2)\mathbf{k}].$$

Example 6. *Show that the points on the helix $\mathbf{r} = a\cos t\mathbf{i} + a\sin t\mathbf{j} + bt\mathbf{k}$, $a > 0$, $b \neq 0$ at which the osculating planes pass through a fixed point are all coplanar.*

Sol. The given helix is

$$\mathbf{r} = a\cos t\mathbf{i} + a\sin t\mathbf{j} + bt\mathbf{k}.$$

$\therefore$
$$\dot{\mathbf{r}} = -a\sin t\mathbf{i} + a\cos t\mathbf{j} + b\mathbf{k}$$
$$\ddot{\mathbf{r}} = -a\cos t\mathbf{i} - a\sin t\mathbf{j} + 0\mathbf{k}$$

The equation of the osculating plane is $[\mathbf{R} - \mathbf{r}\ \ \dot{\mathbf{r}}\ \ \ddot{\mathbf{r}}] = 0$.

Let $\mathbf{R} = X\mathbf{i} + Y\mathbf{j} + Z\mathbf{k}$

$\therefore$ The equation of the osculating plane is

$$\begin{vmatrix} X - a\cos t & Y - a\sin t & Z - bt \\ -a\sin t & a\cos t & b \\ -a\cos t & -a\sin t & 0 \end{vmatrix} = 0$$

$\Rightarrow$ $ab\sin t(X - a\cos t) - ab\cos t(Y - a\sin t) + a^2(Z - bt) = 0$

$\Rightarrow$ $b\sin t\, X - b\cos t\, Y + aZ = abt$

Let the osculating plane at the point $\mathbf{r}$ passes through the fixed point (α, β, γ).

$\therefore$ $(b\sin t)\alpha - (b\cos t)\beta + a\gamma = abt$

$\Rightarrow$ $-b\beta(a\cos t) + b\alpha(a\sin t) - a^2(bt) = -a^2\gamma$

$\Rightarrow$ $b\beta(a\cos t) - b\alpha(a\sin t) + a^2(bt) = a^2\gamma$

$\therefore$ The locus of the point $\mathbf{r}$ (= $(a\cos t, a\sin t, bt)$) is

$b\beta x - b\alpha y + a^2 z = a^2\gamma$, which is a plane.

$\therefore$ The result holds.

Example 7. *Find the equations of characteristic lines and planes to the helix $\mathbf{r} = \cos t\,\mathbf{i} + \sin t\,\mathbf{j} + t\mathbf{k}$ at the point where $t = \pi/2$.*

Sol. We have $\mathbf{r} = \cos t\mathbf{i} + \sin t\mathbf{j} + t\mathbf{k}$.

$\therefore$
$$\mathbf{r}\left(\frac{\pi}{2}\right) = \cos\frac{\pi}{2}\mathbf{i} + \sin\frac{\pi}{2}\mathbf{j} + \frac{\pi}{2}\mathbf{k} = \mathbf{j} + \frac{\pi}{2}\mathbf{k}$$

$\therefore$ The point under consideration is $(0, 1, \pi/2)$.

$$\dot{\mathbf{r}} = -\sin t\mathbf{i} + \cos t\mathbf{j} + \mathbf{k}$$

$\therefore \qquad |\dot{\mathbf{r}}| = \sqrt{\sin^2 t + \cos^2 t + 1} = \sqrt{2}$

$\therefore \qquad \mathbf{t} = \dfrac{\dot{\mathbf{r}}}{|\dot{\mathbf{r}}|} = \dfrac{1}{\sqrt{2}}(-\sin t\mathbf{i} + \cos t\mathbf{j} + \mathbf{k})$

$$\dot{\mathbf{t}} = \frac{1}{\sqrt{2}}(-\cos t\mathbf{i} - \sin t\mathbf{j} + 0\mathbf{k})$$

and $\qquad |\dot{\mathbf{t}}| = \dfrac{1}{\sqrt{2}}\sqrt{\cos^2 t + \sin^2 t} = \dfrac{1}{\sqrt{2}}$

$\therefore \qquad \mathbf{n} = \dfrac{\dot{\mathbf{t}}}{|\dot{\mathbf{t}}|} = \dfrac{1}{\sqrt{2}}(-\cos t\mathbf{i} - \sin t\mathbf{j}) \,.\, \sqrt{2} = -\cos t\mathbf{i} - \sin t\mathbf{j}$

$$\mathbf{b} = \mathbf{t} \times \mathbf{n} = \begin{vmatrix} \mathbf{i} & \mathbf{j} & \mathbf{k} \\ -\dfrac{\sin t}{\sqrt{2}} & \dfrac{\cos t}{\sqrt{2}} & \dfrac{1}{\sqrt{2}} \\ -\cos t & -\sin t & 0 \end{vmatrix} = \frac{\sin t}{\sqrt{2}}\mathbf{i} - \frac{\cos t}{\sqrt{2}}\mathbf{j} + \frac{1}{\sqrt{2}}\mathbf{k}$$

$\therefore \qquad \mathbf{t}(\pi/2) = \dfrac{1}{\sqrt{2}}(-1 \cdot \mathbf{i} + 0 \cdot \mathbf{j} + \mathbf{k}) = -\dfrac{1}{\sqrt{2}}\mathbf{i} + \dfrac{1}{\sqrt{2}}\mathbf{k}$

$$\mathbf{n}(\pi/2) = -0 \cdot \mathbf{i} - 1 \cdot \mathbf{j} = -\mathbf{j}$$

$$\mathbf{b}(\pi/2) = \frac{1}{\sqrt{2}}\mathbf{i} - 0 \cdot \mathbf{j} + \frac{1}{\sqrt{2}}\mathbf{k} = \frac{1}{\sqrt{2}}\mathbf{i} + \frac{1}{\sqrt{2}}\mathbf{k}.$$

Tangent. The equation of the tangent is

$$\mathbf{r} = \mathbf{r}(\pi/2) + \lambda\mathbf{t}(\pi/2) \quad \text{i.e.,} \quad \mathbf{r} = \mathbf{j} + \frac{\pi}{2}\mathbf{k} + \lambda\left(-\frac{1}{\sqrt{2}}\mathbf{i} + \frac{1}{\sqrt{2}}\mathbf{k}\right).$$

Equations in cartesian form are $\dfrac{x-0}{-1} = \dfrac{y-1}{0} = \dfrac{z-\pi/2}{1}$.

Principal normal. The equation of the principal normal is

$$\mathbf{r} = \mathbf{r}(\pi/2) + \lambda\mathbf{n}(\pi/2) \quad \text{i.e.,} \quad \mathbf{r} = \mathbf{j} + \frac{\pi}{2}\mathbf{k} + \lambda(-\mathbf{j}).$$

Equations in the cartesian form are $\dfrac{x-0}{0} = \dfrac{y-1}{-1} = \dfrac{z-\pi/2}{0}$.

Binormal. The equation of the binormal is

$$\mathbf{r} = \mathbf{r}(\pi/2) + \lambda\mathbf{b}(\pi/2) \quad \text{i.e.,} \quad \mathbf{r} = \mathbf{j} + \frac{\pi}{2}\mathbf{k} + \lambda\left(\frac{1}{\sqrt{2}}\mathbf{i} + \frac{1}{\sqrt{2}}\mathbf{k}\right).$$

Equations in the cartesian form are $\dfrac{x-0}{1} = \dfrac{y-1}{0} = \dfrac{z-\pi/2}{1}$.

Normal plane. The equation of the normal plane is $(\mathbf{r} - \mathbf{r}(\pi/2)) \cdot \mathbf{t}(\pi/2) = 0$

i.e., $\qquad \left(\mathbf{r} - \left(\mathbf{j} + \dfrac{\pi}{2}\mathbf{k}\right)\right) . \left(-\dfrac{1}{\sqrt{2}}\mathbf{i} + \dfrac{1}{\sqrt{2}}\mathbf{k}\right) = 0 \quad \text{or} \quad \left(\mathbf{r} - \left(\mathbf{j} + \dfrac{\pi}{2}\mathbf{k}\right)\right) \cdot (-\mathbf{i} + \mathbf{k}) = 0$

or $\qquad \mathbf{r} \cdot (-\mathbf{i} + \mathbf{k}) - \left(0\,(-1) + 1(0) + \dfrac{\pi}{2}(1)\right) = 0 \quad \text{or} \quad \mathbf{r} \cdot (-\mathbf{i} + \mathbf{k}) = \dfrac{\pi}{2}.$

Equation in the cartesian form is $-x+z=\frac{\pi}{2}$ *i.e.*, $x-z+\frac{\pi}{2}=0$.

Rectifying plane. The equation of the rectifying plane is $(\mathbf{r}-\mathbf{r}(\pi/2))\,.\,\mathbf{n}(\pi/2)=0$

i.e., $$\left(\mathbf{r}-\left(\mathbf{j}+\frac{\pi}{2}\mathbf{k}\right)\right)\cdot(-\mathbf{j})=0 \quad\text{or}\quad -\mathbf{r}\cdot\mathbf{j}+1=0 \quad\text{or}\quad \mathbf{r}\cdot\mathbf{j}-1=0.$$

Equation in the cartesian form is $y-1=0$.

Osculating plane. The equation of the osculating plane is

$$(\mathbf{r}-\mathbf{r}(\pi/2))\cdot\mathbf{b}(\pi/2)=0 \quad i.e.,\ \left(\mathbf{r}-\left(\mathbf{j}+\frac{\pi}{2}\mathbf{k}\right)\right).\left(\frac{1}{\sqrt{2}}\mathbf{i}+\frac{1}{\sqrt{2}}\mathbf{k}\right)=0$$

or $$\left(\mathbf{r}-\left(\mathbf{j}+\frac{\pi}{2}\mathbf{k}\right)\right)\cdot(\mathbf{i}+\mathbf{k})=0 \quad\text{or}\quad \mathbf{r}\cdot(\mathbf{i}+\mathbf{k})=\frac{\pi}{2}.$$

Equation in the cartesian form is $x+z=\frac{\pi}{2}$.

Example 8. *For the curve* $\mathbf{r}=(e^{-t}\sin t, e^{-t}\cos t, e^{-t})$, *find the following at the point t:*

(*i*) *the unit tangent vector* **t**

(*ii*) *the equation of the tangent*

(*iii*) *the unit principal normal vector* **n**

(*iv*) *the equation of the normal plane*

(*v*) *the unit binormal vector* **b**

(*vi*) *the equation of the binormal.*

Sol. (*i*) The given curve is

$$\mathbf{r}=e^{-t}\sin t\mathbf{i}+e^{-t}\cos t\mathbf{j}+e^{-t}\mathbf{k}.$$

$$\therefore \quad \dot{\mathbf{r}}=(e^{-t}\cos t-e^{-t}\sin t)\mathbf{i}+(-e^{-t}\sin t-e^{-t}\cos t)\mathbf{j}-e^{-t}\mathbf{k}$$

$$\therefore \quad =e^{-t}\,[(\cos t-\sin t)\mathbf{i}-(\sin t+\cos t)\mathbf{j}-\mathbf{k}]$$

$$\therefore \quad |\,\dot{\mathbf{r}}\,|=e^{-t}\sqrt{(\cos t-\sin t)^2+(\sin t+\cos t)^2+1}=\sqrt{3}\,e^{-t}$$

$$\therefore \quad \mathbf{t}=\frac{\dot{\mathbf{r}}}{|\dot{\mathbf{r}}|}=\frac{1}{\sqrt{3}e^{-t}}\cdot e^{-t}\,[(\cos t-\sin t)\mathbf{i}-(\sin t+\cos t)\mathbf{j}-\mathbf{k}]$$

$$\therefore \quad \mathbf{t}=\frac{1}{\sqrt{3}}\,[(\cos t-\sin t)\mathbf{i}-(\sin t+\cos t)\mathbf{j}-\mathbf{k}]$$

(*ii*) Using **t**, the d.r.'s of the tangent are $\cos t-\sin t$, $-(\sin t+\cos t)$, -1.

$\therefore$ The equations of the tangent at the point t are

$$\frac{x-e^{-t}\sin t}{\cos t-\sin t}=\frac{y-e^{-t}\cos t}{-(\sin t+\cos t)}=\frac{z-e^{-t}}{-1}.$$

(*iii*) $$\dot{\mathbf{t}}=\frac{1}{\sqrt{3}}[(-\sin t-\cos t)\mathbf{i}-(\cos t-\sin t)\mathbf{j}+0\mathbf{k}]$$

$$=\frac{1}{\sqrt{3}}\,[-(\sin t+\cos t)\mathbf{i}+(\sin t-\cos t)\mathbf{j}]$$

$$\therefore \qquad |\dot{\mathbf{t}}| = \frac{1}{\sqrt{3}}\sqrt{(\sin t + \cos t)^2 + (\sin t - \cos t)^2} = \frac{\sqrt{2}}{\sqrt{3}}$$

$$\therefore \qquad \mathbf{n} = \frac{\dot{\mathbf{t}}}{|\dot{\mathbf{t}}|} = \frac{1}{\sqrt{3}}\,[-(\sin t + \cos t)\mathbf{i} + (\sin t - \cos t)\mathbf{j}]\,.\,\frac{\sqrt{3}}{\sqrt{2}}$$

$$\therefore \qquad \mathbf{n} = \frac{1}{\sqrt{2}}\,[-(\sin t + \cos t)\mathbf{i} + (\sin t - \cos t)\mathbf{j}].$$

(*iv*) The equation of the normal plane is $(\mathbf{R} - \mathbf{r}) \cdot \mathbf{t} = 0$.

$$\Rightarrow \quad [(x\mathbf{i} + y\mathbf{j} + z\mathbf{k}) - (e^{-t}\sin t\,\mathbf{i} + e^{-t}\cos t\,\mathbf{j} + e^{-t}\mathbf{k})] \cdot \left[\frac{1}{\sqrt{3}}((\cos t - \sin t)\mathbf{i} - (\sin t + \cos t)\mathbf{j} - \mathbf{k})\right] = 0$$

$$\Rightarrow \quad x(\cos t - \sin t) - y(\sin t + \cos t) - z + e^{-t}(-\sin t\cos t + \sin^2 t + \sin t\cos t + \cos^2 t + 1) = 0$$

$$\Rightarrow \qquad (\cos t - \sin t)x - (\sin t + \cos t)y - z = -2e^{-t}$$

$$\Rightarrow \qquad (\sin t - \cos t)x + (\sin t + \cos t)y + z = 2e^{-t}.$$

(*v*) $\mathbf{b} = \mathbf{t} \times \mathbf{n}$

$$= \begin{vmatrix} \mathbf{i} & \mathbf{j} & \mathbf{k} \\ \frac{1}{\sqrt{3}}(\cos t - \sin t) & -\frac{1}{\sqrt{3}}(\sin t + \cos t) & -\frac{1}{\sqrt{3}} \\ -\frac{1}{\sqrt{2}}(\sin t + \cos t) & \frac{1}{\sqrt{2}}(\sin t - \cos t) & 0 \end{vmatrix}$$

$$= \frac{1}{\sqrt{6}}\begin{vmatrix} \mathbf{i} & \mathbf{j} & \mathbf{k} \\ \sin t - \cos t & \sin t + \cos t & 1 \\ \sin t + \cos t & \cos t - \sin t & 0 \end{vmatrix}$$

$$= \frac{1}{\sqrt{6}}\,[(\sin t - \cos t)\,\mathbf{i} + (\sin t + \cos t)\,\mathbf{j} - 2\mathbf{k}].$$

(*vi*) Using **b**, the d.r.'s of the binormal are $\sin t - \cos t, \sin t + \cos t, -2$.

$\therefore$ The equation of the binormal at the point t are

$$\frac{x - e^{-t}\sin t}{\sin t - \cos t} = \frac{y - e^{-t}\cos t}{\sin t + \cos t} = \frac{z - e^{-t}}{-2}.$$

Example 9. *Find the equation of the osculating plane at a general point on the curve* $\mathbf{r} = (t, t^2, t^3)$*. Show that the osculating planes at three points on this curve meet at a point lying in the plane determined by these three points.*

Sol. The given curve is $\mathbf{r} = (t, t^2, t^3)$. Parametric equations of the curve are

$$x = t, y = t^2, z = t^3.$$

$$\therefore \qquad \dot{x} = 1,\ \dot{y} = 2t,\ \dot{z} = 3t^2,\ \ddot{x} = 0,\ \ddot{y} = 2,\ \ddot{z} = 6t$$

$\therefore$ Equation of the osculating plane at point 't' is

$$\begin{vmatrix} X - t & Y - t^2 & Z - t^3 \\ 1 & 2t & 3t^2 \\ 0 & 2 & 6t \end{vmatrix} = 0$$

$$\Rightarrow \quad 6t^2(X - t) - 6t(Y - t^2) + 2(Z - t^3) = 0$$

$$\Rightarrow \quad 3t^2X - 3tY + Z - t^3 = 0. \quad ...(1)$$

Let t_1, t_2, t_3 be the values of the parameter t at any three points on the curve.

$\therefore$ The equations of the osculating planes at these points are

$$3t_1^2X - 3t_1Y + Z - t_1^3 = 0$$

$$3t_2^2X - 3t_2Y + Z - t_2^3 = 0$$

and

$$3t_3^2X - 3t_3Y + Z - t_3^3 = 0$$

Let these planes intersect at the point (α, β, γ).

$$\therefore \quad 3t_1^2\alpha - 3t_1\beta + \gamma - t_1^3 = 0$$

$$3t_2^2\alpha - 3t_2\beta + \gamma - t_2^3 = 0$$

and

$$3t_3^2\alpha - 3t_3\beta + \gamma - t_3^3 = 0$$

$\therefore$ t_1, t_2, t_3 are the roots of the cubic equation

$$3t^2 \alpha - 3t\beta + \gamma - t^3 = 0$$

or

$$t^3 - 3\alpha t^2 + 3\beta t - \gamma = 0 \quad ...(2)$$

Let the equation of the plane passing through the points t_1, t_2 and t_3 be

$$ax + by + cz + d = 0 \quad ...(3)$$

$$\Rightarrow \quad at + bt^2 + ct^3 + d = 0$$

$$\Rightarrow \quad ct^3 + bt^2 + at + d = 0,$$

where $t = t_1, t_2, t_3$ ...(4)

$\therefore$ Equations (2) and (4) are same.

$$\therefore \quad \frac{1}{c} = \frac{-3\alpha}{b} = \frac{3\beta}{a} = \frac{-\gamma}{d}$$

$$\therefore \quad a = 3\beta c, \; b = -3\alpha c, \; d = -\gamma c$$

$$\therefore \; (3) \Rightarrow \quad 3\beta cx - 3\alpha cy + cz - \gamma c = 0$$

$$\Rightarrow \quad 3\beta x - 3\alpha y + z - \gamma = 0$$

(α, β, γ) lies on this plane if

$$3\beta\alpha - 3\alpha\beta + \gamma - \gamma = 0 \quad \text{if } 0 = 0, \text{ which is true.}$$

$\therefore$ The result holds.

Example 10. *Show that there are three points on the curve $x = at^3 + b$, $y = 3ct^2 + 3dt$, $z = 3et + f$ such that their osculating planes pass through the origin and that the three points lie on the plane $3cex + afy = 0$.*

Sol. The given curve is

$$x = at^3 + b, \; y = 3ct^2 + 3dt, \; z = 3et + f.$$

$$\therefore \quad \dot{x} = 3at^2, \; \dot{y} = 6ct + 3d, \; \dot{z} = 3e, \; \ddot{x} = 6at, \; \ddot{y} = 6c, \; \ddot{z} = 0$$

$\therefore$ Equation of the osculating plane at point 't' is

$$\begin{vmatrix} X - (at^3 + b) & Y - (3ct^2 + 3dt) & Z - (3et + f) \\ 3at^2 & 6ct + 3d & 3e \\ 6at & 6c & 0 \end{vmatrix} = 0$$

$$\Rightarrow \quad -18ce(X - at^3 - b) + 18aet(Y - 3ct^2 - 3dt) + (18act^2 - 36act^2 - 18adt)(Z - 3et - f) = 0$$

$$\Rightarrow \quad ce(X - at^3 - b) - aet(Y - 3ct^2 - 3dt) + (act^2 + adt)(Z - 3et - f) = 0$$

Let this plane pass through the origin.

$\therefore \quad ce(0 - at^3 - b) - aet(0 - 3ct^2 - 3dt) + (act^2 + adt)(0 - 3et - f) = 0$

$\Rightarrow \quad - acet^3 - bce + 3acet^3 + 3adet^2 - 3acet^3 - acft^2 - 3adet^2 - adft = 0$

$\Rightarrow \quad acet^3 + acft^2 + adft + bce = 0 \quad ...(1)$

This is a cubic in t. Let the roots of this equation be t_1, t_2 and t_3.

$\therefore$ There are three points on the given curve whose osculating planes passes through the origin. Let $x_1 = at_1^3 + b, \quad y_1 = 3ct_1^2 + 3dt_1, z_1 = 3et_1 + f$

$$\therefore \quad t_1^3 = \frac{x_1 - b}{a}, t_1 = \frac{z_1 - f}{3e}$$

$$t_1^2 = \frac{y_1 - 3dt_1}{3c} = \frac{1}{3c}\left(y_1 - 3d\left(\frac{z_1 - f}{3e}\right)\right) = \frac{ey_1 - dz_1 + df}{3ce}$$

$$(1) \Rightarrow \quad acet_1^3 + acft_1^2 + adft_1 + bce = 0$$

$$\Rightarrow \quad ace\left(\frac{x_1 - b}{a}\right) + acf\left(\frac{ey_1 - dz_1 + df}{3ce}\right) + adf\left(\frac{z_1 - f}{3e}\right) + bce = 0$$

$$\Rightarrow \quad ce(x_1 - b) + \frac{af}{3e}(ey_1 - dz_1 + df) + \frac{adf}{3e}(z_1 - f) + bce = 0$$

$$\Rightarrow \quad 3ce^2(x_1 - b) + af(ey_1 - dz_1 + df) + adf(z_1 - f) + 3bce^2 = 0$$

$$\Rightarrow \quad 3ce^2x_1 - 3bce^2 + aefy_1 - adf\, z_1 + adf^2 + adf\, z_1 - adf^2 + 3bce^2 = 0$$

$$\Rightarrow \quad 3ce^2x_1 + aefy_1 = 0$$

$$\Rightarrow \quad 3cex_1 + afy_1 = 0$$

$\therefore \quad (x_1, y_1, z_1)$ lies in the plane $3cex + afy = 0$.

Similarly, the points corresponding to t_2 and t_3 also lie on the plane $3cex + afy = 0$.

EXERCISE 1.5

1. Find the intersection of the xy-plane and the tangent lines to the helix $\mathbf{r} = \cos t\mathbf{i} + \sin t\mathbf{j} + t\mathbf{k}$, $(t > 0)$.
2. Find the equation of the osculating plane at any point on the curve $\mathbf{r} = (t, t^2, t^3)$.
3. Find the equation of the osculating plane to the curve $\mathbf{r} = t\mathbf{i} + t^2\mathbf{j} + t^3\mathbf{k}$ at the point for which $t = 1$.
4. Find the equation of the osculating plane to the curve $x = 3t, y = 3t^2, z = 2t^3$ at the points (3, 3, 2), (– 3, 3, – 2) and (6, 12, 16).
5. Find the equation of the osculating plane at the point 't' on the helix $x = a \cos t, y = a \sin t, z = ct$.
6. Show that the osculating plane at the point $t = 1$ of the curve $\mathbf{r} = (3at, 3bt^2, ct^3)$ is $\frac{x}{a} - \frac{y}{b} + \frac{z}{c} = 1$.
7. Find the equation of the osculating plane at the point t of the curve
$$x = a \cosh t, y = a \sinh t, z = bt.$$
8. Find the equation of the osculating plane at the point t of the curve
$$\mathbf{r} = 4a \cos^3 t\mathbf{i} + 4a \sin^3 t\mathbf{j} + 2a \cos 2t\mathbf{k}.$$
9. Find the osculating plane at the point t of the curve $x = a \cos 2t, y = a \sin 2t, z = 2a \sin t$.
10. Find the basic unit vectors $\mathbf{t}$, $\mathbf{n}$ and $\mathbf{b}$ of the curve $\mathbf{r} = (t, t^2, t^3)$ at the point $t = 1$. Find also the equations of the characteristic lines and planes at this point.

Answers

1. $(\cos t + t \sin t, \sin t - t \cos t, 0)$ 2. $3t^2x - 3ty + z = t^3$
3. $3x - 3y + z = 1$
4. $2x - 2y + z = 2,\ 2x + 2y + z = -2,\ 8x - 4y + z = 16$
5. $c(x \sin t - y \cos t - at) + az = 0$ 7. $bx \sinh t - by \cosh t + az = abt$
8. $2x \cos t - 2y \sin t - 3z = 2a \cos 2t$
9. $(\sin 3t + 3 \sin t)x - (\cos 3t + 3 \cos t)y + 4z = 6a \sin t$
10. $\mathbf{t} = \frac{1}{\sqrt{14}}(\mathbf{i} + 2\mathbf{j} + 3\mathbf{k}),\ \mathbf{n} = \frac{1}{\sqrt{266}}(-11\mathbf{i} - 8\mathbf{j} + 9\mathbf{k}),\ \mathbf{b} = \frac{1}{\sqrt{19}}(3\mathbf{i} - 3\mathbf{j} + \mathbf{k})$,

$$\frac{x-1}{1} = \frac{y-1}{2} = \frac{z-1}{3},\quad \frac{x-1}{11} = \frac{y-1}{8} = \frac{z-1}{-9},\quad \frac{x-1}{3} = \frac{y-1}{-3} = \frac{z-1}{1},$$

$x + 2y + 3z = 6,\ 11x + 8y - 9z = 10,\ 3x - 3y + z = 1.$

2

Curvature and Torsion

1. INTRODUCTION

For curves in space, the concepts of curvature and torsion are of fundamental importance. We know that line segments are uniquely determined by their lengths, circles by their radii, triangles by side-angle-side etc. In geometry, we look for geometric quantities which distinguish one figure from another. The importance of curvature and torsion can easily be estimated from the fact that it can be proved that a curve is uniquely determined (except for its position in space) if its curvature and torsion are given as continuous functions of arc length '*s*'.

2. CURVATURE OF A CURVE

Let $\mathbf{r} = \mathbf{r}(s)$ be a regular curve C of class $C^m (m \geq 2)$, where *s* is the parameter 'arc length'.

The vector $\mathbf{r}''(s)$ is called the **curvature vector** on the curve C at the point $\mathbf{r}(s)$ and it is denoted by $\boldsymbol{\kappa}(s)$ (or by $\boldsymbol{\kappa}$). The magnitude of the curvature vector is called the **curvature** of the curve C at the point $\mathbf{r}(s)$ and it is denoted by $\kappa(s)$ (or by κ).

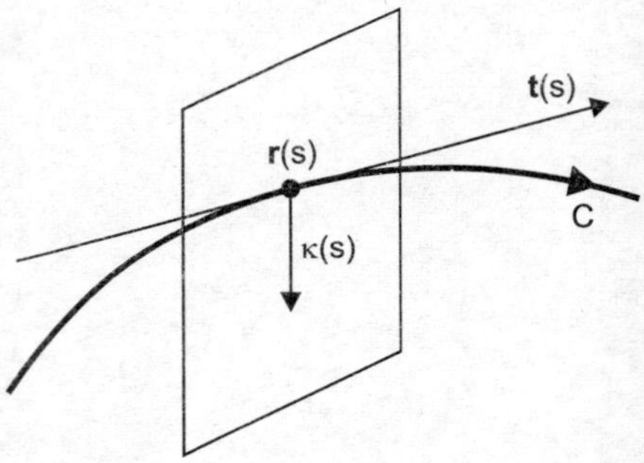

$\therefore \qquad \kappa(s) = |\mathbf{r}''(s)|$

Also $\qquad \mathbf{t}(s) = \mathbf{r}'(s)$, so we have

$$\boldsymbol{\kappa}(s) = \mathbf{t}'(s).$$

We know that $\mathbf{t}(s)$ is a unit vector.

$\Rightarrow \qquad \mathbf{t}(s) \cdot \mathbf{t}(s) = 1$

$\Rightarrow \qquad \mathbf{t}(s) \cdot \mathbf{t}'(s) + \mathbf{t}'(s) \cdot \mathbf{t}(s) = 0$

$\Rightarrow \qquad 2\mathbf{t}'(s) \cdot \mathbf{t}(s) = 0$

$\Rightarrow \qquad \boldsymbol{\kappa}(s) \cdot \mathbf{t}(s) = 0$

$\therefore$ The curvature vector $\boldsymbol{\kappa}(s)$ is orthogonal to $\mathbf{t}(s)$ and hence parallel to the normal plane at $\mathbf{r}(s)$. When $\boldsymbol{\kappa}(s)$ is non-zero, it is in the direction in which the curve is turning.

The reciprocal of the curvature at a point is called the **radius of curvature** at that point and it is denoted by ρ.

$$\therefore \qquad \rho = \frac{1}{\kappa} \qquad \text{(Assuming } \kappa \neq 0)$$

A point on the curve C is called a **point of inflexion** if the curvature κ at that point is zero.

Remark. We have $\qquad \dfrac{\boldsymbol{\kappa}}{\kappa} = \dfrac{\mathbf{r}''}{|\mathbf{r}''|} = \dfrac{\mathbf{t}'}{|\mathbf{t}'|} = \mathbf{n}.$

$$\therefore \qquad \mathbf{n} = \frac{\boldsymbol{\kappa}}{\kappa}.$$

Example 1. *Show that:* $\kappa = |\mathbf{r}' \times \mathbf{r}''|$.

Sol. $$|\mathbf{r}' \times \mathbf{r}''| = |\mathbf{t} \times \mathbf{t}'| = |\mathbf{t}| \cdot |\mathbf{t}'| \sin\frac{\pi}{2} = 1 \cdot \kappa \cdot 1 = \kappa$$

$\therefore$ $$\kappa = |\mathbf{r}' \times \mathbf{r}''|.$$

Example 2. *For the helix* $\mathbf{r} = a \cos t\mathbf{i} + a \sin t\mathbf{j} + bt\mathbf{k}$, $a > 0$, $b \neq 0$, *find the curvature at the point t.*

Sol. We have $$\mathbf{r} = a\cos t\mathbf{i} + a\sin t\mathbf{j} + bt\mathbf{k}.$$

$\therefore$ $$\dot{\mathbf{r}} = -a\sin t\mathbf{i} + a\cos t\mathbf{j} + b\mathbf{k}$$

$\Rightarrow$ $$|\dot{\mathbf{r}}| = \sqrt{a^2\sin^2 t + a^2\cos^2 t + b^2} = \sqrt{a^2 + b^2}$$

$\therefore$ $$\mathbf{t} = \frac{\dot{\mathbf{r}}}{|\dot{\mathbf{r}}|} = \frac{1}{\sqrt{a^2+b^2}}(-a\sin t\mathbf{i} + a\cos t\mathbf{j} + b\mathbf{k})$$

$\therefore$ $$\boldsymbol{\kappa} = \mathbf{t}' = \frac{d\mathbf{t}}{ds} = \frac{d\mathbf{t}}{dt}\cdot\frac{dt}{ds} = \frac{d\mathbf{t}}{dt}\Big/\frac{ds}{dt}$$

$$= \frac{1}{\sqrt{a^2+b^2}}(-a\cos t\mathbf{i} - a\sin t\mathbf{j})\Big/\frac{ds}{dt}$$

$$= -\frac{a}{\sqrt{a^2+b^2}}(\cos t\mathbf{i} + \sin t\mathbf{j})\Big/\sqrt{a^2+b^2} \qquad \left(\because \frac{ds}{dt} = |\dot{\mathbf{r}}|\right)$$

$$= -\frac{a}{a^2+b^2}(\cos t\mathbf{i} + \sin t\mathbf{j})$$

$\therefore$ Curvature, $$\kappa = |\boldsymbol{\kappa}| = \frac{a}{a^2+b^2}\sqrt{(-\cos t)^2 + (-\sin t)^2} = \frac{a}{a^2+b^2}.$$

Example 3. *For the curve* $\mathbf{r} = t\mathbf{i} + \frac{1}{2}t^2\mathbf{j} + \frac{1}{3}t^3\mathbf{k}$, *find the curvature vector and curvature at the point* $t = 1$.

Sol. We have $$\mathbf{r} = t\mathbf{i} + \frac{1}{2}t^2\mathbf{j} + \frac{1}{3}t^3\mathbf{k}.$$

$\therefore$ $$\dot{\mathbf{r}} = \mathbf{i} + t\mathbf{j} + t^2\mathbf{k} \quad \text{and} \quad |\dot{\mathbf{r}}| = \sqrt{1+t^2+t^4}$$

$\therefore$ $$\mathbf{t} = \frac{\dot{\mathbf{r}}}{|\dot{\mathbf{r}}|} = \frac{1}{\sqrt{1+t^2+t^4}}(\mathbf{i} + t\mathbf{j} + t^2\mathbf{k})$$

$$\dot{\mathbf{t}} = \frac{1}{\sqrt{1+t^2+t^4}}(\mathbf{j} + 2t\mathbf{k}) + \left(-\frac{2t+4t^3}{2(1+t^2+t^4)^{3/2}}\right)(\mathbf{i} + t\mathbf{j} + t^2\mathbf{k})$$

$$= \frac{1}{(1+t^2+t^4)^{3/2}} [(1 + t^2 + t^4)(\mathbf{j} + 2t\mathbf{k}) - (t + 2t^3)(\mathbf{i} + t\mathbf{j} + t^2\mathbf{k})]$$

$$= \frac{1}{(1+t^2+t^4)^{3/2}} [-(t + 2t^3)\mathbf{i} + (1 - t^4)\mathbf{j} + (2t + t^3)\mathbf{k}]$$

$$\therefore \qquad \boldsymbol{\kappa} = \mathbf{t}' = \frac{d\mathbf{t}}{ds} = \frac{d\mathbf{t}}{dt}\frac{dt}{ds} = \dot{\mathbf{t}} \Big/ \frac{ds}{dt} = \dot{\mathbf{t}} / |\dot{\mathbf{r}}|$$

$$= \frac{1}{(1+t^2+t^4)^2} [-(t + 2t^3)\mathbf{i} + (1 - t^4)\mathbf{j} + (2t + t^3)\mathbf{k}]$$

$\therefore$ At $t = 1$,

$$\boldsymbol{\kappa} = \frac{1}{(3)^2}[-3\mathbf{i} + 0\mathbf{j} + 3\mathbf{k}] = -\frac{1}{3}\mathbf{i} + \frac{1}{3}\mathbf{k} \text{ and } \kappa = |\boldsymbol{\kappa}| = \sqrt{\frac{1}{9}+\frac{1}{9}} = \frac{\sqrt{2}}{3}.$$

Theorem 1. *Prove that the curvature of a regular curve at a point is equal to the rate of change of direction of the tangent with respect to arc length.*

Proof. Let $\mathbf{r} = \mathbf{r}(s)$ be a regular curve of class $C^m (m \geq 2)$, where s is the parameter 'arc length'. Let $\mathbf{r}(s)$ be any point P on the given curve. Let $\mathbf{r}(s + \delta s)$, $\delta s > 0$ be a neighbouring point Q of $\mathbf{r}(s)$. Let $\mathbf{t}(s)$ and $\mathbf{t}(s + \delta s)$ be the unit tangent vectors at the points $\mathbf{r}(s)$ and $\mathbf{r}(s + \delta s)$ respectively.

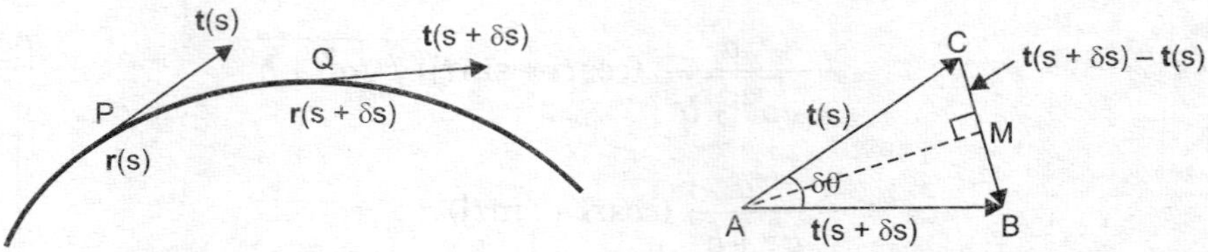

Let $\delta\theta$ denote the angle between the tangent vectors $\mathbf{t}(s)$ and $\mathbf{t}(s + \delta s)$.

By definition,

$$\kappa = |\mathbf{t}'| = \left| \lim_{\delta s \to 0} \frac{\mathbf{t}(s+\delta s) - \mathbf{t}(s)}{\delta s} \right| = \lim_{\delta s \to 0} \left| \frac{\mathbf{t}(s+\delta s) - \mathbf{t}(s)}{\delta s} \right|$$

$$\therefore \qquad \kappa = \lim_{\delta s \to 0} \frac{|\mathbf{t}(s+\delta s) - \mathbf{t}(s)|}{\delta s} \qquad ...(1)$$

Since $\mathbf{t}'$ is a unit vector, we have AC = AB = 1.

In ΔABC, $\qquad$ CB = $|\mathbf{t}(s + \delta s) - \mathbf{t}(s)|$

Also $\qquad$ CB = 2 CM = 2 sin $\angle$CAM = 2 sin $\dfrac{\delta\theta}{2}$

$$= 2\left[\frac{\delta\theta}{2} - \frac{(\delta\theta/2)^3}{3!} + \ldots\ldots\right]$$

(By using Taylor's expansion for the sine function.)

$$= \delta\theta - \frac{(\delta\theta)^3}{24} + \ldots\ldots$$

$$= \delta\theta\left(1 - \frac{(\delta\theta)^2}{24} + \ldots\ldots\right)$$

$\therefore$ (1) $\Rightarrow$
$$\kappa = \lim_{\delta s \to 0} \frac{\delta\theta\left(1 - \frac{(\delta\theta)^2}{24} + \ldots\ldots\right)}{\delta s}$$

$$= \lim_{\delta s \to 0} \frac{\delta\theta}{\delta s} \cdot \lim_{\delta\theta \to 0}\left(1 - \frac{(\delta\theta)^2}{24} + \ldots\ldots\right) \qquad (\because \ \delta\theta \to 0 \text{ as } \delta s \to 0)$$

$$= \frac{d\theta}{ds} \cdot 1 = \frac{d\theta}{ds} = \text{rate of change of } \theta \text{ w.r.t. } s$$

$\therefore$ The curvature at a point is equal to the rate of change of the tangent with respect to the arc length.

Theorem 2. *Prove that a regular curve of class $C^m (m \geq 2)$ is a straight line if and only if its curvature is identically zero.*

Proof. Let $\mathbf{r} = \mathbf{r}(s)$ be a regular curve of class $C^m (m \geq 2)$, where s is the parameter 'arc length'.

Let the curve be a straight line.

Let the curve passes through the point whose position vector is **a** and is parallel to vector **b**.

$\therefore$ $\mathbf{r} = \mathbf{a} + t\mathbf{b}$, where t is a parameter.

$$\Rightarrow \quad \frac{d\mathbf{r}}{dt} = \mathbf{b} \quad \text{and} \quad \left|\frac{d\mathbf{r}}{dt}\right| = |\mathbf{b}|$$

$$\therefore \quad \mathbf{t} = \frac{\dot{\mathbf{r}}}{|\dot{\mathbf{r}}|} = \frac{\mathbf{b}}{|\mathbf{b}|}$$

$$\therefore \quad \kappa = \mathbf{t}' = \frac{d\mathbf{t}}{dt} \cdot \frac{dt}{ds} = \mathbf{0} \cdot \frac{dt}{ds} = \mathbf{0}$$

$\therefore$ $\kappa = |\mathbf{0}| = 0$ *i.e.*, the curvature is identically zero.

Conversely, let the curvature of the curve be identically zero *i.e.*, $\kappa = 0$.

$$\Rightarrow \quad \boldsymbol{\kappa} = \mathbf{0} \quad \Rightarrow \quad \mathbf{t}' = \mathbf{0} \quad \Rightarrow \quad \mathbf{t} = \mathbf{c}, \text{ a constant vector.}$$

$$\Rightarrow \quad \mathbf{r}' = \mathbf{c} \quad \Rightarrow \quad \mathbf{r} = \mathbf{c}s + \mathbf{d}, \text{ where } \mathbf{d} \text{ is a constant vector.}$$

$\therefore$ The curve is a straight line passing through the point whose position vector is **d** and is parallel to the vector **c**.

$\therefore$ The result holds.

Theorem 3. *Let* $\mathbf{r} = \mathbf{r}(t)$ *be a regular curve of class* C^m $(m \geq 2)$, *where t is an arbitrary parameter. Prove that*

$$\kappa = \frac{|\dot{\mathbf{r}} \times \ddot{\mathbf{r}}|}{|\dot{\mathbf{r}}|^3}.$$

Proof. We have $$\dot{\mathbf{r}} = \frac{d\mathbf{r}}{dt} = \frac{d\mathbf{r}}{ds}\frac{ds}{dt} = \mathbf{r}'\dot{s} = \dot{s}\mathbf{r}'$$

and $$\ddot{\mathbf{r}} = \frac{d\dot{\mathbf{r}}}{dt} = \frac{d}{dt}(\dot{s}\mathbf{r}') = \dot{s}\frac{d\mathbf{r}'}{dt} + \ddot{s}\mathbf{r}' = \dot{s}(\mathbf{r}''\dot{s}) + \ddot{s}\mathbf{r}' = \ddot{s}\mathbf{r}' + \dot{s}^2\mathbf{r}''$$

$$\therefore \quad \dot{\mathbf{r}} \times \ddot{\mathbf{r}} = \dot{s}\mathbf{r}' \times (\ddot{s}\mathbf{r}' + \dot{s}^2\mathbf{r}'') = (\dot{s}\,\ddot{s})(\mathbf{r}' \times \mathbf{r}') + (\dot{s}^3)(\mathbf{r}' \times \mathbf{r}'')$$

$$= (\dot{s}\,\ddot{s})\,\mathbf{0} + (\dot{s}^3)(\mathbf{r}' \times \mathbf{r}'') = \dot{s}^3(\mathbf{r}' \times \mathbf{r}'')$$

$$= |\dot{\mathbf{r}}|^3(\mathbf{r}' \times \mathbf{r}'') \qquad (\because \dot{s} = |\dot{\mathbf{r}}|)$$

$\therefore$ $|\dot{\mathbf{r}} \times \ddot{\mathbf{r}}| = |\dot{\mathbf{r}}|^3\,|\mathbf{r}'|\,|\mathbf{r}''|\sin\theta$, where θ is the angle between $\mathbf{r}'$ and $\mathbf{r}''$.

Now $\mathbf{r}' = \mathbf{t}$, $\mathbf{r}'' = \mathbf{t}'$ and $\mathbf{t}$ and $\mathbf{t}'$ are orthogonal.

$\therefore$ $\theta = \pi/2$

Also $|\mathbf{r}'| = |\mathbf{t}| = 1$ and $|\mathbf{r}''| = \kappa$

$\therefore$ $|\dot{\mathbf{r}} \times \ddot{\mathbf{r}}| = |\dot{\mathbf{r}}|^3 .1.\kappa.1$

$$\therefore \quad \kappa = \frac{|\dot{\mathbf{r}} \times \ddot{\mathbf{r}}|}{|\dot{\mathbf{r}}|^3}.$$

Corollary. If $\mathbf{r} = x(t)\mathbf{i} + y(t)\mathbf{j} + z(t)\mathbf{k}$, then

$$|\dot{\mathbf{r}} \times \ddot{\mathbf{r}}| = \sqrt{\Sigma(\dot{y}\ddot{z} - \ddot{y}\dot{z})^2}$$

and $$|\dot{\mathbf{r}}| = \sqrt{\Sigma \dot{x}^2}.$$

$$\therefore \quad \kappa = \frac{|\dot{\mathbf{r}} \times \ddot{\mathbf{r}}|}{|\dot{\mathbf{r}}|^3} = \frac{\sqrt{\Sigma(\dot{y}\ddot{z} - \ddot{y}\dot{z})^2}}{(\Sigma\dot{x}^2)^{3/2}}.$$

Remarks 1. If $\mathbf{r} = \mathbf{r}(s)$, then

$$\kappa = |\mathbf{r}''| = |\mathbf{r}' \times \mathbf{r}''|.$$

2. If $\mathbf{r} = \mathbf{r}(t)$, then $\kappa = \dfrac{|\dot{\mathbf{r}} \times \ddot{\mathbf{r}}|}{|\dot{\mathbf{r}}|^3}$.

Example 4. *For the circle* $\mathbf{r} = a\cos t\mathbf{i} + a\sin t\mathbf{j}$, $a > 0$, *find the radius of curvature at point t.*

Sol. We have $\mathbf{r} = a\cos t\mathbf{i} + a\sin t\mathbf{j}$, $a > 0$.

$\therefore$ $\dot{\mathbf{r}} = -a\sin t\mathbf{i} + a\cos t\mathbf{j}$

$$\Rightarrow \quad |\dot{\mathbf{r}}| = \sqrt{a^2\sin^2 t + a^2\cos^2 t} = a$$

$$\therefore \quad \mathbf{t} = \frac{\dot{\mathbf{r}}}{|\dot{\mathbf{r}}|} = \frac{1}{a}(-a\sin t\mathbf{i} + a\cos t\mathbf{j}) = -\sin t\mathbf{i} + \cos t\mathbf{j}$$

$$\therefore \qquad \boldsymbol{\kappa} = \mathbf{t}' = \frac{d\mathbf{t}}{ds} = \frac{d\mathbf{t}}{dt}\cdot\frac{dt}{ds}$$

$$= (-\cos t\mathbf{i} - \sin t\mathbf{j}) \Big/ \frac{ds}{dt}$$

$$= -(\cos t\mathbf{i} + \sin t\mathbf{j}) / |\dot{\mathbf{r}}|$$

$$= -\frac{1}{a}(\cos t\mathbf{i} + \sin t\mathbf{j})$$

$\therefore$ Curvature, $\kappa = |\boldsymbol{\kappa}| = \dfrac{1}{a}\sqrt{(-\cos t)^2 + (-\sin t)^2} = \dfrac{1}{a}$

$\therefore$ Radius of curvature $= \dfrac{1}{\kappa} = \dfrac{1}{1/a} = a$

$\therefore$ Radius of curvature is equal to the radius of the given circle.

Alternative method

We have $\mathbf{r} = a\cos t\mathbf{i} + a\sin t\mathbf{j},\ a > 0.$

$$\therefore \qquad \dot{\mathbf{r}} = -a\sin t\mathbf{i} + a\cos t\mathbf{j}$$

and

$$\ddot{\mathbf{r}} = -a\cos t\mathbf{i} - a\sin t\mathbf{j}$$

$$\therefore \qquad \dot{\mathbf{r}}\times\ddot{\mathbf{r}} = \begin{vmatrix} \mathbf{i} & \mathbf{j} & \mathbf{k} \\ -a\sin t & a\cos t & 0 \\ -a\cos t & -a\sin t & 0 \end{vmatrix} = (a^2\sin^2 t + a^2\cos^2 t)\mathbf{k} = a^2\mathbf{k}$$

$$\therefore \qquad |\dot{\mathbf{r}}\times\ddot{\mathbf{r}}| = a^2$$

Also $|\dot{\mathbf{r}}|^2 = a^2\sin^2 t + a^2\cos^2 t = a^2 \quad \therefore \quad |\dot{\mathbf{r}}| = a$

$\therefore$ Radius of curvature $= \dfrac{1}{\kappa} = \dfrac{|\dot{\mathbf{r}}|^3}{|\dot{\mathbf{r}}\times\ddot{\mathbf{r}}|} = \dfrac{a^3}{a^2} = a.$

Example 5. *Show that along the plane curve* $\mathbf{r} = x(t)\mathbf{i} + y(t)\mathbf{j}$,

$$\kappa = \frac{|\dot{x}\ddot{y} - \ddot{x}\dot{y}|}{(\dot{x}^2 + \dot{y}^2)^{3/2}}.$$

Sol. We have $\mathbf{r} = x\mathbf{i} + y\mathbf{j}.$ ...(1)

Also, $\kappa = \dfrac{|\dot{\mathbf{r}}\times\ddot{\mathbf{r}}|}{|\dot{\mathbf{r}}|^3}$ ...(2)

(1) $\Rightarrow$ $\dot{\mathbf{r}} = \dot{x}\mathbf{i} + \dot{y}\mathbf{j}$ and $\ddot{\mathbf{r}} = \ddot{x}\mathbf{i} + \ddot{y}\mathbf{j}$

$$\therefore \qquad \dot{\mathbf{r}}\times\ddot{\mathbf{r}} = \begin{vmatrix} \mathbf{i} & \mathbf{j} & \mathbf{k} \\ \dot{x} & \dot{y} & 0 \\ \ddot{x} & \ddot{y} & 0 \end{vmatrix} = (\dot{x}\ddot{y} - \ddot{x}\dot{y})\mathbf{k}$$

$$\therefore \qquad |\dot{\mathbf{r}}\times\ddot{\mathbf{r}}| = |\dot{x}\ddot{y} - \ddot{x}\dot{y}| \quad \text{and} \quad |\dot{\mathbf{r}}| = \left(\dot{x}^2 + \dot{y}^2\right)^{1/2}$$

$\therefore$ (2) $\Rightarrow$ $\kappa = \dfrac{|\dot{x}\ddot{y} - \ddot{x}\dot{y}|}{[(\dot{x}^2 + \dot{y}^2)^{1/2}]^3} = \dfrac{|\dot{x}\ddot{y} - \ddot{x}\dot{y}|}{(\dot{x}^2 + \dot{y}^2)^{3/2}}.$

Example 6. *For the curve*

$$\mathbf{r} = (t - \sin t)\mathbf{i} + (1 - \cos t)\mathbf{j} + t\mathbf{k},$$

find the curvature at point t.

Sol. We have $\mathbf{r} = (t - \sin t)\mathbf{i} + (1 - \cos t)\mathbf{j} + t\mathbf{k}.$

$\therefore$ $\dot{\mathbf{r}} = (1 - \cos t)\mathbf{i} + \sin t\mathbf{j} + \mathbf{k}$

and $\ddot{\mathbf{r}} = \sin t\mathbf{i} + \cos t\mathbf{j} + 0\mathbf{k}$

Now
$$\kappa = \frac{|\dot{\mathbf{r}} \times \ddot{\mathbf{r}}|}{|\dot{\mathbf{r}}|^3} \quad \text{...(1)}$$

$$\dot{\mathbf{r}} \times \ddot{\mathbf{r}} = \begin{vmatrix} \mathbf{i} & \mathbf{j} & \mathbf{k} \\ 1 - \cos t & \sin t & 1 \\ \sin t & \cos t & 0 \end{vmatrix} = -\cos t\mathbf{i} + \sin t\mathbf{j} + (\cos t - 1)\mathbf{k}$$

$\therefore$
$$|\dot{\mathbf{r}} \times \ddot{\mathbf{r}}| = \sqrt{\cos^2 t + \sin^2 + (\cos t - 1)^2}$$

$$= \left(1 + \left(-2\sin^2 \frac{t}{2}\right)^2\right)^{1/2} = \left(1 + 4\sin^4 \frac{t}{2}\right)^{1/2}$$

$$|\dot{\mathbf{r}}| = \sqrt{((1 - \cos t)^2 + \sin^2 t + 1)}$$

$$= (1 + \cos^2 t - 2\cos t + \sin^2 t + 1)^{1/2}$$

$$= (1 + 2(1 - \cos t))^{1/2} = \left(1 + 4\sin^2 \frac{t}{2}\right)^{1/2}$$

$\therefore$ (1) $\Rightarrow$
$$\kappa = \frac{\left(1 + 4\sin^4 \frac{t}{2}\right)^{1/2}}{\left(\left(1 + 4\sin^2 \frac{t}{2}\right)^{1/2}\right)^3} = \frac{\left(1 + 4\sin^4 \frac{t}{2}\right)^{1/2}}{\left(1 + 4\sin^2 \frac{t}{2}\right)^{3/2}}.$$

Example 7. *For the curve $x = 4a\cos^3 t$, $y = 4a\sin^3 t$, $z = 3c\cos 2t$, show that*

$$\kappa = \frac{a}{6(a^2 + c^2)\sin 2t}.$$

Sol. Let $\mathbf{r}$ be the position vector of the point (x, y, z) on the curve.

$\therefore$
$$\mathbf{r} = x\mathbf{i} + y\mathbf{j} + z\mathbf{k} = 4a\cos^3 t\mathbf{i} + 4a\sin^3 t\mathbf{j} + 3c\cos 2t\mathbf{k}$$

$\therefore$
$$\mathbf{t} = \mathbf{r}' = \frac{d\mathbf{r}}{dt}\frac{dt}{ds}$$

$$= (-12a\cos^2 t\sin t\mathbf{i} + 12a\sin^2 t\cos t\mathbf{j} - 6c\sin 2t\mathbf{k})\frac{dt}{ds}$$

$$= (-6a\cos t\sin 2t\mathbf{i} + 6a\sin t\sin 2t\mathbf{j} - 6c\sin 2t\mathbf{k})\frac{dt}{ds}$$

$\therefore$
$$\mathbf{t} = 6\sin 2t(-a\cos t\mathbf{i} + a\sin t\mathbf{j} - c\mathbf{k})\frac{dt}{ds} \quad \text{...(1)}$$

$$\Rightarrow \qquad |\,\mathbf{t}\,| = 6 \sin 2t \,.\, \sqrt{a^2 \cos^2 t + a^2 \sin^2 t + c^2} \left| \frac{dt}{ds} \right|$$

$$\Rightarrow \qquad 1 = 6 \sin 2t \sqrt{a^2 + c^2} \left| \frac{dt}{ds} \right| \qquad (\because \; |\,\mathbf{t}\,| = 1)$$

$$\Rightarrow \qquad \left| \frac{ds}{dt} \right| = 6\sqrt{a^2 + c^2} \sin 2t$$

$$\Rightarrow \qquad \frac{ds}{dt} = 6\sqrt{a^2 + c^2} \sin 2t \qquad \text{(Assuming } \sin 2t > 0\text{)}$$

$$\therefore \quad (1) \Rightarrow \qquad \mathbf{t} = 6 \sin 2t\,(-a \cos t\mathbf{i} + a \sin t\mathbf{j} - c\mathbf{k}) \,.\, \frac{1}{6\sqrt{a^2 + c^2} \sin 2t}$$

$$= \frac{1}{\sqrt{a^2 + c^2}} (-a \cos t\mathbf{i} + a \sin t\mathbf{j} - c\mathbf{k})$$

Now

$$\mathbf{t}' = \frac{d\mathbf{t}}{ds} = \frac{d\mathbf{t}}{dt}\frac{dt}{ds}$$

$$= \frac{1}{\sqrt{a^2 + c^2}} (a \sin t\mathbf{i} + a \cos t\mathbf{j} - 0\mathbf{k}) \frac{1}{6\sqrt{a^2 + c^2} \sin 2t}$$

$$= \frac{a}{6(a^2 + c^2) \sin 2t} (\sin t\mathbf{i} + \cos t\mathbf{j})$$

$$\therefore \qquad \kappa = |\,\mathbf{t}'\,| = \frac{a}{6(a^2 + c^2) \sin 2t} \sqrt{\sin^2 t + \cos^2 t} = \frac{a}{6(a^2 + c^2) \sin 2t}.$$

Example 8. *Find the radius of curvature at any point of the curve*

$$x^2 + y^2 = a^2,\; x^2 - y^2 = az.$$

Sol. The given curve is

$$x^2 + y^2 = a^2 \qquad ...(1) \qquad x^2 - y^2 = az \qquad ...(2)$$

Let $x = a \cos t,\; y = a \sin t$

$\therefore$ (1) is satisfied.

(2) $\Rightarrow$ $a^2 \cos^2 t - a^2 \sin^2 t = az$

$\Rightarrow$ $z = a \cos 2t$

$\therefore$ The given curve is

$$x = a \cos t,\; y = a \sin t,\; z = a \cos 2t.$$

Let $\mathbf{r}$ be the position vector of the point (x, y, z) on the curve.

$$\therefore \qquad \mathbf{r} = x\mathbf{i} + y\mathbf{j} + z\mathbf{k} = a \cos t\mathbf{i} + a \sin t\mathbf{j} + a \cos 2t\,\mathbf{k}$$

$$\therefore \qquad \dot{\mathbf{r}} = -a \sin t\mathbf{i} + a \cos t\mathbf{j} - 2a \sin 2t\mathbf{k}$$

and

$$\ddot{\mathbf{r}} = -a \cos t\mathbf{i} - a \sin t\mathbf{j} - 4a \cos 2t\mathbf{k}$$

$$\therefore \quad \dot{\mathbf{r}} \times \ddot{\mathbf{r}} = \begin{vmatrix} \mathbf{i} & \mathbf{j} & \mathbf{k} \\ -a\sin t & a\cos t & -2a\sin 2t \\ -a\cos t & -a\sin t & -4a\cos 2t \end{vmatrix}$$

$$= a^2 \begin{vmatrix} \mathbf{i} & \mathbf{j} & \mathbf{k} \\ \sin t & -\cos t & 2\sin 2t \\ \cos t & \sin t & 4\cos 2t \end{vmatrix}$$

$$= a^2 [(-4\cos t\cos 2t - 2\sin t\sin 2t)\mathbf{i} - (4\sin t\cos 2t - 2\cos t\sin 2t)\mathbf{j} + (\sin^2 t + \cos^2 t)\mathbf{k}]$$

$$= a^2 [(-2\cos t\cos 2t - 2\cos(t - 2t))\mathbf{i} + (-2\sin t\cos 2t + 2\sin(2t - t))\mathbf{j} + \mathbf{k}]$$

$$= a^2[-4\cos^3 t\mathbf{i} + 4\sin^3 t\mathbf{j} + \mathbf{k}]$$

$$\therefore \quad |\dot{\mathbf{r}}|^2 = a^2\sin^2 t + a^2\cos^2 t + 4a^2\sin^2 2t$$

$$= a^2(1 + 4\sin^2 2t) = a^2(5 - 4\cos^2 2t)$$

$$= a^2\left(5 - \frac{4z^2}{a^2}\right) = 5a^2 - 4z^2$$

$$\therefore \quad |\dot{\mathbf{r}}| = \sqrt{5a^2 - 4z^2}$$

$$|\dot{\mathbf{r}} \times \ddot{\mathbf{r}}|^2 = a^4(16\cos^6 t + 16\sin^6 t + 1)$$

$$= a^4[16\{(\cos^2 t + \sin^2 t)^3 - 3\cos^2 t\sin^2 t(\cos^2 t + \sin^2 t)\} + 1]$$

$$= a^4[16\{1^3 - 3\cos^2 t\sin^2 t \,.\, 1\} + 1] = a^4[17 - 48\cos^2 t\sin^2 t]$$

$$= a^4[17 - 12\sin^2 2t] = a^4[5 + 12\cos^2 2t]$$

$$= a^4\left[5 + \frac{12z^2}{a^2}\right] = a^2(5a^2 + 12z^2)$$

$$\therefore \quad |\dot{\mathbf{r}} \times \ddot{\mathbf{r}}| = a\sqrt{5a^2 + 12z^2}$$

$$\therefore \quad \rho = \frac{1}{\kappa} = \frac{|\dot{\mathbf{r}}|^3}{|\dot{\mathbf{r}} \times \ddot{\mathbf{r}}|} = \frac{(5a^2 - 4z^2)^{3/2}}{a\sqrt{5a^2 + 12z^2}}.$$

Example 9. *Find the equation of the osculating plane and curvature at point t of the curve x = a cos 2t, y = a sin 2t, z = 2a sin t.*

Sol. Let **r** be the position vector of the point (x, y, z) on the curve.

$$\therefore \quad \mathbf{r} = x\mathbf{i} + y\mathbf{j} + z\mathbf{k} = a\cos 2t\mathbf{i} + a\sin 2t\mathbf{j} + 2a\sin t\mathbf{k}$$

$$\therefore \quad \dot{\mathbf{r}} = -2a\sin 2t\mathbf{i} + 2a\cos 2t\mathbf{j} + 2a\cos t\mathbf{k}$$

and
$$\ddot{\mathbf{r}} = -4a\cos 2t\mathbf{i} - 4a\sin 2t\mathbf{j} - 2a\sin t\,\mathbf{k}$$

$$\therefore \quad \dot{\mathbf{r}} \times \ddot{\mathbf{r}} = \begin{vmatrix} \mathbf{i} & \mathbf{j} & \mathbf{k} \\ -2a\sin 2t & 2a\cos 2t & 2a\cos t \\ -4a\cos 2t & -4a\sin 2t & -2a\sin t \end{vmatrix}$$

$$= -4a^2 \begin{vmatrix} \mathbf{i} & \mathbf{j} & \mathbf{k} \\ -\sin 2t & \cos 2t & \cos t \\ 2\cos 2t & 2\sin 2t & \sin t \end{vmatrix}$$

$$= -4a^2[(\cos 2t \sin t - 2 \sin 2t \cos t)\mathbf{i} - (-\sin 2t \sin t - 2 \cos 2t \cos t)\mathbf{j} + (-2 \sin^2 2t - 2 \cos^2 2t)\mathbf{k}]$$

$$= -4a^2[(\sin (t - 2t) - \sin 2t \cos t)\mathbf{i} + (\cos (2t - t) + \cos 2t \cos t)\mathbf{j} - 2\mathbf{k}]$$

$$= 4a^2[(\sin t + \sin 2t \cos t)\mathbf{i} - (\cos t + \cos 2t \cos t)\mathbf{j} + 2\mathbf{k}]$$

Equation of the osculating plane is $[\mathbf{R} - \mathbf{r}\ \dot{\mathbf{r}}\ \ddot{\mathbf{r}}] = 0$.

$$\Rightarrow \quad (\mathbf{R} - \mathbf{r}) \cdot (\dot{\mathbf{r}} \times \ddot{\mathbf{r}}) = 0 \qquad \text{...(1)}$$

Let $\quad \mathbf{R} = x\mathbf{i} + y\mathbf{j} + z\mathbf{k}$

$\therefore$ (1) $\Rightarrow$ $[(x - a \cos 2t)\mathbf{i} + (y - a \sin 2t)\mathbf{j} + (z - 2a \sin t)\mathbf{k}] \cdot 4a^2[(\sin t + \sin 2t \cos t)\mathbf{i} - (\cos t + \cos 2t \cos t)\mathbf{j} + 2\mathbf{k}] = 0$

$\Rightarrow$ $(x - a \cos 2t) (\sin t + \sin 2t \cos t) - (y - a \sin 2t) (\cos t + \cos 2t \cos t) + (z - 2a \sin t)2 = 0$

$\Rightarrow$ $(\sin t + \sin 2t \cos t)x - (\cos t \cdot 2 \cos^2 t)y + 2z$

$= a \cos 2t \sin t + a \cos 2t \sin 2t \cos t - a \sin 2t \cos t - a \sin 2t \cos 2t \cos t + 4a \sin t$

$= a \sin (t - 2t) + 4a \sin t$

$= 3a \sin t$

$\therefore$ $(\sin t + \sin 2t \cos t)x - 2 \cos^3 t\, y + 2z = 3a \sin t$

This is the equation of the osculating plane.

$|\dot{\mathbf{r}} \times \ddot{\mathbf{r}}|^2 = 16a^4[(\sin t + \sin 2t \cos t)^2 + (\cos t + \cos 2t \cos t)^2 + 4]$

$= 16a^4[\sin^2 t + \sin^2 2t \cos^2 t + 2 \sin t \sin 2t \cos t + \cos^2 t + \cos^2 2t \cos^2 t + 2 \cos t \cos 2t \cos t + 4]$

$= 16a^4[1 + (\sin^2 2t + \cos^2 2t) \cos^2 t + 2 \cos t (\sin t \sin 2t + \cos t \cos 2t) + 4]$

$= 16a^4[1 + \cos^2 t + 2 \cos t \cos (t - 2t) + 4]$

$= 16a^4[5 + 3 \cos^2 t]$

$\therefore$ $|\dot{\mathbf{r}} \times \ddot{\mathbf{r}}| = 4a^2 \sqrt{5 + 3 \cos^2 t}$

Also, $|\dot{\mathbf{r}}|^2 = 4a^2 \sin^2 2t + 4a^2 \cos^2 2t + 4a^2 \cos^2 t$

$= 4a^2 + 4a^2 \cos^2 t = 4a^2(1 + \cos^2 t)$

$\therefore$ $|\dot{\mathbf{r}}|^3 = 8a^3(1 + \cos^2 t)^{3/2}$

$$\therefore \quad \kappa = \frac{|\dot{\mathbf{r}} \times \ddot{\mathbf{r}}|}{|\dot{\mathbf{r}}|^3} = \frac{4a^2 \sqrt{5 + 3 \cos^2 t}}{8a^3(1 + \cos^2 t)^{3/2}} = \frac{\sqrt{5 + 3 \cos^2 t}}{2a\,(1 + \cos^2 t)^{3/2}}.$$

Example 10. *Show that a curve* $\mathbf{r} = \mathbf{r}(s)$ *of class* $C^m (m \geq 2)$ *is a straight line if all tangent lines are concurrent.*

Sol. The equation of the tangent line at the point $\mathbf{r}(s)$ is

$$\mathbf{R}(s) = \mathbf{r}(s) + \lambda(s)\mathbf{t}(s),$$

where $\mathbf{R}(s)$ is a general point on the tangent line and $\lambda(s)$ is a parameter.

Let all tangent lines intersect at the point $\mathbf{r}_0(s)$.

$\therefore$ $\mathbf{r}_0(s) = \mathbf{r}(s) + \lambda_0(s)\mathbf{t}(s)$ for some value $\lambda_0(s)$ of $\lambda(s)$

Differentiating w.r.t. s, we get

$$0 = \mathbf{r}'(s) + \lambda_0(s)\mathbf{t}'(s) + \lambda_0'(s)\mathbf{t}(s)$$

$$\Rightarrow \quad 0 = \mathbf{t}(s) + \lambda_0(s)\mathbf{t}'(s) + \lambda_0'(s)\mathbf{t}(s)$$

$$\Rightarrow \quad 0 = (1 + \lambda_0'(s))\mathbf{t}(s) + \lambda_0(s)\mathbf{t}'(s)$$

Multiplying by $\mathbf{t}'(s)$, we get

$$\Rightarrow \quad 0 = (1 + \lambda_0'(s))\,(\mathbf{t}(s) \cdot \mathbf{t}'(s)) + \lambda_0(s)(\mathbf{t}'(s) \,.\, \mathbf{t}'(s))$$

$$\Rightarrow \quad 0 = (1 + \lambda_0'(s)) \cdot 0 + \lambda_0(s)\, |\mathbf{t}'(s)|^2$$

$$\Rightarrow \quad \lambda_0(s)\, |\mathbf{t}'(s)|^2 = 0 \Rightarrow |\mathbf{t}'(s)| = 0 \Rightarrow \mathbf{t}'(s) = 0 \qquad (\text{Assuming } \lambda_0(s) \neq 0)$$

$$\Rightarrow \quad \mathbf{t}(s) = \mathbf{c}, \text{ a constant vector}$$

$$\Rightarrow \quad \mathbf{r}'(s) = \mathbf{c} \Rightarrow \mathbf{r}(s) = s\mathbf{c} + \mathbf{d}, \text{ where } \mathbf{d} \text{ is a constant vector}$$

$\therefore$ The curve $\mathbf{r} = \mathbf{r}(s)$ is a straight line passing through the point whose position vector is **d** and is parallel to the vector **c**.

WORKING RULES FOR SOLVING PROBLEMS

Rule I. (*i*) $\boldsymbol{\kappa} = \mathbf{r}'' = \mathbf{t}'$ (*ii*) $\kappa = |\mathbf{r}''| = |\mathbf{t}'|$

Rule II. *Radius of curvature,* $\rho = \dfrac{1}{\kappa}$

Rule III. (*i*) $\kappa = |\mathbf{r}' \times \mathbf{r}''|$ (*ii*) $\kappa = \dfrac{|\dot{\mathbf{r}} \times \ddot{\mathbf{r}}|}{|\dot{\mathbf{r}}|^3}$.

Rule IV. *Curve is a straight line if and only if its curvature is identically zero.*

EXERCISE 2.1

1. For the curve $\mathbf{r} = a \cos t\mathbf{i} + b \sin t\mathbf{j}$, $a, b > 0$, find the curvature at point t.
2. For the curve $\mathbf{r} = \cosh t\mathbf{i} + \sinh t\mathbf{j}$, find the curvature at point t'.
3. For the curve $\mathbf{r} = t\mathbf{i} + t^{3/2}\mathbf{j}$, $t > 0$, find the curvature at point t.
4. Show that a curve $\mathbf{r} = \mathbf{r}(t)$ of class C^m $(m \geq 2)$, where t is an arbitrary parameter, is a straight line if $\dot{\mathbf{r}}(t)$ and $\ddot{\mathbf{r}}(t)$ are linearly dependent for all t.
5. For the curve $\mathbf{r} = t\mathbf{i} + t^2\mathbf{j} + t^3\mathbf{k}$, find the curvature at the point (0, 0, 0).
6. Show that the curvature of a circle of radius a is equal to $1/a$.
7. Let $\mathbf{r} = \mathbf{r}(t)$ be a regular curve of class $C^m(m \geq 2)$, where t is an arbitrary parameter. Show that:
$$\kappa = \frac{\sqrt{(\dot{\mathbf{r}}\,.\,\dot{\mathbf{r}})\,(\ddot{\mathbf{r}}\,.\,\ddot{\mathbf{r}}) - (\dot{\mathbf{r}}\,.\,\ddot{\mathbf{r}})^2}}{(\dot{\mathbf{r}}\,.\,\dot{\mathbf{r}})^{3/2}}.$$
8. Show that for a curve $y = y(x)$ in the xy-plane:
$$\kappa(x) = \frac{|y''|}{(1 + y'^2)^{3/2}}.$$
9. For the following curves in the xy-plane, find curvature: (*i*) $y = x^2$ (*ii*) $xy = \lambda$.
10. For the curve $x = a(3t - t^3)$, $y = 3at^2$, $z = a(3t + t^3)$, show that:
$$\kappa = \frac{1}{3a(1 + t^2)^2}.$$
11. For the curve $x = t$, $y = t^2$, $z = t^3$, show that:
$$\kappa^2 = \frac{4(9t^4 + 9t^2 + 1)}{(9t^4 + 4t^2 + 1)^3}.$$

12. For the helix $x = a \cos t$, $y = a \sin t$, $z = at \cot \alpha$, show that: $\kappa = \frac{1}{a} \sin^2 \alpha$.

13. Find the curvature of the curve given by $\mathbf{r} = a(t - \sin t)\mathbf{i} + a(1 - \cos t)\mathbf{j} + bt\mathbf{k}$.

14. For the curve $x = 3t$, $y = 3t^2$, $z = 2t^3$, show that: $\rho = \frac{3}{2}(1 + 2t^2)^2$.

Answers

1. $\dfrac{ab}{(a^2 \sin^2 t + b^2 \cos^2 t)^{3/2}}$ **2.** $\dfrac{1}{\cosh^2 2t}$ **3.** $\dfrac{6}{\sqrt{t}\,(4 + 9t)^{3/2}}$

5. 2 **9.** (*i*) $\dfrac{2}{(1 + 4x^2)^{3/2}}$ (*ii*) $\dfrac{2\lambda x^3}{(x^4 + \lambda^2)^{3/2}}$

13. $\dfrac{a\left(b^2 + 4a^2 \sin^4 \dfrac{t}{2}\right)^{1/2}}{\left(b^2 + 4a^2 \sin^2 \dfrac{t}{2}\right)^{3/2}}$.

Hints

4. Let $\dot{\mathbf{r}}(t) = \lambda \ddot{\mathbf{r}}(t)$.

7. $(\dot{\mathbf{r}} \cdot \dot{\mathbf{r}})(\ddot{\mathbf{r}} \cdot \ddot{\mathbf{r}}) - (\dot{\mathbf{r}} \cdot \ddot{\mathbf{r}})^2 = |\dot{\mathbf{r}}|^2 |\ddot{\mathbf{r}}|^2 - (|\dot{\mathbf{r}}||\ddot{\mathbf{r}}| \cos \theta)^2 = |\dot{\mathbf{r}}|^2 |\ddot{\mathbf{r}}|^2 (1 - \cos^2 \theta) = |\dot{\mathbf{r}} \times \ddot{\mathbf{r}}|^2$.

3. TORSION OF A CURVE

Let $\mathbf{r} = \mathbf{r}(s)$ be a regular curve C of class $C^m (m \geq 3)$ and $\mathbf{r}''(s) \neq 0$, where s is the parameter 'arc length'.

$$\mathbf{r}''(s) \neq \mathbf{0} \quad \Rightarrow \mathbf{n}(s) = \frac{\mathbf{t}'(s)}{|\mathbf{t}'(s)|} = \frac{\mathbf{t}'(s)}{|\mathbf{r}''(s)|} \quad \Rightarrow \mathbf{n}(s) \text{ is defined.}$$

Also,
$$\mathbf{b}(s) = \mathbf{t}(s) \times \mathbf{n}(s) = \mathbf{r}'(s) \times \frac{\mathbf{r}''(s)}{|\mathbf{r}''(s)|} = \frac{\mathbf{r}'(s) \times \mathbf{r}''(s)}{|\mathbf{r}''(s)|} \quad \Rightarrow \mathbf{b}(s) \text{ is defined.}$$

∴ $\mathbf{n}(s)$ and $\mathbf{b}(s)$ exist at the point $\mathbf{r}(s)$.

Since $\mathbf{r}'''(s)$ exists, the binormal $\mathbf{b}(s)$ is differentiable w.r.t. s.

∴ $\mathbf{b}'(s)$ exists.

The scalar quantity $-\mathbf{n}(s) \cdot \mathbf{b}'(s)$ is called the **torsion** of the curve C at the point $\mathbf{r}(s)$ and it is denoted by $\tau(s)$ (or by τ).

The reciprocal of the torsion is called the **radius of torsion** at that point and it is denoted by σ.

∴
$$\sigma = \frac{1}{\tau}. \qquad \text{(Assuming } \tau \neq 0\text{)}$$

Remark. It can be proved that a curve is uniquely determined (except for its position in space) if we are given its curvature $\kappa(\neq 0)$ and torsion τ as continuous functions of arc length s. This result shows the importance of curvature and torsion in the study of differential geometry of space curves.

Example 1. *For the helix* $\mathbf{r} = a \cos t\mathbf{i} + a \sin t\mathbf{j} + bt\mathbf{k}$, $a > 0$, $b \neq 0$, *find the torsion at the point t.*

Sol. We have
$$\mathbf{r} = a \cos t\mathbf{i} + a \sin t\mathbf{j} + bt\mathbf{k}$$

$\therefore$
$$\dot{\mathbf{r}} = -a \sin t\mathbf{i} + a \cos t\mathbf{j} + b\mathbf{k}$$

$\Rightarrow$
$$|\dot{\mathbf{r}}| = \sqrt{a^2 \sin^2 t + a^2 \cos^2 t + b^2} = \sqrt{a^2 + b^2}$$

$\therefore$
$$\mathbf{t} = \frac{\dot{\mathbf{r}}}{|\dot{\mathbf{r}}|} = \frac{1}{\sqrt{a^2 + b^2}}(-a \sin t\mathbf{i} + a \cos t\mathbf{j} + b\mathbf{k})$$

$$\mathbf{t}' = \frac{d\mathbf{t}}{ds} = \frac{d\mathbf{t}}{dt}\frac{dt}{ds} = \frac{d\mathbf{t}}{dt}\Big/\frac{ds}{dt}$$

$$= \frac{1}{\sqrt{a^2 + b^2}}(-a \cos t\mathbf{i} - a \sin t\mathbf{j}) \;/|\dot{\mathbf{r}}|$$

$$= -\frac{a}{\sqrt{a^2 + b^2}}(\cos t\mathbf{i} + \sin t\mathbf{j}) \Big/ \sqrt{a^2 + b^2}$$

$$= -\frac{a}{a^2 + b^2}(\cos t\mathbf{i} + \sin t\mathbf{j})$$

$\therefore$
$$|\mathbf{t}'| = \frac{a}{a^2 + b^2}(\cos^2 t + \sin^2 t)^{1/2} = \frac{a}{a^2 + b^2}$$

$\therefore$
$$\mathbf{n} = \frac{\mathbf{t}'}{|\mathbf{t}'|} = -\frac{a}{a^2 + b^2}(\cos t\mathbf{i} + \sin t\mathbf{j}) \cdot \frac{a^2 + b^2}{a} = -(\cos t\mathbf{i} + \sin t\mathbf{j})$$

$\therefore$
$$\mathbf{b} = \mathbf{t} \times \mathbf{n} = \begin{vmatrix} \mathbf{i} & \mathbf{j} & \mathbf{k} \\ -\frac{a}{\sqrt{a^2 + b^2}}\sin t & \frac{a}{\sqrt{a^2 + b^2}}\cos t & \frac{b}{\sqrt{a^2 + b^2}} \\ -\cos t & -\sin t & 0 \end{vmatrix}$$

$$= -\frac{1}{\sqrt{a^2 + b^2}}\begin{vmatrix} \mathbf{i} & \mathbf{j} & \mathbf{k} \\ -a \sin t & a \cos t & b \\ \cos t & \sin t & 0 \end{vmatrix}$$

$$= -\frac{1}{\sqrt{a^2 + b^2}}[-b \sin t\mathbf{i} + b \cos t\mathbf{j} - a\mathbf{k}]$$

$\therefore$
$$\mathbf{b}' = \frac{d\mathbf{b}}{ds} = \frac{d\mathbf{b}}{dt}\frac{dt}{ds} = \frac{d\mathbf{b}}{dt}\Big/\frac{ds}{dt}$$

$$= -\frac{1}{\sqrt{a^2 + b^2}}(-b \cos t\mathbf{i} - b \sin t\mathbf{j} - 0\mathbf{k}) \;/|\dot{\mathbf{r}}|$$

$$= \frac{1}{\sqrt{a^2+b^2}} (b \cos t\mathbf{i} + b \sin t\mathbf{j}) \Big/ \sqrt{a^2+b^2}$$

$$= \frac{b}{a^2+b^2} (\cos t\mathbf{i} + \sin t\mathbf{j})$$

$\therefore$ Torsion, $\tau = -\mathbf{n} \cdot \mathbf{b}'$

$$= (\cos t\mathbf{i} + \sin t\mathbf{j}) \cdot \frac{b}{a^2+b^2} (\cos t\mathbf{i} + \sin t\mathbf{j})$$

$$= \frac{b}{a^2+b^2} \cdot (\cos^2 t + \sin^2 t) = \frac{b}{a^2+b^2}.$$

Note. Torsion at each point of a helix is always constant.

Theorem 1. *(Serret-Frenet formulae). Let $r = r(s)$ be a regular curve of class $C^m (m \geq 3)$, where s is the parameter 'arc length' and $r''(s) \neq 0$. Prove that*

(i) $t' = \kappa n$ *(ii)* $n' = -\kappa t + \tau b$ *(iii)* $b' = -\tau n$.

Proof. *(i)* $\kappa\mathbf{n} = |\mathbf{r}''|\mathbf{n} = |\mathbf{t}'| \left(\frac{\mathbf{t}'}{|\mathbf{t}'|}\right) = \mathbf{t}'$

$\therefore$ $\mathbf{t}' = \kappa\mathbf{n}$.

(iii) $(\mathbf{b} \cdot \mathbf{t})' = \mathbf{b} \cdot \mathbf{t}' + \mathbf{b}' \cdot \mathbf{t} = \mathbf{b} \cdot (\kappa\mathbf{n}) + \mathbf{b}' \cdot \mathbf{t}$

$= \kappa(\mathbf{b} \cdot \mathbf{n}) + \mathbf{b}' \cdot \mathbf{t} = 0 + \mathbf{b}' \cdot \mathbf{t} = \mathbf{b}' \cdot \mathbf{t}$

Also, $(\mathbf{b} \,.\, \mathbf{t})' = 0' = 0$

$\Rightarrow$ $\mathbf{b}' \,.\, \mathbf{t} = 0$ $\Rightarrow$ $\mathbf{b}'$ is perpendicular to $\mathbf{t}$.

Also, $|\mathbf{b}| = 1$ $\Rightarrow$ $\mathbf{b} \cdot \mathbf{b} = 1$ $\Rightarrow$ $\mathbf{b} \cdot \mathbf{b}' + \mathbf{b}' \cdot \mathbf{b} = 0$ $\Rightarrow$ $\mathbf{b}' \cdot \mathbf{b} = 0$

$\Rightarrow$ $\mathbf{b}'$ is perpendicular to $\mathbf{b}$.

$\therefore$ $\mathbf{b}'$ is perpendicular to the plane determined by $\mathbf{t}$ and $\mathbf{b}$.

$\therefore$ $\mathbf{b}'$ is parallel to $\mathbf{n}$.

Let $\mathbf{b}' = \lambda\mathbf{n}$.

$\Rightarrow$ $\mathbf{n} \,.\, \mathbf{b}' = \mathbf{n} \cdot (\lambda\mathbf{n}) = \lambda(\mathbf{n} \cdot \mathbf{n}) = \lambda \cdot 1 = \lambda$

$\Rightarrow$ $-\tau = \lambda$ $(\because \ \tau = -\mathbf{n} \,.\, \mathbf{b}')$

$\therefore$ $\mathbf{b}' = -\tau\mathbf{n}$.

(ii) We have $\mathbf{n} = \mathbf{b} \times \mathbf{t}$.

$\therefore$ $\mathbf{n}' = (\mathbf{b} \times \mathbf{t})' = \mathbf{b} \times \mathbf{t}' + \mathbf{b}' \times \mathbf{t} = \mathbf{b} \times (\kappa\mathbf{n}) + (-\tau\mathbf{n}) \times \mathbf{t}$

(Using *(i)* and *(iii)*)

$= \kappa(\mathbf{b} \times \mathbf{n}) - \tau(\mathbf{n} \times \mathbf{t}) = \kappa(-\mathbf{t}) - \tau(-\mathbf{b}) = -\kappa\mathbf{t} + \tau\mathbf{b}$

$\therefore$ $\mathbf{n}' = -\kappa\mathbf{t} + \tau\mathbf{b}$.

Remark 1. Serret-Frenet equations shows that we can express the vectors $\mathbf{t}'$, $\mathbf{n}'$, $\mathbf{b}'$ as linear combinations of the vectors $\mathbf{t}$, $\mathbf{n}$, $\mathbf{b}$.

Remark 2. We have proved equation *(iii)* before proving equation *(ii)* because the result of *(iii)* is used in proving *(ii)*.

Remark 3. The Serret-Frenet equations can also be written as

$$\mathbf{t}' = 0\mathbf{t} + \kappa\mathbf{n} + 0\mathbf{b}$$
$$\mathbf{n}' = -\kappa\mathbf{t} + 0\mathbf{n} + \tau\mathbf{b}$$
$$\mathbf{b}' = 0\mathbf{t} - \tau\mathbf{n} + 0\mathbf{b}.$$

In the above equations, the coefficients of $\mathbf{t}$, $\mathbf{n}$ and $\mathbf{b}$ form the matrix $\begin{bmatrix} 0 & \kappa & 0 \\ -\kappa & 0 & \tau \\ 0 & -\tau & 0 \end{bmatrix}$.

Remark 4. In the above discussion, **t**, **n**, **b**, **t**′, **n**′, **b**′, κ, τ are all functions of the parameter s. For the sake of simplicity, we have written **t**(s) as **t** etc.

Example 2. *Show that along the curve* **r** = **r**(*s*), κτ = | **t**′ · **b**′ |.

Sol. We have $\mathbf{t}' \cdot \mathbf{b}' = (\kappa \mathbf{n}) \cdot (-\tau \mathbf{n}) = -\kappa\tau(\mathbf{n} . \mathbf{n}) = -\kappa\tau(1) = -\kappa\tau$

$$\therefore \quad |\mathbf{t}' \cdot \mathbf{b}'| = |-\kappa\tau| = \kappa\tau.$$

Example 3. *Show that along the curve* **r** = **r**(*s*), τ = [**t n n**′], *provided* κ ≠ 0.

Sol.

$$\mathbf{n} \times \mathbf{n}' = \mathbf{n} \times (-\kappa\mathbf{t} + \tau\mathbf{b}) = -\kappa(\mathbf{n} \times \mathbf{t}) + \tau(\mathbf{n} \times \mathbf{b})$$

$$= -\kappa(-\mathbf{b}) + \tau(\mathbf{t}) = \kappa\mathbf{b} + \tau\mathbf{t}$$

$$\therefore \quad [\mathbf{t}\ \mathbf{n}\ \mathbf{n}'] = \mathbf{t} \cdot (\mathbf{n} \times \mathbf{n}') = \mathbf{t} \cdot (\kappa\mathbf{b} + \tau\mathbf{t})$$

$$= \kappa(\mathbf{t} \cdot \mathbf{b}) + \tau(\mathbf{t} . \mathbf{t}) = \kappa(0) + \tau(1) = \tau$$

$$\therefore \quad \tau = [\mathbf{t}\ \mathbf{n}\ \mathbf{n}'].$$

Example 4. *Show that along the curve* **r** = **r**(*s*), $\tau = \dfrac{[\mathbf{r}'\ \mathbf{r}''\ \mathbf{r}''']}{\kappa^2}$, *provided* κ ≠ 0.

Sol. We have $\mathbf{r}' = \mathbf{t}$ and $\mathbf{r}'' = \mathbf{t}' = \kappa\mathbf{n}$

$$\therefore \quad \mathbf{r}''' = \frac{d\mathbf{r}''}{ds} = \frac{d\mathbf{t}'}{ds} = \frac{d}{ds}(\kappa\mathbf{n}) = \kappa\mathbf{n}' + \kappa'\mathbf{n}$$

$$= \kappa(-\kappa\mathbf{t} + \tau\mathbf{b}) + \kappa'\mathbf{n} \qquad (\text{Using } \mathbf{n}' = -\kappa\mathbf{t} + \tau\mathbf{b})$$

$$= -\kappa^2\mathbf{t} + \kappa\tau\mathbf{b} + \kappa'\mathbf{n}$$

$$\therefore \quad \mathbf{r}'' \times \mathbf{r}''' = \kappa\mathbf{n} \times (-\kappa^2\mathbf{t} + \kappa\tau\mathbf{b} + \kappa'\mathbf{n})$$

$$= -\kappa^3(\mathbf{n} \times \mathbf{t}) + \kappa^2\tau(\mathbf{n} \times \mathbf{b}) + \kappa\kappa'(\mathbf{n} \times \mathbf{n})$$

$$= -\kappa^3(-\mathbf{b}) + \kappa^2\tau\mathbf{t} + \kappa\kappa'\mathbf{0} = \kappa^3\mathbf{b} + \kappa^2\tau\mathbf{t}$$

$$\therefore \quad [\mathbf{r}'\ \mathbf{r}''\ \mathbf{r}'''] = \mathbf{r}' \cdot (\mathbf{r}'' \times \mathbf{r}''') = \mathbf{t} \cdot (\kappa^3\mathbf{b} + \kappa^2\tau\mathbf{t})$$

$$= \kappa^3(\mathbf{t} \cdot \mathbf{b}) + \kappa^2\tau(\mathbf{t} \cdot \mathbf{t}) = 0 + \kappa^2\tau \cdot 1 = \kappa^2\tau$$

$$\therefore \quad \tau = \frac{[\mathbf{r}'\ \mathbf{r}''\ \mathbf{r}''']}{\kappa^2}.$$

Theorem 2. *Let* **r** = **r**(*t*) *be a regular curve of class* $C^m (m \geq 3)$, *where t is an arbitrary parameter. Prove that*

$$\tau = \frac{[\dot{\mathbf{r}}\ \ddot{\mathbf{r}}\ \dddot{\mathbf{r}}]}{|\dot{\mathbf{r}} \times \ddot{\mathbf{r}}|^2}, \textit{ provided } |\dot{\mathbf{r}} \times \ddot{\mathbf{r}}| \neq 0.$$

Proof. We have

$$\dot{\mathbf{r}} = \frac{d\mathbf{r}}{dt} = \frac{d\mathbf{r}}{ds}\frac{ds}{dt} = \mathbf{r}'\dot{s} = \dot{s}\mathbf{t}$$

$$\ddot{\mathbf{r}} = \frac{d\dot{\mathbf{r}}}{dt} = \frac{d}{dt}(\dot{s}\mathbf{t}) = \dot{s}\frac{d\mathbf{t}}{dt} + \ddot{s}\mathbf{t} = \dot{s}(\mathbf{t}'\dot{s}) + \ddot{s}\mathbf{t}$$

$$= \dot{s}^2(\kappa\mathbf{n}) + \ddot{s}\mathbf{t} = \ddot{s}\mathbf{t} + \kappa\dot{s}^2\mathbf{n}$$

$$\therefore \quad \dot{\mathbf{r}} \times \ddot{\mathbf{r}} = (\dot{s}\mathbf{t}) \times (\ddot{s}\mathbf{t} + \kappa\dot{s}^2\mathbf{n}) = \dot{s}\ddot{s}(\mathbf{t} \times \mathbf{t}) + \kappa\dot{s}^3(\mathbf{t} \times \mathbf{n})$$

$$= \dot{s}\ddot{s}(\mathbf{0}) + \kappa\dot{s}^3\mathbf{b} = \kappa\dot{s}^3\mathbf{b}$$

Differentiating w.r.t. t, we get

$$\dot{\mathbf{r}} \times \dddot{\mathbf{r}} + \ddot{\mathbf{r}} \times \ddot{\mathbf{r}} = \kappa'\dot{s}^3\mathbf{b} + \kappa 3\dot{s}^2\ddot{s}\mathbf{b} + \kappa\dot{s}^3(\mathbf{b}'\dot{s})$$

$$\Rightarrow \qquad \dot{\mathbf{r}} \times \dddot{\mathbf{r}} + \mathbf{0} = (\kappa' \dot{s}^3 + 3\kappa \dot{s}^2 \ddot{s})\mathbf{b} + \kappa \dot{s}^4(-\tau \mathbf{n})$$

$$\Rightarrow \qquad \dot{\mathbf{r}} \times \dddot{\mathbf{r}} = (\kappa' \dot{s}^3 + 3\kappa \dot{s}^2 \ddot{s})\mathbf{b} - \kappa \tau \dot{s}^4 \mathbf{n}$$

$$\therefore \qquad \ddot{\mathbf{r}} . (\dot{\mathbf{r}} \times \dddot{\mathbf{r}}) = (\ddot{s}\mathbf{t} + \kappa \dot{s}^2 \mathbf{n}) \cdot \left[(\kappa' \dot{s}^3 + 3\kappa \dot{s}^2 \ddot{s})\mathbf{b} - \kappa \tau \dot{s}^4 \mathbf{n}\right]$$

$$= -\kappa^2 \dot{s}^6 \tau (\mathbf{n} \cdot \mathbf{n}) \qquad \text{(Using } \mathbf{t} \cdot \mathbf{b} = 0, \mathbf{t} \cdot \mathbf{n} = 0, \mathbf{n} \cdot \mathbf{b} = 0\text{)}$$

$$\Rightarrow \qquad [\ddot{\mathbf{r}} \ \dot{\mathbf{r}} \ \dddot{\mathbf{r}}] = -\kappa^2 \dot{s}^6 \tau$$

$$\Rightarrow \qquad -[\dot{\mathbf{r}} \ \ddot{\mathbf{r}} \ \dddot{\mathbf{r}}] = -\kappa^2 \dot{s}^6 \tau$$

$$\Rightarrow \qquad [\dot{\mathbf{r}} \ \ddot{\mathbf{r}} \ \dddot{\mathbf{r}}] = \kappa^2 \dot{s}^6 \tau \qquad \text{...(1)}$$

Also $\quad |\dot{\mathbf{r}} \times \ddot{\mathbf{r}}| = \kappa \dot{s}^3 |\mathbf{b}| = \kappa \dot{s}^3$

$$\therefore \ (1) \Rightarrow \qquad [\dot{\mathbf{r}} \ \ddot{\mathbf{r}} \ \dddot{\mathbf{r}}] = |\dot{\mathbf{r}} \times \ddot{\mathbf{r}}|^2 \tau$$

$$\therefore \qquad \tau = \frac{[\dot{\mathbf{r}} \ \ddot{\mathbf{r}} \ \dddot{\mathbf{r}}]}{|\dot{\mathbf{r}} \times \ddot{\mathbf{r}}|^2}.$$

Corollary. **If** $\mathbf{r} = x(t)\mathbf{i} + y(t)\mathbf{j} + z(t)\mathbf{k}$**,** then

$$[\dot{\mathbf{r}} \ \ddot{\mathbf{r}} \ \dddot{\mathbf{r}}] = \begin{vmatrix} \dot{x} & \dot{y} & \dot{z} \\ \ddot{x} & \ddot{y} & \ddot{z} \\ \dddot{x} & \dddot{y} & \dddot{z} \end{vmatrix} \quad \text{and} \quad |\dot{\mathbf{r}} \times \ddot{\mathbf{r}}|^2 = \Sigma(\dot{y}\ddot{z} - \ddot{y}\dot{z})^2.$$

$$\therefore \qquad \tau = \frac{[\dot{\mathbf{r}} \ddot{\mathbf{r}} \dddot{\mathbf{r}}]}{|\dot{\mathbf{r}} \times \ddot{\mathbf{r}}|^2} = \begin{vmatrix} \dot{x} & \dot{y} & \dot{z} \\ \ddot{x} & \ddot{y} & \ddot{z} \\ \dddot{x} & \dddot{y} & \dddot{z} \end{vmatrix} \div \Sigma(\dot{y}\ddot{z} - \ddot{y}\dot{z})^2.$$

Remarks 1. If $\mathbf{r} = \mathbf{r}(s)$, then

$$\tau = \frac{[\mathbf{r}' \ \mathbf{r}'' \ \mathbf{r}''']}{|\mathbf{r}' \times \mathbf{r}''|^2}. \qquad \text{(Using } \kappa = |\mathbf{r}' \times \mathbf{r}''|\text{)}$$

2. If $\mathbf{r} = \mathbf{r}(t)$, then

$$\tau = \frac{[\dot{\mathbf{r}} \ \ddot{\mathbf{r}} \ \dddot{\mathbf{r}}]}{|\dot{\mathbf{r}} \times \ddot{\mathbf{r}}|^2}.$$

Theorem 3. *Prove that a regular curve of class $C^m (m \geq 3)$ is a plane curve if and only if its torsion is identically zero.*

Proof. Let $\mathbf{r} = \mathbf{r}(s)$ be a regular curve of class $C^m (m \geq 3)$, where s is the parameter 'arc length'.

Let the curve be a plane curve.

Let the plane of the curve be normal to the vector $\mathbf{a}$.

$\therefore \quad \mathbf{r} \cdot \mathbf{a} = \lambda$, where λ is some constant and $\mathbf{r} = \mathbf{r}(s)$.

$$\Rightarrow \qquad \frac{d}{ds}(\mathbf{r} \cdot \mathbf{a}) = 0$$

$$\Rightarrow \qquad \mathbf{r}' \cdot \mathbf{a} = 0 \ \Rightarrow \ \mathbf{t} \cdot \mathbf{a} = 0 \qquad \text{...(1)}$$

$$(1) \Rightarrow \quad \frac{d}{ds}(\mathbf{t} \,.\, \mathbf{a}) = 0 \quad \Rightarrow \quad \mathbf{t}' \,.\, \mathbf{a} = 0$$

$$\Rightarrow \quad \frac{\mathbf{t}'}{|\mathbf{t}'|} \,.\, \mathbf{a} = 0 \quad \Rightarrow \quad \mathbf{n} \,.\, \mathbf{a} = 0 \qquad ...(2)$$

(1) and (2) imply that **a** is perpendicular to the plane of **t** and **n.**

$\therefore$ **a** is parallel to the unit binormal vector **b.**

Let $\mathbf{b} = \mu\, \mathbf{a}.$

$$\therefore \quad \mathbf{b}' = \frac{d\mathbf{b}}{ds} = \frac{d}{ds}(\mu\, \mathbf{a}) = \mathbf{0}$$

$\therefore \quad \tau = -\,\mathbf{n} \,.\, \mathbf{b}' = -\,\mathbf{n} \,.\, \mathbf{0} = 0$ *i.e.*, the torsion is identically zero.

Conversely, let the torsion of the curve be identically zero.

By **Serret-Frenet equation,** $\mathbf{b}' = -\,\tau\mathbf{n}.$

$\therefore \quad \mathbf{b}' = 0\mathbf{n} = \mathbf{0} \quad \therefore \quad \mathbf{b} = \mathbf{b}_0$, a constant vector.

Let **r** be any point on the curve.

$$\therefore \quad \frac{d}{ds}(\mathbf{r} \,.\, \mathbf{b}_0) = \mathbf{r}' .\, \mathbf{b}_0 = \mathbf{t} \,.\, \mathbf{b}_0 = 0 \qquad (\because \mathbf{t} \,.\, \mathbf{b} = 0)$$

$\therefore \quad \mathbf{r} \,.\, \mathbf{b}_0 = \text{constant}.$

This equation represents a plane.

$\therefore$ The curve $\mathbf{r} = \mathbf{r}(s)$ lies on a plane.

$\therefore$ The result holds.

Remark. For a curve to lie in a plane it is sufficient to show that its unit binormal vector **b** is a constant vector.

Example 5. *Show that the curve* $\mathbf{r} = a \cos t\, \mathbf{i} + a \sin t\, \mathbf{j} + b\mathbf{k}$, $a > 0$, $b \neq 0$ *is a plane curve.*

Sol. We have $\mathbf{r} = a \cos t\, \mathbf{i} + a \sin t\, \mathbf{j} + b\mathbf{k}$

$$\therefore \quad \dot{\mathbf{r}} = -\,a \sin t\, \mathbf{i} + a \cos t\, \mathbf{j} + 0\mathbf{k}$$

$$|\,\dot{\mathbf{r}}\,| = \sqrt{a^2 \sin^2 t + a^2 \cos^2 t} = a$$

$$\therefore \quad \mathbf{t} = \frac{\dot{\mathbf{r}}}{|\,\dot{\mathbf{r}}\,|} = \frac{1}{a}(-\,a \sin t\, \mathbf{i} + a \cos t\, \mathbf{j})$$

$$= -\sin t\, \mathbf{i} + \cos t\, \mathbf{j}$$

$$\therefore \quad \dot{\mathbf{t}} = -\cos t\, \mathbf{i} - \sin t\, \mathbf{j} \quad \text{and} \quad |\,\dot{\mathbf{t}}\,| = \sqrt{\cos^2 t + \sin^2 t} = 1$$

$$\therefore \quad \mathbf{n} = \frac{\dot{\mathbf{t}}}{|\,\dot{\mathbf{t}}\,|} = -\frac{1}{1}(\cos t\mathbf{i} + \sin t\,\mathbf{j}) = -\,(\cos t\, \mathbf{i} + \sin t\, \mathbf{j})$$

$$\therefore \quad \mathbf{b} = \mathbf{t} \times \mathbf{n} = \begin{vmatrix} \mathbf{i} & \mathbf{j} & \mathbf{k} \\ -\sin t & \cos t & 0 \\ -\cos t & -\sin t & 0 \end{vmatrix} = \mathbf{k}$$

$\therefore$ **b** is a constant vector.

$$\Rightarrow \qquad \mathbf{b}' = \mathbf{0} \;\Rightarrow\; \tau = -\,\mathbf{n}\,.\,\mathbf{b}' = -\,\mathbf{n}\,.\,\mathbf{0} = 0.$$

∴ The given curve is a plane curve.

Alternative method

$$[\dot{\mathbf{r}}\ \ddot{\mathbf{r}}\ \dddot{\mathbf{r}}] = \begin{vmatrix} -a\sin t & a\cos t & 0 \\ -a\cos t & -a\sin t & 0 \\ a\sin t & -a\cos t & 0 \end{vmatrix} = 0$$

$$\therefore \qquad \tau = \frac{[\dot{\mathbf{r}}\ \ddot{\mathbf{r}}\ \dddot{\mathbf{r}}]}{|\dot{\mathbf{r}}\times\ddot{\mathbf{r}}|^2} = \frac{0}{|\dot{\mathbf{r}}\times\ddot{\mathbf{r}}|^2} = 0\,.$$

∴ The given curve is a plane curve.

Example 6. *Show that the curve* $\mathbf{r} = \left(t, \dfrac{1+t}{t}, \dfrac{1-t^2}{t}\right)$ *lies on a plane.*

Sol. The given curve is

$$\mathbf{r} = t\mathbf{i} + \frac{1+t}{t}\mathbf{j} + \frac{1-t^2}{t}\mathbf{k}$$

$$\therefore \qquad \dot{\mathbf{r}} = \mathbf{i} + \left(-\frac{1}{t^2}\right)\mathbf{j} + \left(-\frac{1}{t^2} - 1\right)\mathbf{k}$$

$$\ddot{\mathbf{r}} = \frac{2}{t^3}\mathbf{j} + \frac{2}{t^3}\mathbf{k}$$

and

$$\dddot{\mathbf{r}} = -\frac{6}{t^4}\mathbf{j} - \frac{6}{t^4}\mathbf{k}$$

$$\therefore \qquad [\dot{\mathbf{r}}\ \ddot{\mathbf{r}}\ \dddot{\mathbf{r}}] = \begin{vmatrix} 1 & -\dfrac{1}{t^2} & -\dfrac{1}{t^2}-1 \\ 0 & \dfrac{2}{t^3} & \dfrac{2}{t^3} \\ 0 & -\dfrac{6}{t^4} & -\dfrac{6}{t^4} \end{vmatrix} = -\frac{12}{t^7} + \frac{12}{t^7} = 0$$

$$\therefore \qquad \tau = \frac{[\dot{\mathbf{r}}\ \ddot{\mathbf{r}}\ \dddot{\mathbf{r}}]}{|\dot{\mathbf{r}}\times\ddot{\mathbf{r}}|^2} = \frac{0}{|\dot{\mathbf{r}}\times\ddot{\mathbf{r}}|^2} = 0\,.$$

∴ The given curve lies on a plane.

Example 7. *Find the torsion of the curve* $\mathbf{r} = t\mathbf{i} + t^2\mathbf{j} + t^3\mathbf{k}$ at the point '*t*'.

Sol. We have $\qquad \mathbf{r} = t\mathbf{i} + t^2\mathbf{j} + t^3\mathbf{k}.$

$$\therefore \qquad \dot{\mathbf{r}} = \mathbf{i} + 2t\mathbf{j} + 3t^2\mathbf{k},$$

$$\ddot{\mathbf{r}} = 2\mathbf{j} + 6t\mathbf{k} \quad \text{and} \quad \dddot{\mathbf{r}} = 6\mathbf{k}$$

Now,

$$\tau = \frac{[\dot{\mathbf{r}}\ \ddot{\mathbf{r}}\ \dddot{\mathbf{r}}]}{|\dot{\mathbf{r}}\times\ddot{\mathbf{r}}|^2} \qquad \text{...(1)}$$

$$\dot{\mathbf{r}}\times\ddot{\mathbf{r}} = \begin{vmatrix} \mathbf{i} & \mathbf{j} & \mathbf{k} \\ 1 & 2t & 3t^2 \\ 0 & 2 & 6t \end{vmatrix} = 6t^2\mathbf{i} - 6t\mathbf{j} + 2\mathbf{k}$$

$$\therefore \quad |\dot{\mathbf{r}} \times \ddot{\mathbf{r}}|^2 = 36t^4 + 36t^2 + 4 = 4(1 + 9t^2 + 9t^4)$$

$$[\dot{\mathbf{r}}\ \ddot{\mathbf{r}}\ \dddot{\mathbf{r}}] = \begin{vmatrix} 1 & 2t & 3t^2 \\ 0 & 2 & 6t \\ 0 & 0 & 6 \end{vmatrix} = 12$$

$$\therefore \quad (1) \Rightarrow \qquad \tau = \frac{12}{4(1+9t^2+9t^4)} = \frac{3}{1+9t^2+9t^4}.$$

Example 8. *For the curve* $\mathbf{r} = (at - a \sin t)\mathbf{i} + (a - a \cos t)\mathbf{j} + bt\mathbf{k}$, *find the torsion at the point 't'.*

Sol. We have $\mathbf{r} = (at - a \sin t)\mathbf{i} + (a - a \cos t)\mathbf{j} + bt\mathbf{k}$

$$\therefore \quad \dot{\mathbf{r}} = (a - a \cos t)\mathbf{i} + a \sin t\mathbf{j} + b\mathbf{k}$$

$$\ddot{\mathbf{r}} = a \sin t\ \mathbf{i} + a \cos t\mathbf{j} + 0\mathbf{k} \quad \text{and} \quad \dddot{\mathbf{r}} = a \cos t\mathbf{i} - a \sin t\mathbf{j} + 0\mathbf{k}$$

Now,
$$\tau = \frac{[\dot{\mathbf{r}}\ \ddot{\mathbf{r}}\ \dddot{\mathbf{r}}]}{|\dot{\mathbf{r}} \times \ddot{\mathbf{r}}|^2} \qquad ...(1)$$

$$\dot{\mathbf{r}} \times \ddot{\mathbf{r}} = \begin{vmatrix} \mathbf{i} & \mathbf{j} & \mathbf{k} \\ a - a\cos t & a \sin t & b \\ a \sin t & a \cos t & 0 \end{vmatrix}$$

$$= -ab \cos t\ \mathbf{i} + ab \sin t\ \mathbf{j} + (a^2 \cos t - a^2)\mathbf{k}$$

$$\therefore \quad |\dot{\mathbf{r}} \times \ddot{\mathbf{r}}|^2 = a^2b^2 \cos^2 t + a^2b^2 \sin^2 t + a^4 (\cos t - 1)^2$$

$$= a^2b^2 + a^4\left(-2\sin^2\frac{t}{2}\right)^2 = a^2\left[b^2 + 4a^2 \sin^4 \frac{t}{2}\right]$$

Also,
$$[\dot{\mathbf{r}}\ \ddot{\mathbf{r}}\ \dddot{\mathbf{r}}] = \begin{vmatrix} a - a\cos t & a \sin t & b \\ a \sin t & a \cos t & 0 \\ a \cos t & -a \sin t & 0 \end{vmatrix} = b[-a^2\sin^2 t - a^2\cos^2 t] = -a^2b$$

$$\therefore \quad (1) \Rightarrow \qquad \tau = \frac{-a^2b}{a^2\left(b^2 + 4a^2 \sin^4 \frac{t}{2}\right)} = -\frac{b}{b^2 + 4a^2 \sin^4 \frac{t}{2}}.$$

Example 9. *For the curve* $x = a \tan t$, $y = a \cot t$, $z = \sqrt{2}\ a \log \tan t$, *prove that*

$$\rho = \sigma = \frac{2\sqrt{2}\ a}{\sin^2 2t}.$$

Sol. The given curve is

$$x = a \tan t,\ y = a \cot t,\ z = \sqrt{2}\ a \log \tan t.$$

Let $\mathbf{r}$ be the position vector of the point (x, y, z) on the curve.

$$\therefore \quad \mathbf{r} = x\mathbf{i} + y\mathbf{j} + z\mathbf{k} = a \tan t\mathbf{i} + a \cot t\mathbf{j} + \sqrt{2} a \log \tan t\mathbf{k}$$

$$\therefore \quad \mathbf{t} = \mathbf{r}' = \dot{\mathbf{r}}\frac{dt}{ds}$$

$$= \left(a \sec^2 t\mathbf{i} - a \operatorname{cosec}^2 t\mathbf{j} + \sqrt{2}\ a \frac{\sec^2 t}{\tan t}\mathbf{k}\right)\frac{dt}{ds}$$

$$\therefore \quad \mathbf{t} = a\left(\sec^2 t\mathbf{i} - \text{cosec}^2 t\mathbf{j} + \frac{\sqrt{2}}{\sin t \cos t}\mathbf{k}\right)\frac{dt}{ds} \quad \ldots(1)$$

$$\Rightarrow \quad |\,\mathbf{t}\,| = a\left(\sec^4 t + \text{cosec}^4 t + \frac{2}{\sin^2 t \cos^2 t}\right)^{1/2}\frac{dt}{ds}$$

$$\Rightarrow \quad \frac{ds}{dt}\,.\,1 = a\left(\frac{\sin^4 t + \cos^4 t + 2\sin^2 t\cos^2 t}{\sin^4 t \cos^4 t}\right)^{1/2} = \frac{a}{\sin^2 t \cos^2 t}$$

$$\therefore \quad \frac{dt}{ds} = \frac{\sin^2 t \cos^2 t}{a}$$

$$\therefore \ (1) \Rightarrow \quad \mathbf{t} = a\left(\sec^2 t\mathbf{i} - \text{cosec}^2 t\mathbf{j} + \frac{\sqrt{2}}{\sin t \cos t}\mathbf{k}\right).\ \frac{\sin^2 t\cos^2 t}{a}$$

or $$\mathbf{t} = (\sin^2 t\mathbf{i} - \cos^2 t\mathbf{j} + \sqrt{2}\,\sin t \cos t\mathbf{k})$$

$$\therefore \quad \mathbf{t}' = \dot{\mathbf{t}}\,\frac{dt}{ds} = (2\sin t\cos t\mathbf{i} + 2\cos t \sin t\mathbf{j} + \sqrt{2}\,\cos 2t\mathbf{k})\,.\,\frac{\sin^2 t\cos^2 t}{a}$$

$$\Rightarrow \quad \kappa\mathbf{n} = \frac{\sin^2 t\cos^2 t}{a}(\sin 2t\mathbf{i} + \sin 2t\mathbf{j} + \sqrt{2}\,\cos 2t\mathbf{k}) \quad \ldots(2)$$

(Using $\mathbf{t}' = \kappa\mathbf{n}$)

$$\Rightarrow \quad \kappa|\,\mathbf{n}\,| = \frac{\sin^2 t\cos^2 t}{a}(\sin^2 2t + \sin^2 2t + 2\cos^2 2t)^{1/2} = \frac{\sqrt{2}\sin^2 t\cos^2 t}{a}$$

$$\therefore \quad \kappa = \frac{\sqrt{2}\sin^2 t\cos^2 t}{a} \quad \text{and} \quad \rho = \frac{1}{\kappa} = \frac{a}{\sqrt{2}\sin^2 t\cos^2 t}$$

$$\therefore \ (2) \Rightarrow \quad \mathbf{n} = \frac{1}{\sqrt{2}}(\sin 2t\mathbf{i} + \sin 2t\mathbf{j} + \sqrt{2}\,\cos 2t\mathbf{k})$$

$$\Rightarrow \quad \mathbf{n}' = \dot{\mathbf{n}}\,\frac{dt}{ds} = \frac{1}{\sqrt{2}}(2\cos 2t\mathbf{i} + 2\cos 2t\mathbf{j} - 2\sqrt{2}\,\sin 2t\mathbf{k})\,\frac{\sin^2 t\cos^2 t}{a}$$

$$\Rightarrow \quad \tau\mathbf{b} - \kappa\mathbf{t} = \frac{\sqrt{2}\sin^2 t\cos^2 t}{a}(\cos 2t\mathbf{i} + \cos 2t\mathbf{j} - \sqrt{2}\,\sin 2t\mathbf{k})$$

(Using $\mathbf{n}' = \tau\mathbf{b} - \kappa\mathbf{t}$)

$$\Rightarrow \quad (\tau\mathbf{b} - \kappa\mathbf{t})\cdot(\tau\mathbf{b} - \kappa\mathbf{t})$$

$$= \frac{2\sin^4 t\cos^4 t}{a^2}(\cos^2 2t + \cos^2 2t + 2\sin^2 2t)$$

$$\Rightarrow \quad \tau^2 + \kappa^2 = \frac{4\sin^4 t\cos^4 t}{a^2}$$

$$\Rightarrow \quad \tau^2 = \frac{4\sin^4 t\cos^4 t}{a^2} - \frac{2\sin^4 t\cos^4 t}{a^2} = \frac{2\sin^4 t\cos^4 t}{a^2}$$

$$\therefore \quad \tau = \frac{\sqrt{2}\sin^2 t\cos^2 t}{a} \quad \text{and} \quad \sigma = \frac{1}{\tau} = \frac{a}{\sqrt{2}\sin^2 t\cos^2 t}$$

$$\therefore \quad \rho = \sigma = \frac{a}{\sqrt{2}\sin^2 t\cos^2 t} = \frac{2\sqrt{2}\,a}{\sin^2 2t}.$$

Example 10. *Prove that for the curve of intersection of the surfaces $x^2 + y^2 = z^2$ and $z = a\tan^{-1}\dfrac{y}{x}$:*

$$\rho = \frac{a(2+\theta^2)^{3/2}}{(8+5\theta^2+\theta^4)^{1/2}} \quad \textit{and} \quad \sigma = \frac{a(8+5\theta^2+\theta^4)}{6+\theta^2}, \quad \textit{where} \quad y = x\tan\theta.$$

Sol. Given surfaces are

$$x^2 + y^2 = z^2 \qquad ...(1) \qquad \text{and} \qquad z = a\tan^{-1}\frac{y}{x} \qquad ...(2)$$

Let $y = x\tan\theta \quad \therefore \quad (2) \Rightarrow z = a\tan^{-1}(\tan\theta) = a\theta$

(1) $\Rightarrow \quad x^2 + x^2\tan^2\theta = a^2\theta^2 \quad \Rightarrow x^2\sec^2\theta = a^2\theta^2 \quad \Rightarrow x = a\theta\cos\theta$

$\therefore \quad y = (a\theta\cos\theta)\tan\theta = a\theta\sin\theta$

$\therefore$ The parametric equations of the given curve are

$$x = a\theta\cos\theta,\ y = a\theta\sin\theta,\ z = a\theta.$$

Let **r** be the position vector of the point (x, y, z) on the curve.

$$\therefore \quad \mathbf{r} = x\mathbf{i} + y\mathbf{j} + z\mathbf{k} = a\theta\cos\theta\mathbf{i} + a\theta\sin\theta\mathbf{j} + a\theta\mathbf{k}$$

$$\therefore \quad \dot{\mathbf{r}} = a\,[(\cos\theta - \theta\sin\theta)\mathbf{i} + (\sin\theta + \theta\cos\theta)\mathbf{j} + \mathbf{k}]$$

$$\ddot{\mathbf{r}} = a\,[(-2\sin\theta - \theta\cos\theta)\mathbf{i} + (2\cos\theta - \theta\sin\theta)\mathbf{j}]$$

$$\dddot{\mathbf{r}} = a\,[(-3\cos\theta + \theta\sin\theta)\mathbf{i} + (-3\sin\theta - \theta\cos\theta)\mathbf{j}]$$

$$\therefore \quad \dot{\mathbf{r}} \times \ddot{\mathbf{r}} = a^2 \begin{vmatrix} \mathbf{i} & \mathbf{j} & \mathbf{k} \\ \cos\theta - \theta\sin\theta & \sin\theta + \theta\cos\theta & 1 \\ -2\sin\theta - \theta\cos\theta & 2\cos\theta - \theta\sin\theta & 0 \end{vmatrix}$$

$$= a^2\,[(-2\cos\theta + \theta\sin\theta)\mathbf{i} + (-2\sin\theta - \theta\cos\theta)\mathbf{j}$$
$$+ (2\cos^2\theta - \theta\sin\theta\cos\theta - 2\theta\sin\theta\cos\theta + \theta^2\sin^2\theta$$
$$+ 2\sin^2\theta + 2\theta\sin\theta\cos\theta + \theta\sin\theta\cos\theta + \theta^2\cos^2\theta)\mathbf{k}]$$
$$= a^2\,[(-2\cos\theta + \theta\sin\theta)\mathbf{i} - (2\sin\theta + \theta\cos\theta)\mathbf{j} + (2 + \theta^2)\mathbf{k}]$$

$$[\dot{\mathbf{r}}\ \ddot{\mathbf{r}}\ \dddot{\mathbf{r}}] = \dot{\mathbf{r}}\,.\,(\ddot{\mathbf{r}} \times \dddot{\mathbf{r}}) = (\dot{\mathbf{r}} \times \ddot{\mathbf{r}})\,.\,\dddot{\mathbf{r}}$$
$$= a^2\,[(-2\cos\theta + \theta\sin\theta)\mathbf{i} - (2\sin\theta + \theta\cos\theta)\mathbf{j}$$
$$+ (2+\theta^2)\mathbf{k}]\,.\,a\,[(-3\cos\theta + \theta\sin\theta)\mathbf{i} - (3\sin\theta + \theta\cos\theta)\mathbf{j}]$$
$$= a^3\,[(6\cos^2\theta + \theta^2\sin^2\theta - 5\theta\sin\theta\cos\theta)$$
$$+ (6\sin^2\theta + \theta^2\cos^2\theta + 5\theta\sin\theta\cos\theta)]$$
$$= a^3(6 + \theta^2)$$

$$|\dot{\mathbf{r}}|^2 = a^2 [(\cos^2\theta + \theta^2 \sin^2\theta - 2\theta\cos\theta\sin\theta) + \sin^2\theta + \theta^2\cos^2\theta + 2\theta\sin\theta\cos\theta + 1]$$

$$= a^2 (2 + \theta^2)$$

$$|\dot{\mathbf{r}} \times \ddot{\mathbf{r}}|^2 = a^4[(4\cos^2\theta + \theta^2\sin^2\theta - 4\theta\cos\theta\sin\theta) + (4\sin^2\theta + \theta^2\cos^2\theta + 4\theta\sin\theta\cos\theta) + (2+\theta^2)^2]$$

$$= a^4 [4 + \theta^2 + 4 + \theta^4 + 4\theta^2] = a^4 [8 + 5\theta^2 + \theta^4]$$

$$\therefore \quad \rho = \frac{1}{\kappa} = \frac{|\dot{\mathbf{r}}|^3}{|\dot{\mathbf{r}} \times \ddot{\mathbf{r}}|} = \frac{a^3(2+\theta^2)^{3/2}}{a^2(8+5\theta^2+\theta^4)^{1/2}} = \frac{a(2+\theta^2)^{3/2}}{(8+5\theta^2+\theta^4)^{1/2}}$$

and

$$\sigma = \frac{1}{\tau} = \frac{|\dot{\mathbf{r}} \times \ddot{\mathbf{r}}|^2}{[\dot{\mathbf{r}}\,\ddot{\mathbf{r}}\,\dddot{\mathbf{r}}]} = \frac{a^4(8+5\theta^2+\theta^4)}{a^3(6+\theta^2)} = \frac{a(8+5\theta^2+\theta^4)}{6+\theta^2}.$$

Example 11. *For a point on the curve of intersection of the surfaces $x^2 - y^2 = c^2$ and $y = x \tanh \frac{z}{c}$, show that $\rho = \sigma = \frac{2x^2}{c}$.*

Sol. Given curve is

$$x^2 - y^2 = c^2 \quad \ldots(1) \qquad y = x \tanh \frac{z}{c} \quad \ldots(2)$$

Let $\quad x = c\cosh t, \quad y = c\sinh t.$

$\therefore$ (1) is satisfied.

(2) $\Rightarrow \quad c\sinh t = c\cosh t \tanh \frac{z}{c}$

$\Rightarrow \quad \tanh \frac{z}{c} = \tanh t \;\Rightarrow\; z = ct$

$\therefore$ The parametric equations of the given curve are

$$x = c\cosh t, y = c\sinh t, z = ct.$$

Let $\mathbf{r}$ be the position vector of the point (x, y, z) on the curve.

$$\therefore \quad \mathbf{r} = x\mathbf{i} + y\mathbf{j} + z\mathbf{k} = c\cosh t\,\mathbf{i} + c\sinh t\,\mathbf{j} + ct\mathbf{k}$$

$$\therefore \quad \dot{\mathbf{r}} = c\sinh t\,\mathbf{i} + c\cosh t\,\mathbf{j} + c\mathbf{k}$$

$$\ddot{\mathbf{r}} = c\cosh t\,\mathbf{i} + c\sinh t\,\mathbf{j}$$

and

$$\dddot{\mathbf{r}} = c\sinh t\,\mathbf{i} + c\cosh t\,\mathbf{j}$$

$$\therefore \quad |\dot{\mathbf{r}}| = c(\sinh^2 t + \cosh^2 t + 1)^{1/2} = c(2\cosh^2 t)^{1/2} = \sqrt{2}\,c\cosh t$$

$$\dot{\mathbf{r}} \times \ddot{\mathbf{r}} = \begin{vmatrix} \mathbf{i} & \mathbf{j} & \mathbf{k} \\ c\sinh t & c\cosh t & c \\ c\cosh t & c\sinh t & 0 \end{vmatrix} = c^2 \begin{vmatrix} \mathbf{i} & \mathbf{j} & \mathbf{k} \\ \sinh t & \cosh t & 1 \\ \cosh t & \sinh t & 0 \end{vmatrix}$$

$$= c^2 [-\sinh t\mathbf{i} + \cosh t\mathbf{j} - \mathbf{k}]$$

$$\therefore \quad |\dot{\mathbf{r}} \times \ddot{\mathbf{r}}| = c^2 (\sinh^2 t + \cosh^2 t + 1)^{1/2} = c^2(2\cosh^2 t)^{1/2} = \sqrt{2}\,c^2\cosh t$$

$$\therefore \quad \kappa = \frac{|\dot{\mathbf{r}} \times \ddot{\mathbf{r}}|}{|\dot{\mathbf{r}}|^3} = \frac{\sqrt{2}\,c^2\cosh t}{(\sqrt{2}\,c\cosh t)^3} = \frac{1}{2c\cosh^2 t}$$

$$\therefore \qquad \rho = \frac{1}{\kappa} = 2c\cosh^2 t = 2c\left(\frac{x}{c}\right)^2 = \frac{2x^2}{c}.$$

$$[\dot{\mathbf{r}}\ \ddot{\mathbf{r}}\ \dddot{\mathbf{r}}] = \begin{vmatrix} c\sinh t & c\cosh t & c \\ c\cosh t & c\sinh t & 0 \\ c\sinh t & c\cosh t & 0 \end{vmatrix} = c\,(c^2\cosh^2 t - c^2\sinh^2 t) = c^3$$

$$\therefore \qquad \tau = \frac{[\dot{\mathbf{r}}\,\ddot{\mathbf{r}}\,\dddot{\mathbf{r}}]}{|\dot{\mathbf{r}}\times\ddot{\mathbf{r}}|^2} = \frac{c^3}{(\sqrt{2}\,c^2\cosh t)^2} = \frac{1}{2c\cosh^2 t}$$

$$\therefore \qquad \sigma = \frac{1}{\tau} = 2c\cosh^2 t = 2c\left(\frac{x}{c}\right)^2 = \frac{2x^2}{c}.$$

$$\therefore \qquad \rho = \sigma = \frac{2x^2}{c}.$$

Example 12. *Determine the function f(u) so that the curve given by* $\mathbf{r} = (a\cos u, a\sin u, f(u))$ *should be a plane curve.*

Sol. The given curve is

$$\mathbf{r} = a\cos u\mathbf{i} + a\sin u\mathbf{j} + f(u)\mathbf{k}.$$

Given curve is a plane curve iff $\tau = 0$

$$\text{iff} \qquad \frac{[\dot{\mathbf{r}}\ \ddot{\mathbf{r}}\ \dddot{\mathbf{r}}]}{|\dot{\mathbf{r}}\times\ddot{\mathbf{r}}|^2} = 0 \quad \text{iff} \quad [\dot{\mathbf{r}}\ \ddot{\mathbf{r}}\ \dddot{\mathbf{r}}] = 0.$$

$\therefore$ Given curve is a plane curve iff $[\dot{\mathbf{r}}\ \ddot{\mathbf{r}}\ \dddot{\mathbf{r}}] = 0$.

We shall choose $f(u)$ so that we may have $[\dot{\mathbf{r}}\ \ddot{\mathbf{r}}\ \dddot{\mathbf{r}}] = 0$.

$$\dot{\mathbf{r}} = -a\sin u\mathbf{i} + a\cos u\mathbf{j} + \dot{f}(u)\mathbf{k}$$

$$\ddot{\mathbf{r}} = -a\cos u\mathbf{i} - a\sin u\mathbf{j} + \ddot{f}(u)\mathbf{k}$$

$$\dddot{\mathbf{r}} = a\sin u\mathbf{i} - a\cos u\mathbf{j} + \dddot{f}(u)\mathbf{k}$$

$$[\dot{\mathbf{r}}\ \ddot{\mathbf{r}}\ \dddot{\mathbf{r}}] = \begin{vmatrix} -a\sin u & a\cos u & \dot{f}(u) \\ -a\cos u & -a\sin u & \ddot{f}(u) \\ a\sin u & -a\cos u & \dddot{f}(u) \end{vmatrix}$$

$$= \begin{vmatrix} 0 & 0 & \dot{f}(u)+\dddot{f}(u) \\ -a\cos u & -a\sin u & \ddot{f}(u) \\ a\sin u & -a\cos u & \dddot{f}(u) \end{vmatrix} \qquad (\text{Operating } R_1 \to R_1 + R_3)$$

$$= (\dot{f}(u)+\dddot{f}(u))\,(a^2\cos^2 u + a^2\sin^2 u) = a^2(\dot{f}(u)+\dddot{f}(u))$$

$$\therefore \qquad a^2(\dot{f}(u)+\dddot{f}(u)) = 0 \qquad (\because\ [\dot{\mathbf{r}}\,\ddot{\mathbf{r}}\,\dddot{\mathbf{r}}] = 0)$$

$$\Rightarrow \qquad \dot{f}(u)+\dddot{f}(u) = 0$$

Integrating, we get $f(u) + \ddot{f}(u) = c_1$

$$\Rightarrow \qquad \ddot{f}(u) = c_1 - f(u) \quad \Rightarrow \quad 2\dot{f}(u)\,\ddot{f}(u) = 2(c_1 - f(u))\,\dot{f}(u)$$

Integrating, we get

$$(\dot{f}(u))^2 = -(c_1 - f(u))^2 + c_2$$

$$\Rightarrow \qquad \dot{f}(u) = \sqrt{c_2 - (c_1 - f(u))^2}$$

$$\Rightarrow \qquad \frac{d(f(u))}{\sqrt{c_2 - (c_1 - f(u))^2}} = du$$

Integrating, we get

$$-\sin^{-1}\frac{c_1 - f(u)}{\sqrt{c_2}} = u + c_3$$

$$\Rightarrow \qquad \sin(-(u + c_3)) = \frac{c_1 - f(u)}{\sqrt{c_2}}$$

$$\Rightarrow \qquad -\sqrt{c_2}\,\sin(u + c_3) = c_1 - f(u)$$

$$\Rightarrow \qquad f(u) = \sqrt{c_2}\,\sin(u + c_3) + c_1$$

$$= \sqrt{c_2}\,(\sin u \cos c_3 + \cos u \sin c_3) + c_1$$

$$= \sqrt{c_2}\,\cos c_3 \sin u + \sqrt{c_2}\,\sin c_3 \cos u + c_1$$

$$\therefore \qquad f(u) = \text{A} \sin u + \text{B} \cos u + \text{C},$$

where $\text{A} = \sqrt{c_2}\,\cos c_3$, $\text{B} = \sqrt{c_2}\,\sin c_3$, $\text{C} = c_1$ are arbitrary constants.

Example 13. *If the tangent and binormal at a point of a curve make angles* θ *and* ϕ *with a fixed direction, show that:* $\dfrac{\sin\theta}{\sin\phi}\dfrac{d\theta}{d\phi} = -\dfrac{\kappa}{\tau}$.

Sol. Let the equation of the curve be $\mathbf{r} = \mathbf{r}(s)$, where the parameter '$s$' is arc length.

Let the tangent and binormal at the point P of the curve make angles θ and ϕ with vector $\mathbf{a}$ which is along the given fixed direction.

$\therefore$ Angle between $\mathbf{t}$ and $\mathbf{a}$ is θ and angle between $\mathbf{b}$ and $\mathbf{a}$ is ϕ.

$$\therefore \qquad \mathbf{t} \cdot \mathbf{a} = a \cos\theta \qquad \text{...(1)} \qquad (\because \ |\mathbf{t}| = 1)$$

$$\text{and} \qquad \mathbf{b} \cdot \mathbf{a} = a \cos\phi \qquad \text{...(2)} \qquad (\because \ |\mathbf{b}| = 1)$$

Differentiating (1) w.r.t. s, we get

$$\mathbf{t}' \cdot \mathbf{a} = -a \sin\theta \frac{d\theta}{ds}$$

$$\Rightarrow \qquad \kappa\mathbf{n} \cdot \mathbf{a} = -a \sin\theta \frac{d\theta}{ds} \qquad \text{...(3)}$$

Differentiating (2) w.r.t. s, we get

$$\mathbf{b}' \cdot \mathbf{a} = -a \sin\phi \frac{d\phi}{ds}$$

$$\Rightarrow \qquad -\tau\mathbf{n} \cdot \mathbf{a} = -a \sin\phi \frac{d\phi}{ds} \qquad \text{...(4)}$$

Dividing (3) by (4), we get

$$\frac{\kappa(\mathbf{n}.\mathbf{a})}{-\tau(\mathbf{n}.\mathbf{a})} = \frac{-a\sin\theta\dfrac{d\theta}{ds}}{-a\sin\phi\dfrac{d\phi}{ds}}$$

$$\Rightarrow \qquad -\frac{\kappa}{\tau} = \frac{\sin\theta}{\sin\phi}\frac{d\theta}{d\phi}.$$

Example 14. *For the curve* $\mathbf{r} = \mathbf{r}(s)$, *if*

$$\frac{d\mathbf{t}}{ds} = \mathbf{w}\times\mathbf{t},\quad \frac{d\mathbf{n}}{ds} = \mathbf{w}\times\mathbf{n} \text{ and } \frac{d\mathbf{b}}{ds} = \mathbf{w}\times\mathbf{b}, \text{ find the vector } \mathbf{w}.$$

Sol. Given equations are

$$\frac{d\mathbf{t}}{ds} = \mathbf{w}\times\mathbf{t} \quad ...(1) \qquad \frac{d\mathbf{n}}{ds} = \mathbf{w}\times\mathbf{n} \quad ...(2) \qquad \frac{d\mathbf{b}}{ds} = \mathbf{w}\times\mathbf{b} \quad ...(3)$$

By *Frenet formula*, $\dfrac{d\mathbf{t}}{ds} = \kappa\mathbf{n}$.

$$\Rightarrow \qquad \frac{d\mathbf{t}}{ds} = \mathbf{0} + \kappa\mathbf{n} = \tau(\mathbf{t}\times\mathbf{t}) + \kappa(\mathbf{b}\times\mathbf{t}) = (\tau\mathbf{t} + \kappa\mathbf{b})\times\mathbf{t}$$

$$\therefore \qquad \frac{d\mathbf{t}}{ds} = (\tau\mathbf{t} + \kappa\mathbf{b})\times\mathbf{t} \quad ...(4)$$

By *Frenet formula*, $\dfrac{d\mathbf{n}}{ds} = -\kappa\mathbf{t} + \tau\mathbf{b}$.

$$\Rightarrow \qquad \frac{d\mathbf{n}}{ds} = -\kappa(\mathbf{n}\times\mathbf{b}) + \tau(\mathbf{t}\times\mathbf{n}) = \kappa(\mathbf{b}\times\mathbf{n}) + \tau(\mathbf{t}\times\mathbf{n})$$

$$= (\kappa\mathbf{b} + \tau\mathbf{t})\times\mathbf{n} = (\tau\mathbf{t} + \kappa\mathbf{b})\times\mathbf{n}$$

$$\therefore \qquad \frac{d\mathbf{n}}{ds} = (\tau\mathbf{t} + \kappa\mathbf{b})\times\mathbf{n} \quad ...(5)$$

By *Frenet formula*, $\dfrac{d\mathbf{b}}{ds} = -\tau\mathbf{n}$.

$$\Rightarrow \qquad \frac{d\mathbf{b}}{ds} = -\tau(\mathbf{b}\times\mathbf{t}) + \mathbf{0} = \tau(\mathbf{t}\times\mathbf{b}) + \kappa(\mathbf{b}\times\bar{\mathbf{b}}) = (\tau\mathbf{t} + \kappa\mathbf{b})\times\mathbf{b}$$

$$\therefore \qquad \frac{d\mathbf{b}}{ds} = (\tau\mathbf{t} + \kappa\mathbf{b})\times\mathbf{b} \quad ...(6)$$

If $\mathbf{w} = \tau\mathbf{t} + \kappa\mathbf{b}$, then given equations (1), (2) and (3) are satisfied.

Note. The vector $\mathbf{w} = \tau\mathbf{t} + \kappa\mathbf{b}$ is called the **Darboux vector** for the curve $\mathbf{r} = \mathbf{r}(s)$.

Example 15. *Using Serret-Frenet formula, find the direction cosines of the unit principal normal vector and the unit binormal vector at the point 's' for the curve* $\mathbf{r} = \mathbf{r}(s)$.

Sol. Given curve is $\quad \mathbf{r} = \mathbf{r}(s)$.

Let $\mathbf{r}$ be the position vector of the point (x, y, z) on the curve.

$$\therefore \qquad \mathbf{r} = x\mathbf{i} + y\mathbf{j} + z\mathbf{k}$$

We have $\mathbf{t}' = \kappa\mathbf{n}$.

$$\Rightarrow \quad \mathbf{n} = \frac{\mathbf{t}'}{\kappa} = \frac{(\mathbf{r}')'}{\kappa} = \frac{\mathbf{r}''}{\kappa} = \frac{1}{\kappa}(x''\mathbf{i} + y''\mathbf{j} + z''\mathbf{k}) = \frac{x''}{\kappa}\mathbf{i} + \frac{y''}{\kappa}\mathbf{j} + \frac{z''}{\kappa}\mathbf{k}$$

$$|\,\mathbf{n}\,| = 1 \Rightarrow \left(\frac{x''^2}{\kappa^2} + \frac{y''^2}{\kappa^2} + \frac{z''^2}{\kappa^2}\right)^{1/2} = 1 \Rightarrow \kappa^2 = x''^2 + y''^2 + z''^2$$

$$\therefore \quad \kappa = \sqrt{x''^2 + y''^2 + z''^2}$$

$$\therefore \quad \mathbf{n} = \frac{x''}{\sqrt{x''^2 + y''^2 + z''^2}}\mathbf{i} + \frac{y''}{\sqrt{x''^2 + y''^2 + z''^2}}\mathbf{j} + \frac{z''}{\sqrt{x''^2 + y''^2 + z''^2}}\mathbf{k}$$

Since **n** is a unit vector, the d.c.'s of **n** are

$$\frac{x''}{\sqrt{x''^2 + y''^2 + z''^2}}, \frac{y''}{\sqrt{x''^2 + y''^2 + z''^2}}, \frac{z''}{\sqrt{x''^2 + y''^2 + z''^2}}.$$

$$\mathbf{b} = \mathbf{t} \times \mathbf{n} = \mathbf{r}' \times \frac{\mathbf{t}'}{\kappa} = \frac{1}{\kappa}(\mathbf{r}' \times \mathbf{r}'') = \frac{1}{\kappa}\begin{vmatrix} \mathbf{i} & \mathbf{j} & \mathbf{k} \\ x' & y' & z' \\ x'' & y'' & z'' \end{vmatrix}$$

$$\left(\kappa = \mathbf{t}' \Rightarrow \kappa = |\,\mathbf{t}'\,| \Rightarrow \mathbf{n} = \frac{\mathbf{t}'}{|\,\mathbf{t}'\,|} = \frac{\mathbf{t}'}{\kappa}\right)$$

$$= \frac{1}{\kappa}\,[(y'z'' - y''z')\mathbf{i} + (z'x'' - z''x')\mathbf{j} + (x'y'' - x''y')\mathbf{k}]$$

$$= \frac{y'z'' - y''z'}{\sqrt{x''^2 + y''^2 + z''^2}}\mathbf{i} + \frac{z'x'' - z''x'}{\sqrt{x''^2 + y''^2 + z''^2}}\mathbf{j} + \frac{x'y'' - x''y'}{\sqrt{x''^2 + y''^2 + z''^2}}\mathbf{k}$$

Since **b** is a unit vector, the d.c.'s of **b** are

$$\frac{y'z'' - y''z'}{\sqrt{x''^2 + y''^2 + z''^2}}, \frac{z'x'' - z''x'}{\sqrt{x''^2 + y''^2 + z''^2}}, \frac{x'y'' - x''y'}{\sqrt{x''^2 + y''^2 + z''^2}}.$$

Example 16. *Let* $\mathbf{r} = \mathbf{r}(t)$ *be a curve. Prove that:*

(i) $\dot{\mathbf{r}} = \dot{s}\mathbf{t}$ *(ii)* $\ddot{\mathbf{r}} = \ddot{s}\mathbf{t} + \kappa\dot{s}^2\,\mathbf{n}$

(iii) $\dddot{\mathbf{r}} = (\dddot{s} - \kappa^2\dot{s}^3)\,\mathbf{t} + \dot{s}\,(3\kappa\ddot{s} - \dot{\kappa}\dot{s})\,\mathbf{n} + \kappa\tau\dot{s}^3\mathbf{b}$.

Hence deduce that:

(a) $\mathbf{n} = \dfrac{\dot{s}\,\ddot{\mathbf{r}} - \ddot{s}\,\dot{\mathbf{r}}}{\kappa\dot{s}^3}$ *(b)* $\mathbf{b} = \dfrac{\dot{\mathbf{r}} \times \ddot{\mathbf{r}}}{\kappa\dot{s}^3}$

(c) $\kappa^2 = \dfrac{|\,\ddot{\mathbf{r}}\,|^2 - \ddot{s}^2}{\dot{s}^4}$ *(d)* $\tau = \dfrac{[\dot{\mathbf{r}}\,\ddot{\mathbf{r}}\,\dddot{\mathbf{r}}]}{\kappa^2\dot{s}^6}$.

Sol. *(i)*

$$\dot{\mathbf{r}} = \frac{d\mathbf{r}}{dt} = \frac{d\mathbf{r}}{ds}\frac{ds}{dt} = \mathbf{r}'\dot{s} = \dot{s}\mathbf{t}.$$

(ii)

$$\ddot{\mathbf{r}} = \frac{d}{dt}(\dot{\mathbf{r}}) = \frac{d}{dt}(\dot{s}\mathbf{t}) = \dot{s}\frac{d\mathbf{t}}{dt} + \ddot{s}\mathbf{t}$$

$$= \dot{s}\,(\mathbf{t}'\,\dot{s}) + \ddot{s}\mathbf{t} = \dot{s}^2\,(\kappa\mathbf{n}) + \ddot{s}\mathbf{t} = \ddot{s}\,\mathbf{t} + \kappa\dot{s}^2\,\mathbf{n}.$$

(*iii*) $$\dddot{\mathbf{r}} = \frac{d}{dt}(\ddot{\mathbf{r}}) = \frac{d}{dt}(\ddot{s}\mathbf{t} + \kappa\dot{s}^2\mathbf{n})$$

$$= \left(\ddot{s}\frac{d\mathbf{t}}{dt} + \dddot{s}\mathbf{t}\right) + \left(\dot{\kappa}\dot{s}^2\mathbf{n} + \kappa 2\dot{s}\,\ddot{s}\,\mathbf{n} + \kappa\dot{s}^2\frac{d\mathbf{n}}{dt}\right)$$

$$= \ddot{s}\,(\mathbf{t}'\dot{s}) + \dddot{s}\mathbf{t} + \dot{\kappa}\dot{s}^2\mathbf{n} + 2\kappa\dot{s}\ddot{s}\,\mathbf{n} + \kappa\dot{s}^2\mathbf{n}'\dot{s}$$

$$= \dot{s}\ddot{s}\,(\kappa\mathbf{n}) + \dddot{s}\mathbf{t} + \dot{\kappa}\dot{s}^2\mathbf{n} + 2\kappa\dot{s}\ddot{s}\,\mathbf{n} + \kappa\dot{s}^3(-\kappa\mathbf{t} + \tau\mathbf{b})$$

$$= (\dddot{s} - \kappa^2\dot{s}^3)\mathbf{t} + \dot{s}(3\kappa\ddot{s} + \dot{\kappa}\dot{s})\mathbf{n} + \kappa\tau\,\dot{s}^3\mathbf{b}.$$

(*a*) $$\dot{s}\ddot{\mathbf{r}} - \ddot{s}\dot{\mathbf{r}} = \dot{s}(\ddot{s}\mathbf{t} + \kappa\dot{s}^2\mathbf{n}) - \ddot{s}\,(\dot{s}\mathbf{t}) = \kappa\dot{s}^3\mathbf{n}$$

$\therefore$ $$\mathbf{n} = \frac{\dot{s}\ddot{\mathbf{r}} - \ddot{s}\dot{\mathbf{r}}}{\kappa\dot{s}^3}.$$

(*b*) $$\dot{\mathbf{r}}\times\ddot{\mathbf{r}} = (\dot{s}\mathbf{t})\times(\ddot{s}\mathbf{t} + \kappa\dot{s}^2\mathbf{n}) = \dot{s}\,\ddot{s}\,(\mathbf{t}\times\mathbf{t}) + \kappa\dot{s}^3(\mathbf{t}\times\mathbf{n})$$

$$= \dot{s}\,\ddot{s}\,(\mathbf{0}) + \kappa\dot{s}^3\mathbf{b} = \kappa\dot{s}^3\,\mathbf{b}$$

$\therefore$ $$\mathbf{b} = \frac{\dot{\mathbf{r}}\times\ddot{\mathbf{r}}}{\kappa\dot{s}^3}.$$

(*c*) $$|\ddot{\mathbf{r}}|^2 = \ddot{\mathbf{r}}\,.\,\ddot{\mathbf{r}} = (\ddot{s}\mathbf{t} + \kappa\dot{s}^2\mathbf{n})\,.\,(\ddot{s}\mathbf{t} + \kappa\dot{s}^2\mathbf{n})$$

$$= \ddot{s}^2\,(\mathbf{t}\,.\,\mathbf{t}) + \kappa\dot{s}^2\ddot{s}\,(\mathbf{t}\,.\,\mathbf{n}) + \kappa\dot{s}^2\ddot{s}\,(\mathbf{n}\,.\,\mathbf{t}) + \kappa^2\dot{s}^4\,(\mathbf{n}\,.\,\mathbf{n})$$

$$= \ddot{s}^2\,(1) + 0 + 0 + \kappa^2\dot{s}^4(1) = \ddot{s}^2 + \kappa^2\dot{s}^4$$

$\therefore$ $$|\ddot{\mathbf{r}}|^2 - \ddot{s}^2 = \kappa^2\dot{s}^4$$

$\therefore$ $$\kappa^2 = \frac{|\ddot{\mathbf{r}}|^2 - \ddot{s}^2}{\dot{s}^4}.$$

(*d*) $$[\dot{\mathbf{r}}\,\ddot{\mathbf{r}}\,\dddot{\mathbf{r}}] = \dot{\mathbf{r}}\,.(\ddot{\mathbf{r}}\times\dddot{\mathbf{r}}) = (\dot{\mathbf{r}}\times\ddot{\mathbf{r}})\,.\,\dddot{\mathbf{r}}$$

$$= (\kappa\dot{s}^3\mathbf{b})\,.\,[(\dddot{s} - \kappa^2\dot{s}^3)\mathbf{t} + \dot{s}\,(3\kappa\ddot{s} + \dot{\kappa}\dot{s})\mathbf{n} + \kappa\tau\dot{s}^3\mathbf{b}]$$ (Using (*iii*) and (*b*))

$$= \kappa^2\dot{s}^6\tau\,(\mathbf{b}\,.\,\mathbf{b}) = \kappa^2\dot{s}^6\tau$$ (Using $\mathbf{b}\,.\,\mathbf{t} = 0$, $\mathbf{b}\,.\,\mathbf{n} = 0$)

$\therefore$ $$\tau = \frac{[\dot{\mathbf{r}}\,\ddot{\mathbf{r}}\,\dddot{\mathbf{r}}]}{\kappa^2\dot{s}^6}.$$

Example 17. *Let* $\mathbf{r} = \mathbf{r}(s)$ *be a curve. Prove that:*

(*i*) $\mathbf{r}'\,.\,\mathbf{r}'' = 0$ (*ii*) $\mathbf{r}''' = -\kappa^2\mathbf{t} + \kappa'\mathbf{n} + \kappa\tau\mathbf{b}$

(*iii*) $\mathbf{r}'\,.\,\mathbf{r}''' = -\kappa^2$ (*iv*) $\mathbf{r}''\,.\,\mathbf{r}''' = \kappa\kappa'$

(*v*) $\mathbf{r}'''' = -3\kappa\kappa'\mathbf{t} + (\kappa'' - \kappa^3 - \kappa\tau^2)\mathbf{n} + (2\kappa'\tau + \tau'\kappa)\mathbf{b}$

(*vi*) $\mathbf{r}'\,.\,\mathbf{r}'''' = -3\kappa\kappa'$ (*vii*) $\mathbf{r}''\,.\,\mathbf{r}'''' = \kappa(\kappa'' - \kappa^3 - \kappa\tau^2)$

(*viii*) $\mathbf{r}'''\,.\,\mathbf{r}'''' = \kappa'\kappa'' + 2\kappa^3\kappa' + \kappa^2\tau\tau' + \kappa\kappa'\tau^2$

(*ix*) $[\mathbf{t}'\,\mathbf{t}''\,\mathbf{t}'''] = \kappa^3(\kappa\tau' - \kappa'\tau) = \kappa^5\dfrac{d}{ds}\left(\dfrac{\tau}{\kappa}\right)$ (*x*) $[\mathbf{b}'\,\mathbf{b}''\,\mathbf{b}'''] = \tau^3(\kappa'\tau - \kappa\tau') = \tau^5\dfrac{d}{ds}\left(\dfrac{\kappa}{\tau}\right)$.

Sol. (*i*) $\mathbf{r}' \cdot \mathbf{r}'' = \mathbf{t} \cdot \mathbf{t}' = \mathbf{t} \cdot (\kappa\mathbf{n}) = \kappa\,(\mathbf{t} \cdot \mathbf{n}) = \kappa \cdot 0 = 0.$ $\quad (\because\ \mathbf{r}' = \mathbf{t}$ and $\mathbf{t}' = \kappa\mathbf{n})$

(*ii*) $\mathbf{r}''' = (\mathbf{r}')'' = \mathbf{t}'' = (\mathbf{t}')' = (\kappa\mathbf{n})' = \kappa\mathbf{n}' + \kappa'\mathbf{n} = \kappa(-\kappa\mathbf{t} + \tau\mathbf{b}) + \kappa'\mathbf{n}$

$= -\kappa^2\mathbf{t} + \kappa'\mathbf{n} + \kappa\tau\mathbf{b}.$ $\quad (\because\ \mathbf{n}' = -\kappa\mathbf{t} + \tau\mathbf{b})$

(*iii*) $\mathbf{r}' \cdot \mathbf{r}''' = \mathbf{t} \cdot (-\kappa^2\mathbf{t} + \kappa'\mathbf{n} + \kappa\tau\mathbf{b})$ (Using (*iii*))

$= -\kappa^2\,(\mathbf{t} \cdot \mathbf{t}) + \kappa'(\mathbf{t} \cdot \mathbf{n}) + \kappa\tau(\mathbf{t} \cdot \mathbf{b}) = -\kappa^2 \cdot 1 + 0 + 0 = -\kappa^2.$

(*iv*) $\mathbf{r}'' \cdot \mathbf{r}''' = (\mathbf{r}')' \cdot \mathbf{r}''' = \mathbf{t}' \cdot \mathbf{r}'''$

$= (\kappa\mathbf{n}) \cdot (-\kappa^2\mathbf{t} + \kappa'\mathbf{n} + \kappa\tau\mathbf{b})$ (Using (*iii*))

$= -\kappa^3(\mathbf{n} \cdot \mathbf{t}) + \kappa\kappa'(\mathbf{n} \cdot \mathbf{n}) + \kappa^2\tau(\mathbf{n} \cdot \mathbf{b}) = 0 + \kappa\kappa' \cdot 1 + 0 = \kappa\kappa'.$

(*v*) $r'''' = (r''')' = (-\kappa^2\mathbf{t} + \kappa'\mathbf{n} + \kappa\tau\mathbf{b})'$ (Using (*iii*))

$= -(\kappa^2\mathbf{t}' + 2\kappa\kappa'\mathbf{t}) + (\kappa'\mathbf{n}' + \kappa''\mathbf{n}) + (\kappa'\tau\mathbf{b} + \kappa\tau'\mathbf{b} + \kappa\tau\mathbf{b}')$

$= -\kappa^2\,(\kappa\mathbf{n}) - 2\kappa\kappa'\mathbf{t} + \kappa'\,(-\kappa\mathbf{t} + \tau\mathbf{b}) + \kappa''\mathbf{n} + \kappa'\tau\mathbf{b} + \kappa\tau'\mathbf{b} + \kappa\tau\,(-\tau\mathbf{n})$

$= -\kappa^3\mathbf{n} - 2\kappa\kappa'\mathbf{t} - \kappa\kappa'\mathbf{t} + \kappa'\tau\mathbf{b} + \kappa''\mathbf{n} + \kappa'\tau\mathbf{b} + \kappa\tau'\mathbf{b} - \kappa\tau^2\mathbf{n}.$

$= -3\kappa\kappa'\mathbf{t} + (\kappa'' - \kappa^3 - \kappa\tau^2)\mathbf{n} + (2\kappa'\tau + \tau'\kappa)\,\mathbf{b}.$

(*vi*) $\mathbf{r}' \cdot \mathbf{r}'''' = \mathbf{t}\,.\,[-3\kappa\kappa'\mathbf{t} + (\kappa'' - \kappa^3 - \kappa\tau^2)\mathbf{n} + (2\kappa'\tau + \tau'\kappa)\mathbf{b}]$ (Using (*v*))

$= -3\kappa\kappa'\,(\mathbf{t} \cdot \mathbf{t}) + (\kappa'' - \kappa^3 - \kappa\tau^2)(\mathbf{t}\,.\,\mathbf{n}) + (2\kappa'\tau + \tau'\kappa)\,(\mathbf{t}\,.\,\mathbf{b})$

$= -3\kappa\kappa' \cdot 1 + 0 + 0 = -3\kappa\kappa'.$

(*vii*) $\mathbf{r}'' \cdot \mathbf{r}'''' = \mathbf{t}' \cdot r''''$

$= (\kappa\mathbf{n}) \cdot (-3\kappa\kappa'\mathbf{t} + (\kappa'' - \kappa^3 - \kappa\tau^2)\mathbf{n} + (2\kappa'\tau + \tau'\,\kappa)\mathbf{b}]$

$= -3\kappa^2\kappa'\,(\mathbf{n} \cdot \mathbf{t}) + \kappa(\kappa'' - \kappa^3 - \kappa\tau^2)\,(\mathbf{n} \cdot \mathbf{n}) + \kappa(2\kappa'\tau + \tau'\kappa)\,(\mathbf{n} \cdot \mathbf{b})$

$= 0 + \kappa(\kappa'' - \kappa^3 - \kappa\tau^2) \cdot 1 + 0 = \kappa(\kappa'' - \kappa^3 - \kappa\tau^2).$

(*viii*) $\mathbf{r}''' \cdot \mathbf{r}'''' = (-\kappa^2\mathbf{t} + \kappa'\mathbf{n} + \kappa\tau\mathbf{b}).\,(-3\kappa\kappa'\mathbf{t} + (\kappa'' - \kappa^3 - \kappa\tau^2)\,\mathbf{n} + (2\kappa'\tau + \tau'\kappa)\mathbf{b})$

(Using (*ii*) and (*v*))

$= 3\kappa^3\kappa'\,(\mathbf{t} \cdot \mathbf{t}) + \kappa'\,(\kappa'' - \kappa^3 - \kappa\tau^2)\,(\mathbf{n} \cdot \mathbf{n}) + \kappa\tau(2\kappa'\tau + \tau'\kappa)\,(\mathbf{b} \cdot \mathbf{b})$

$= 3\kappa^3\kappa' + \kappa'\kappa'' - \kappa'\kappa^3 - \kappa\kappa'\tau^2 + 2\kappa\kappa'\tau^2 + \kappa^2\tau\tau'$

$= \kappa'\kappa'' + 2\kappa^3\kappa' + \kappa^2\tau\tau' + \kappa\kappa'\tau^2.$

(*ix*) $[\mathbf{t}'\ \mathbf{t}''\ \mathbf{t}'''] = [\mathbf{r}''\ \mathbf{r}'''\ \mathbf{r}'''']$

$$= \begin{vmatrix} 0 & \kappa & 0 \\ -\kappa^2 & \kappa' & \kappa\tau \\ -3\kappa\kappa' & \kappa'' - \kappa^3 - \kappa\tau^2 & 2\kappa'\tau + \tau'\kappa \end{vmatrix}$$

$(\because\ \mathbf{r}'' = \mathbf{t}' = \kappa\mathbf{n}$ and the vectors $\mathbf{t}, \mathbf{n}, \mathbf{b}$ form a right handed triad)

$= -\kappa[-\kappa^2(2\kappa'\tau + \tau'\kappa) + 3\kappa\kappa'(\kappa\tau)] = 2\kappa^3\kappa'\tau + \kappa^4\tau' - 3\kappa^3\kappa'\tau$

$= \kappa^4\,\tau' - \kappa^3\,\kappa'\,\tau = \kappa^3(\kappa\tau' - \kappa'\,\tau)$

$$= \kappa^5 \cdot \left(\frac{\kappa\tau' - \kappa'\tau}{\kappa^2}\right) = \kappa^5\,\frac{d}{ds}\left(\frac{\tau}{\kappa}\right).$$

(*x*) We have $\mathbf{b}' = -\tau\mathbf{n} = 0\mathbf{t} - \tau\mathbf{n} + 0\mathbf{b}$

$\mathbf{b}'' = (\mathbf{b}')' = (-\tau\mathbf{n})' = -\tau\mathbf{n}' - \tau'\mathbf{n}$

$= -\tau(-\kappa\mathbf{t} + \tau\mathbf{b}) - \tau'\mathbf{n} = \tau\kappa\mathbf{t} - \tau'\mathbf{n} - \tau^2\mathbf{b}$

$\mathbf{b}''' = (\mathbf{b}'')' = (\tau\kappa\mathbf{t} - \tau'\mathbf{n} - \tau^2\mathbf{b})'$

$= (\tau'\kappa\mathbf{t} + \tau\kappa'\mathbf{t} + \tau\kappa\mathbf{t}') - (\tau''\mathbf{n} + \tau'\mathbf{n}') - (2\tau\tau'\mathbf{b} + \tau^2\mathbf{b}')$

$= \tau'\kappa\mathbf{t} + \tau\kappa'\mathbf{t} + \tau k(\kappa\mathbf{n}) - \tau''\mathbf{n} - \tau'(-\kappa\mathbf{t} + \tau\mathbf{b}) - 2\tau\tau'\mathbf{b} - \tau^2\,(-\tau\mathbf{n})$

$= \tau'\kappa\mathbf{t} + \tau\kappa'\mathbf{t} + \tau\kappa^2\mathbf{n} - \tau''\mathbf{n} + \tau'\kappa\mathbf{t} - \tau'\tau\,\mathbf{b} - 2\tau\tau'\mathbf{b} + \tau^3\mathbf{n}$

$= (2\tau'\kappa + \tau\kappa')\,\mathbf{t} + (\tau\kappa^2 - \tau'' + \tau^3)\,\mathbf{n} - 3\tau\tau'\mathbf{b}$

$$\therefore \quad [\mathbf{b}' \ \mathbf{b}'' \ \mathbf{b}'''] = \begin{vmatrix} 0 & -\tau & 0 \\ \tau\kappa & -\tau' & -\tau^2 \\ 2\tau'\kappa + \tau\kappa' & \tau\kappa^2 - \tau'' + \tau^3 & -3\tau\tau' \end{vmatrix}$$

($\because$ **t**, **n**, **b** form a right handed triad)

$$= \tau\,[\tau\kappa\,(-3\tau\tau') + (2\tau'\kappa + \tau\kappa')\tau^2]$$
$$= \tau\,[-3\tau^2\tau'\kappa + 2\tau^2\tau'\kappa + \tau^3\kappa']$$
$$= \tau[-\tau^2\tau'\kappa + \tau^3\kappa'] = \tau^3(\kappa'\tau - \kappa\tau')$$
$$= \tau^5\left(\frac{\tau\kappa' - \kappa\tau'}{\tau^2}\right) = \tau^5\,\frac{d}{ds}\left(\frac{\kappa}{\tau}\right).$$

WORKING RULES FOR SOLVING PROBLEMS

Rule I. $\tau = -\mathbf{n}\cdot\mathbf{b}'$

Rule II. *Radius of torsion,* $\sigma = \dfrac{1}{\tau}$

Rule III. *Serret-Frenet Formula:*

(*i*) $\mathbf{t}' = \kappa\mathbf{n}$ (*ii*) $\mathbf{n}' = -\kappa\mathbf{t} + \tau\mathbf{b}$ (*iii*) $\mathbf{b}' = -\tau\mathbf{n}$

Rule IV. (*i*) $\tau = \dfrac{[\mathbf{r}'\ \mathbf{r}''\ \mathbf{r}''']}{|\mathbf{r}' \times \mathbf{r}''|^2}$ (*ii*) $\tau = \dfrac{[\dot{\mathbf{r}}\ \ddot{\mathbf{r}}\ \dddot{\mathbf{r}}]}{|\dot{\mathbf{r}} \times \ddot{\mathbf{r}}|^2}$

Rule V. *Curve is a plane curve if and only if its torsion is identically zero.*

EXERCISE 2.2

1. Find the torsion of the curve $\mathbf{r} = t\mathbf{i} + t^2\mathbf{j} + t^3\mathbf{k}$ at the point where $t = 2$.
2. Show that the torsion of a plane curve (with $\kappa > 0$) is identically zero.
3. Find the torsion of the curve $\mathbf{r} = (3t - t^3)\mathbf{i} + 3t^2\mathbf{j} + (3t + t^3)\mathbf{k}$ at point t.
4. Find the torsion of the curve $\mathbf{r} = (t - \sin t)\mathbf{i} + (1 - \cos t)\mathbf{j} + t\mathbf{k}$ at point t.
5. For the curve $x = a(3t - t^3)$, $y = 3at^2$, $z = a(3t + t^3)$, show that curvature κ and torsion τ each is equal to $\dfrac{1}{3a(1+t^2)^2}$.
6. Find the torsion of the helix $x = a\cos t$, $y = a\sin t$, $z = at\tan\alpha$ at point t.
7. For the curve $x = 3t$, $y = 3t^2$, $z = 2t^3$, show that:
$$\kappa = \tau = \frac{2}{3(1+2t^2)^2}.$$
8. For a point on the curve of intersection of the surfaces $x^2 + y^2 = a^2$, $x^2 - y^2 = az$, find the torsion.
9. Find the torsion at any point t of the curve $x = a\cos 2t$, $y = a\sin 2t$, $z = 2a\sin t$.
10. Let $\mathbf{r} = \mathbf{r}(t)$ be a regular curve of class C^m ($m \geq 3$), where t is an arbitrary parameter. Prove that
$$\tau = \frac{[\dot{\mathbf{r}}\ \ddot{\mathbf{r}}\ \dddot{\mathbf{r}}]}{(\dot{\mathbf{r}}\,.\,\dot{\mathbf{r}})\,(\ddot{\mathbf{r}}\,.\,\ddot{\mathbf{r}}) - (\dot{\mathbf{r}}\,.\,\ddot{\mathbf{r}})^2},\ \text{provided } \kappa \neq 0.$$
11. Show that the curve $\mathbf{r} = \mathbf{r}(t)$ is a plane curve if and only if $[\dot{\mathbf{r}}\ \ddot{\mathbf{r}}\ \dddot{\mathbf{r}}] = 0$.

12. Is $x = a \cos t$, $y = a \sin t$, $z = ct$, $a > 0$, $c \neq 0$ a plane curve ? Calculate the curvature and torsion of the above curve at point 't'.

13. Show that the position vector of the current point on the curve $\mathbf{r} = \mathbf{r}(s)$ satisfies the differential equation:

$$\frac{d}{ds}\left[\sigma \frac{d}{ds}\left(\rho \frac{d^2\mathbf{r}}{ds^2}\right)\right] + \frac{d}{ds}\left[\frac{\sigma}{\rho}\frac{d\mathbf{r}}{ds}\right] + \frac{\rho}{\sigma}\frac{d^2\mathbf{r}}{ds^2} = \mathbf{0}.$$

Answers

1. $\dfrac{3}{181}$ **3.** $\dfrac{2}{3(1+t^2)^2}$ **4.** $-\dfrac{1}{1+4\sin^4\dfrac{t}{2}}$

6. $a \sec^2 \alpha$ **8.** $\dfrac{6\sqrt{a^2 - z^2}}{5a^2 + 12z^2}$ **9.** $\dfrac{3}{a\,(5 \sec t + 3 \cos t)}$

12. $\dfrac{a}{a^2 + c^2}, \dfrac{c}{a^2 + c^2}$.

Hint

13. Using $\rho = \dfrac{1}{\kappa}$, $\sigma = \dfrac{1}{\tau}$, $\dfrac{d\mathbf{r}}{ds} = \mathbf{r}' = \mathbf{t}$ and $\dfrac{d^2\mathbf{r}}{ds^2} = \mathbf{r}'' = \mathbf{t}' = \kappa\mathbf{n}$, we get

$$\frac{d}{ds}\left[\sigma \frac{d}{ds}\left(\rho \frac{d^2\mathbf{r}}{ds^2}\right)\right] + \frac{d}{ds}\left[\frac{\sigma}{\rho}\frac{d\mathbf{r}}{ds}\right] + \frac{\rho}{\sigma}\frac{d^2\mathbf{r}}{ds^2}$$

$$= \left[\frac{1}{\tau}\left(\frac{1}{\kappa}.\kappa\mathbf{n}\right)'\right]' + \left[\frac{\kappa}{\tau}\mathbf{t}\right]' + \frac{\tau}{\kappa}(\kappa\mathbf{n}) = \left[\frac{1}{\tau}(-\kappa\mathbf{t} + \tau\mathbf{b})\right]' + \left[\frac{\kappa}{\tau}\mathbf{t}\right]' + \tau\mathbf{n}$$

$$= -\left[\frac{\kappa}{\tau}\mathbf{t}\right]' + \mathbf{b}' + \left[\frac{\kappa}{\tau}\mathbf{t}\right]' + \tau\mathbf{n} = -\tau\mathbf{n} + \tau\mathbf{n} = \mathbf{0}.$$

---

4. CONTACT OF A CURVE WITH A SURFACE

In this section, we shall study the degree of contact of a curve with a surface.

A curve $\mathbf{r} = x(t)\mathbf{i} + y(t)\mathbf{j} + z(t)\mathbf{k}$ of sufficiently high class is said to have n-**point contact** (or **contact of** $(n-1)^{\text{th}}$ **order**) with a surface $F(x, y, z) = 0$ at the point corresponding to t_0 if the function $f(t) = F(x(t), y(t), z(t))$ satisfies :

$$f(t_0) = \dot{f}(t_0) = \ddot{f}(t_0) = \ldots\ldots = f^{(n-1)}(t_0) = 0 \quad \text{and} \quad f^{(n)}(t_0) \neq 0.$$

Example 1. *Show that the curve* $\mathbf{r} = t\mathbf{i} + t^2\mathbf{j} + t^3\mathbf{k}$ *has 6-point contact with the paraboloid* $x^2 + z^2 - y = 0$ *at the origin.*

Sol. Given curve is $\mathbf{r} = t\mathbf{i} + t^2\mathbf{j} + t^3\mathbf{k}$.

$$t = 0,\ t^2 = 0,\ t^3 = 0 \ \Rightarrow\ t = 0$$

$\therefore$ The origin corresponds to $t = 0$.

Given surface is $x^2 + z^2 - y = 0$.

Let $$f(t) = t^2 + (t^3)^2 - t^2 = t^6$$

$\therefore$ $$\dot{f}(t) = 6t^5,\ \ddot{f}(t) = 30t^4,\ \dddot{f}(t) = 120t^3, f^{(4)}(t) = 360t^2, f^{(5)}(t) = 720t, f^{(6)}(t) = 720$$

$\therefore$ $$\dot{f}(0) = 0,\ \ddot{f}(0) = 0,\ \dddot{f}(0) = 0, f^{(4)}(0) = 0, f^{(5)}(0) = 0 \text{ and } f^{(6)}(0) \neq 0.$$

$\therefore$ The given curve has 6-point contact with the given paraboloid at the origin.

Example 2. *If the circle $lx + my + nz = 0$, $x^2 + y^2 + z^2 = 2cz$ has 3-point contact with the paraboloid $ax^2 + by^2 = 2z$ at the origin then show that:* $c = \dfrac{l^2 + m^2}{bl^2 + am^2}$.

Sol. Given circle is $lx + my + nz = 0$ ...(1) $x^2 + y^2 + z^2 = 2cz$ ...(2)

Let the parametric equations of this circle be $x = \phi_1(t)$, $y = \phi_2(t)$, $z = \phi_3(t)$.

Substituting the values of x, y, z in (1) and (2) and differentiating w.r.t. t, we get

$$l\dot{x} + m\dot{y} + n\dot{z} = 0 \quad ...(3)$$

and $$2x\dot{x} + 2y\dot{y} + 2z\dot{z} = 2c\dot{z}$$

or $$x\dot{x} + y\dot{y} + z\dot{z} = c\dot{z} \quad ...(4)$$

The circle passes through the origin.

$\therefore$ (4) $\Rightarrow$ $0\dot{x} + 0\dot{y} + 0\dot{z} = c\dot{z}$ $\Rightarrow$ $c\dot{z} = 0$ $\Rightarrow$ $\dot{z} = 0$

$\therefore$ (3) $\Rightarrow$ $l\dot{x} + m\dot{y} + n(0) = 0$ $\Rightarrow$ $l\dot{x} + m\dot{y} = 0$

$$\Rightarrow \quad \frac{\dot{x}}{m} = \frac{\dot{y}}{-l} = \lambda, \text{ say} \quad ...(5)$$

The paraboloid is $ax^2 + by^2 = 2z$. After substituting the values of x, y, z in terms of t, let

$$f(t) = ax^2 + by^2 - 2z \quad ...(6)$$

$\therefore$ $$\dot{f}(t) = 2ax\dot{x} + 2by\dot{y} - 2\dot{z} \quad ...(7)$$

$$\ddot{f}(t) = 2a[\dot{x}^2 + x\ddot{x}] + 2b\,[\dot{y}^2 + y\ddot{y}] - 2\ddot{z}$$

or $$\ddot{f}(t) = 2a\dot{x}^2 + 2ax\ddot{x} + 2b\dot{y}^2 + 2by\ddot{y} - 2\ddot{z} \quad ...(8)$$

Since the circle has 3-point contact with the paraboloid at the origin, we have

$$f(t) = \dot{f}(t) = \ddot{f}(t) = 0,\ \dddot{f}(t) \neq 0 \text{ at the origin.}$$

$\therefore$ (6) $\Rightarrow$ $a(0)^2 + b(0)^2 - 2(0) = 0$ ...(9)

(7) $\Rightarrow$ $2a(0)\dot{x} + 2b(0)\dot{y} - 2\dot{z} = 0$ ...(10)

(8) $\Rightarrow$ $2a\dot{x}^2 + 2a(0)\ddot{x} + 2b\dot{y}^2 + 2b(0)\ddot{y} - 2\ddot{z} = 0$ ...(11)

(9) $\Rightarrow$ $0 = 0$, which is always true.

(10) $\Rightarrow$ $\dot{z} = 0$, which is also true.

(11) $\Rightarrow$ $2a\dot{x}^2 + 2b\dot{y}^2 - 2\ddot{z} = 0$ $\Rightarrow$ $a\dot{x}^2 + b\dot{y}^2 = \ddot{z}$ ...(12)

Differentiating (4) w.r.t. t, we get

$$\dot{x}^2 + x\ddot{x} + \dot{y}^2 + y\ddot{y} + \dot{z}^2 + z\ddot{z} = c\ddot{z}.$$

Since the circle passes through the origin, we have

$$\dot{x}^2 + 0\ddot{x} + \dot{y}^2 + 0\ddot{y} + \dot{z}^2 + 0\ddot{z} = c\ddot{z}.$$

$$\Rightarrow \qquad \dot{x}^2 + \dot{y}^2 + \dot{z}^2 = c\ddot{z}$$

$$\Rightarrow \qquad \dot{x}^2 + \dot{y}^2 = c\ddot{z} \qquad \text{...(13)} (\because \ \dot{z} = 0)$$

Dividing (13) by (12), we get

$$c = \frac{\dot{x}^2 + \dot{y}^2}{a\dot{x}^2 + b\dot{y}^2} = \frac{m^2\lambda^2 + l^2\lambda^2}{am^2\lambda^2 + bl^2\lambda^2} = \frac{m^2 + l^2}{am^2 + bl^2}. \qquad \text{(Using (5))}$$

$$\therefore \qquad c = \frac{l^2 + m^2}{bl^2 + am^2}.$$

Example 3. *Find the equation of the plane that has 3-point contact with the curve $x = t^4 - 1, y = t^3 - 1, z = t^2 - 1$ at the origin.*

Sol. Given curve is

$$x = t^4 - 1, y = t^3 - 1, z = t^2 - 1.$$

$$t^4 - 1 = 0, t^3 - 1 = 0, t^2 - 1 = 0 \ \Rightarrow \ t = 1$$

$\therefore$ The origin corresponds to $t = 1$.

Let the equation of the required plane through the origin be $ax + by + cz = 0$.

Let $\qquad f(t) = a(t^4 - 1) + b(t^3 - 1) + c(t^2 - 1)$

$\therefore \qquad \dot{f}(t) = 4at^3 + 3bt^2 + 2ct$

$$\ddot{f}(t) = 12at^2 + 6bt + 2c$$

$$\dddot{f}(t) = 24at + 6b$$

Since the plane has 3-point contact at $t = 1$, we have

$$f(1) = 0, \dot{f}(1) = 0, \ddot{f}(1) = 0 \text{ and } \dddot{f}(1) \neq 0.$$

$$f(1) = 0 \quad \Rightarrow \quad 0 = 0$$

$$\dot{f}(1) = 0 \quad \Rightarrow \quad 4a + 3b + 2c = 0 \qquad \text{...(1)}$$

$$\ddot{f}(1) = 0 \quad \Rightarrow \quad 12a + 6b + 2c = 0 \qquad \text{...(2)}$$

(1) and (2) $\Rightarrow \quad \dfrac{a}{6-12} = \dfrac{b}{24-8} = \dfrac{c}{24-36}$

$$\Rightarrow \qquad \frac{a}{-6} = \frac{b}{16} = \frac{c}{-12} \quad \Rightarrow \quad \frac{a}{3} = \frac{b}{-8} = \frac{c}{6}$$

Let $\qquad a = 3, b = -8, c = 6.$

Also $\qquad \dddot{f}(1) = 24a + 6b = 24(3) + 6(-8) = 24 \neq 0$

$\therefore$ The equation of the required plane is $3x - 8y + 6z = 0$.

Example 4. *Find the lines that have 4-point contact with the surface $x^4 + 3xyz + x^2 - y^2 - z^2 + 2yz - 3xy - 2y + 2z - 1 = 0$ at the point (0, 0, 1).*

Sol. Let $$\frac{x-0}{a} = \frac{y-0}{b} = \frac{z-1}{c} = t \qquad ...(1)$$

be a line passing through (0, 0, 1).

$\therefore$ $$x = at, y = bt, z = ct + 1$$

$\therefore$ The point (0, 0, 1) corresponds to the value $t = 0$.

Given surface is

$$x^4 + 3xyz + x^2 - y^2 - z^2 + 2yz - 3xy - 2y + 2z - 1 = 0.$$

Let $f(t) = (at)^4 + 3(at)(bt)(ct + 1) + (at)^2 - (bt)^2 - (ct + 1)^2 + 2(bt)(ct + 1)$
$- 3(at)(bt) - 2(bt) + 2(ct + 1) - 1$

$\therefore$ $f(t) = a^4t^4 + 3abt^2(ct + 1) + a^2t^2 - b^2t^2 - (ct + 1)^2 + 2bt(ct + 1) - 3abt^2 - 2bt + 2ct + 1$

$\therefore$ $\dot{f}(t) = 4a^3t^3 + 9abct^2 + 6abt + 2a^2t - 2b^2t - 2(ct + 1)c + 4bct + 2b - 6abt - 2b + 2c$

$\ddot{f}(t) = 12a^3t^2 + 18abct + 6ab + 2a^2 - 2b^2 - 2c^2 + 4bc - 6ab$

$\dddot{f}(t) = 24a^3t + 18abc$

$\ddddot{f}(t) = 24a^3.$

Let the line (1) has 4-point contact with the given surface at (0, 0, 1).

$\therefore$ $f(0) = 0, \dot{f}(0) = 0, \ddot{f}(0) = 0, \dddot{f}(0) = 0, \ddddot{f}(0) \neq 0.$

$f(0) = 0 \quad \Rightarrow \quad -1 + 1 = 0,$ which is true.

$\dot{f}(0) = 0 \quad \Rightarrow \quad -2c + 2b - 2b + 2c = 0,$ which is true.

$\ddot{f}(0) = 0 \quad \Rightarrow \quad 6ab + 2a^2 - 2b^2 - 2c^2 + 4bc - 6ab = 0$

$$\Rightarrow \quad a^2 - b^2 - c^2 + 2bc = 0 \qquad ...(2)$$

$\dddot{f}(0) = 0 \Rightarrow 18abc = 0 \Rightarrow abc = 0 \Rightarrow a = 0$ or $b = 0$ or $c = 0$

Case I. a = 0

(2) $\Rightarrow b^2 + c^2 - 2bc = 0 \Rightarrow b = c$

$\therefore$ The line is $\frac{x}{0} = \frac{y}{c} = \frac{z-1}{c}$ or $\frac{x}{0} = \frac{y}{1} = \frac{z-1}{1}$.

Case II. b = 0

(2) $\Rightarrow a^2 - c^2 = 0 \Rightarrow a = \pm c$

$\therefore$ The lines are $\frac{x}{\pm c} = \frac{y}{0} = \frac{z-1}{c}$ or $\frac{x}{\pm 1} = \frac{y}{0} = \frac{z-1}{1}$.

Case III. c = 0

(2) $\Rightarrow a^2 - b^2 = 0 \Rightarrow a = \pm b$

$\therefore$ The lines are $\frac{x}{\pm b} = \frac{y}{b} = \frac{z-1}{0}$ or $\frac{x}{\pm 1} = \frac{y}{1} = \frac{z-1}{0}$.

$\therefore$ There are five possible lines.

Theorem 1. *Let $r = r(s)$ be any curve and $P(s_0)$ be any point on the curve. Prove that the curve $r = r(s)$ has at least 2-point contact with a plane through P at the point P iff the plane contains the tangent line at P.*

Proof. Let the equation of a plane through $P(s_0)$ be $(\mathbf{r} - \mathbf{r}_0) . \mathbf{N} = 0$, where $\mathbf{r_0} = \mathbf{r}(s_0)$ *i.e.,* the position vector of P and $\mathbf{N}$ is a unit vector perpendicular to the plane.

Let $$f(s) = (\mathbf{r}(s) - \mathbf{r}_0) . \mathbf{N}$$

$\therefore$ $$f'(s) = \mathbf{r}'(s) . \mathbf{N} = \mathbf{t}(s) . \mathbf{N}$$

Now $$f(s_0) = (\mathbf{r}_0 - \mathbf{r}_0) . \mathbf{N} = 0 \quad \text{and} \quad f'(s_0) = \mathbf{t}(s_0) . \mathbf{N}$$

$\therefore$ $$f'(s_0) = 0 \text{ iff } \mathbf{t}(s_0) . \mathbf{N} = 0$$

iff $\mathbf{N}$ is orthogonal to $\mathbf{t}(s_0)$ iff the plane $(\mathbf{r} - \mathbf{r}_0) . \mathbf{N} = 0$ contains the tangent line at P.

$\therefore$ The curve $\mathbf{r} = \mathbf{r}(s)$ has at least 2-point contact with a plane through P at the point P on the curve iff the plane contains the tangent line at P.

Theorem 2. *Let $r = r(s)$ be any curve and $P(s_0)$ be any non-inflexional point on the curve. Prove that the curve $r = r(s)$ has at least 3-point contact with a plane through P at the point P iff the plane is the osculating plane at P.*

Proof. Let the equation of a plane through P be $(\mathbf{r} - \mathbf{r}_0) . \mathbf{N} = 0$, where $\mathbf{r}_0 = \mathbf{r}(s_0)$ *i.e.,* the position vector of P and $\mathbf{N}$ is a unit vector perpendicular to the plane.

Let $$f(s) = (\mathbf{r}(s) - r_0) . \mathbf{N}$$

$\therefore$ $$f'(s) = \mathbf{r}'(s) . \mathbf{N} = \mathbf{t}(s) . \mathbf{N}$$

and $$f''(s) = \mathbf{t}'(s) . \mathbf{N} = \kappa(s)\, \mathbf{n}(s) . \mathbf{N}$$

Now $$f(s_0) = (\mathbf{r}_0 - \mathbf{r}_0) . \mathbf{N} = 0, \mathbf{f}'(s_0) = \mathbf{t}(s_0) . \mathbf{N} \quad \text{and} \quad f''(s_0) = \kappa(s_0)\, \mathbf{n}(s_0) . \mathbf{N}.$$

$\therefore$ $$f'(s_0) = 0, \quad f''(s_0) = 0 \text{ iff } \quad \mathbf{t}(s_0) . \mathbf{N} = 0, \kappa(s_0)\, \mathbf{n}(s_0) . \mathbf{N} = 0$$

iff $$\mathbf{t}(s_0) . \mathrm{N} = 0 \quad \mathbf{n}(s_0) . \mathbf{N} = 0 \qquad (\because \quad \kappa(s_0) \neq 0)$$

iff $\mathbf{N}$ is orthogonal to $\mathbf{t}(s_0)$ and $\mathbf{n}(s_0)$ iff the plane $(\mathbf{r} - \mathbf{r}_0) . \mathbf{N} = 0$ is the osculating plane at P.

$\therefore$ The curve $\mathbf{r} = \mathbf{r}(s)$ has at least 3-point contact with a plane through P at the point P on the curve iff the plane is the osculating plane at P.

Remark. If P is an inflexional point, then $f''(s_0) = \kappa(s_0)\, \mathbf{n}(s_0) . \mathbf{N} = 0$ even if the plane contains only the tangent line at P and is not the osculating plane at P.

Example 5. *Show that the osculating plane has at least 4-point contact with a curve at P iff either the curvature or the torsion vanishes at P.*

Sol. Let the equation of the curve be $\mathbf{r} = \mathbf{r}(s)$. Let $\mathbf{r}_0 = \mathbf{r}(s_0)$ be the position vector of the point P on the curve.

The equation of the osculating plane at P is $(\mathbf{r} - \mathbf{r}_0) . \mathbf{b}_0 = 0$.

Let $$f(s) = (\mathbf{r}(s) - \mathbf{r}_0) . \mathbf{b}_0$$

$\therefore$ $$f'(s) = \mathbf{r}'(s) . \mathbf{b}_0 = \mathbf{t} . \mathbf{b}_0,$$

$$f''(s) = \mathbf{t}' . \mathbf{b}_0 = (\kappa\mathbf{n}) . \mathbf{b}_0 = \kappa(\mathbf{n} . \mathbf{b}_0)$$

and $f'''(s) = \kappa'(\mathbf{n} \,.\, \mathbf{b}_0) + \kappa(\mathbf{n}' \,.\, \mathbf{b}_0)$

$= \kappa'(\mathbf{n} \,.\, \mathbf{b}_0) + \kappa(-\kappa\mathbf{t} + \tau\mathbf{b}) \,.\, \mathbf{b}_0$

$= \kappa'(\mathbf{n} \,.\, \mathbf{b}_0) - \kappa^2 (\mathbf{t} \,.\, \mathbf{b}_0) + \kappa\tau (\mathbf{b} \,.\mathbf{b}_0)$

$\therefore$ $f(s_0) = (\mathbf{r}_0 - \mathbf{r}_0) \,.\, \mathbf{b}_0 = 0,$

$f'(s_0) = \mathbf{t}_0 \,.\, \mathbf{b}_0 = 0,$

$f''(s_0) = \kappa_0 (\mathbf{n}_0 \,.\, \mathbf{b}_0) = 0$

and $f'''(s_0) = \kappa_0'(\mathbf{n}_0 \,.\, \mathbf{b}_0) - \kappa_0^2(\mathbf{t}_0 \,.\, \mathbf{b}_0) + \kappa_0\tau_0 (\mathbf{b}_0 \,.\, \mathbf{b}_0) = \kappa_0\tau_0$

$\therefore$ $f'''(s_0) = 0$ iff $\kappa_0 = 0$ or $\tau_0 = 0$.

$\therefore$ **The osculating plane at P has at least 4-point contact with the curve at P iff either the curvature or the torsion vanishes at P.**

EXERCISE 2.3

1. Show that the curve $\mathbf{r} = t\mathbf{i} + t^2\mathbf{j} + t^3\mathbf{k}$ has 2-point contact with the paraboloid $x^2 + y^2 = z$ at the origin.

2. Find the equation of the plane that has 3-point contact with the curve $x = 2t + 1$, $y = 3t^2 + 2$, $z = 4t^3 + 3$ at the point (3, 5, 7).

3. Let $\mathbf{r} = \mathbf{r}(s)$ be any curve and $P(s_0)$ be any point of inflexion on the curve. Prove that the curve $\mathbf{r} = \mathbf{r}(s)$ has at least 3-point contact with a plane through P at the point P iff the plane contains the tangent line at P.

4. Show that the osculating plane at P has 3-point contact with a curve at P iff neither the curvature nor the torsion vanishes at P.

5. Show that the osculating plane at P has at least 3-point contact with a curve at P.

Answer

2. $6x - 4y + z = 5$.

---

5. CONTACT OF A CURVE WITH A CURVE

A curve $\mathbf{r} = x(t)\mathbf{i} + y(t)\mathbf{j} + z(t)\mathbf{k}$ of sufficiently high class is said to have ***n*-point contact** (or **contact of** $(n - 1)$th **order**) with a curve $F(x, y, z) = 0$, $G(x, y, z) = 0$ at the point corresponding to t_0 if the functions $f(t) = F(x(t), y(t), z(t))$ and $g(t) = G(x(t), y(t), z(t))$ satisfy:

$$f(t_0) = \dot{f}(t_0) = \ddot{f}(t_0) = \ldots = f^{(n-1)}(t_0) = 0$$

$$g(t_0) = \dot{g}(t_0) = \ddot{g}(t_0) = \ldots = g^{(n-1)}(t_0) = 0$$

and either $f^{(n)}(t_0) \neq 0$ or $g^{(n)}(t_0) \neq 0$.

Thus, the curve $\mathbf{r} = x(t)\mathbf{i} + y(t)\mathbf{j} + z(t)\mathbf{k}$ has n-point contact with the curve $F(x, y, z) = 0$, $G(x, y, z) = 0$ if and only if the curve $\mathbf{r} = x(t)\mathbf{i} + y(t)\mathbf{j} + z(t)\mathbf{k}$ has n-point contact with one of the surfaces $F(x, y, z) = 0$ and $G(x, y, z) = 0$ and at least n-point contact with the other surface.

6. OSCULATING CIRCLE TO A CURVE

A circle having at least 3-point contact with a given curve C at a point P on the curve is called the **osculating circle** to the curve C at the point P.

The centre of the osculating circle at point P is called the **centre of curvature** of the curve C at the point P.

By the definition of contact between curves, the osculating circle to the curve C at point P can be considered as the intersection of a sphere with at least 3-point contact with the curve C at point P and a plane with at least 3-point contact with C at P. If $\kappa \neq 0$ at P, then the osculating plane at P is the unique plane having at least 3-point contact with the curve C at P. In particular, if $\tau \neq 0$ in addition to $\kappa \neq 0$ at P, then the osculating plane is the unique plane having exactly 3-point contact with C at P.

Therefore the osculating circle to a curve at a point always lies on the osculating plane to the curve at that point, provided $\kappa \neq 0$ at the point under consideration.

Thus, the osculating circle to a curve at a point can be considered as the intersection of a sphere with at least 3-point contact with the curve at that point and the osculating plane to the curve at the point under consideration, provided $\kappa \neq 0$.

7. EQUATION OF OSCULATING CIRCLE

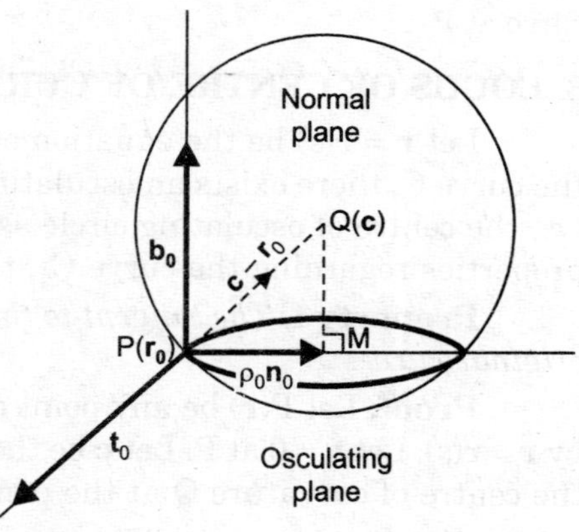

Let $\mathbf{r} = \mathbf{r}(s)$ be the equation of a curve C, where s is the parameter 'arc length'. Let P be any point on the curve C for the value s_0 of s. Let $\mathbf{r_0} = \mathbf{r}(s_0)$. Let curvature $\kappa_0 (= \kappa(s_0))$ be non-zero at P.

The equation of the osculating plane at P is $(\mathbf{r} - \mathbf{r_0}) \cdot \mathbf{b_0} = 0$, where $\mathbf{b_0} = \mathbf{b}(s_0)$.

Let the osculating circle at P be the intersection of the osculating plane at P and the sphere

$$|\mathbf{r} - \mathbf{c}|^2 = a^2$$

with centre at Q(**c**) and passing through P and having at least 3-point contact with the curve $\mathbf{r} = \mathbf{r}(s)$ at P.

$$\therefore \quad |\mathbf{r_0} - \mathbf{c}|^2 = a^2$$

Let $\quad f(s) = |\mathbf{r}(s) - \mathbf{c}|^2 - a^2.$

$$\therefore \quad f(s) = (\mathbf{r}(s) - \mathbf{c}) \cdot (\mathbf{r}(s) - \mathbf{c}) - a^2$$

$$\therefore \quad f'(s) = (\mathbf{r}(s) - \mathbf{c}) \cdot (\mathbf{r}'(s)) + \mathbf{r}'(s) \cdot (\mathbf{r}(s) - \mathbf{c}) - 0 = 2(\mathbf{r}(s) - \mathbf{c}) \cdot \mathbf{t}(s)$$

$$f''(s) = 2(\mathbf{r}(s) - \mathbf{c}) \cdot \mathbf{t}'(s) + 2(\mathbf{t}(s)) \cdot \mathbf{t}(s)$$

$$= 2(\mathbf{r}(s) - \mathbf{c}) \cdot \kappa(s)\, \mathbf{n}(s) + 2(1) = 2\kappa(s)\, (\mathbf{r}(s) - \mathbf{c}) \cdot \mathbf{n}(s) + 2$$

Since the sphere has at least 3-point of contact at P, we have $f(s_0) = f'(s_0) = f''(s_0) = 0$.

$f(s_0) = 0 \quad \Rightarrow \quad (\mathbf{r}_0 - \mathbf{c}) \cdot (\mathbf{r}_0 - \mathbf{c}) - a^2 = 0 \Rightarrow |\mathbf{r}_0 - \mathbf{c}|^2 - a^2 = 0.$

$\Rightarrow \quad a^2 - a^2 = 0$, which is true.

$f'(s_0) = 0 \quad \Rightarrow \quad 2(\mathbf{r}_0 - \mathbf{c}) \cdot \mathbf{t}_0 = 0 \Rightarrow (\mathbf{c} - \mathbf{r_0}) \cdot \mathbf{t}_0 = 0$

$\Rightarrow \quad \mathbf{c} - \mathbf{r}_0$ lies in the normal plane at P

$\Rightarrow$ centre (Q) of the sphere is in the normal plane at P.

$f''(s_0) = 0 \quad \Rightarrow \quad 2\kappa_0(\mathbf{r_0} - \mathbf{c}) \cdot \mathbf{n_0} + 2 = 0$

$$\Rightarrow \quad (\mathbf{r_0} - \mathbf{c}) \cdot \mathbf{n_0} = -\frac{1}{\kappa_0} = -\rho_0 \qquad \text{(Using } \kappa_0 \neq 0\text{)}$$

$$\Rightarrow \quad (\mathbf{c} - \mathbf{r_0}) \cdot \mathbf{n_0} = \rho_0$$

$\therefore$ Projection of $\mathbf{c} - \mathbf{r_0}$ on $\mathbf{n_0} = \rho_0$ ...(1)

Let QM be perpendicular to the principal normal at P.

$\therefore$ M is the centre of the osculating circle at P.

Also, (1) $\Rightarrow$ $\mathbf{PM} = \rho_0 \mathbf{n_0}$ $\Rightarrow$ PM $= \rho_0$ $\Rightarrow$ M is the centre of curvature at P.

Also, the centre of curvature (M) lies on the principal normal at P and at a distance ρ_0 from P.

$\therefore$ The radius of the osculating circle at P is ρ_0 which is also equal to the radius of curvature of the given curve at P.

Also, the position vector of the centre of the osculating circle at P($\mathbf{r_0}$)

$$= \text{P.V. of M} = \mathbf{r_0} + \rho_0 \mathbf{n_0}$$

and it lies on the principal normal of the given curve at P.

Remark. If $\kappa_0 = 0$, then $f''(s_0) = 0 \Rightarrow 0 + 2 = 0$, which is impossible.

$\therefore$ If $\kappa_0 = 0$, then there does not exist any sphere having at least 3-point of contact with the given curve at P.

8. LOCUS OF CENTRE OF CURVATURE

Let $\mathbf{r} = \mathbf{r}(s)$ be the equation of a curve C. For each point P with non-zero curvature, on the curve C, there exists an osculating circle. Let C_1 denote the locus of the centre of curvature *i.e.,* the centre of osculating circle as the point P moves along the curve C. We shall prove two properties regarding the curve C_1, the locus of centre of curvature.

Property I. *The tangent to the locus of centre of curvature lies in the normal plane of the original curve.*

Proof. Let P($\mathbf{r}$) be any point on a curve C given by $\mathbf{r} = \mathbf{r}(s)$. Let $\kappa \neq 0$ at P. Let $\mathbf{c}$ be the position vector of the centre of curvature Q at the point P.

$$\therefore \quad \mathbf{c} = \mathbf{r} + \rho \mathbf{n} \quad \text{...(1)}$$

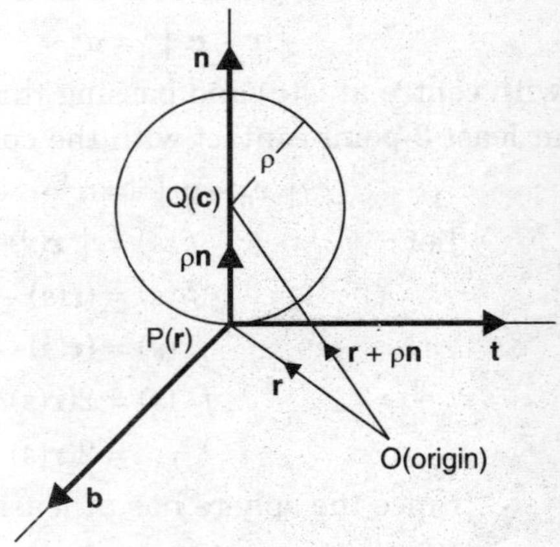

We shall use suffix '1' with quantities corresponding to the curve C_1 of the locus of centre of curvature.

Differentiating (1) w.r.t. s_1, we get

$$\frac{d\mathbf{c}}{ds_1} = \frac{d}{ds}(\mathbf{r} + \rho\mathbf{n}) \frac{ds}{ds_1}.$$

$$\Rightarrow \quad \mathbf{t_1} = (\mathbf{r}' + \rho\mathbf{n}' + \rho'\mathbf{n}) \frac{ds}{ds_1}$$

$$\Rightarrow \quad \mathbf{t_1} = (\mathbf{r}' + \rho(-\kappa\mathbf{t} + \tau\mathbf{b}) + \rho'\mathbf{n}) \frac{ds}{ds_1}$$

$$\Rightarrow \quad \mathbf{t_1} = (\mathbf{t} - \mathbf{t} + \rho\tau\mathbf{b} + \rho'\mathbf{n}) \frac{ds}{ds_1},$$

$$\Rightarrow \quad \mathbf{t_1} = (\rho\tau\mathbf{b} + \rho'\mathbf{n}) \frac{ds}{ds_1}$$

$\therefore$ $\mathbf{t_1}$ lies in the plane of $\mathbf{b}$ and $\mathbf{n}$.

$\therefore$ The tangent at a point to the curve C_1 lies in the corresponding normal plane of curve C.

$\therefore$ The tangent to the locus of the centre of curvature lies in the normal plane of the original curve.

Remark. Let α be the angle between the tangent to the locus of the centre of curvature at the centre of curvature at point P and the principal normal at P.

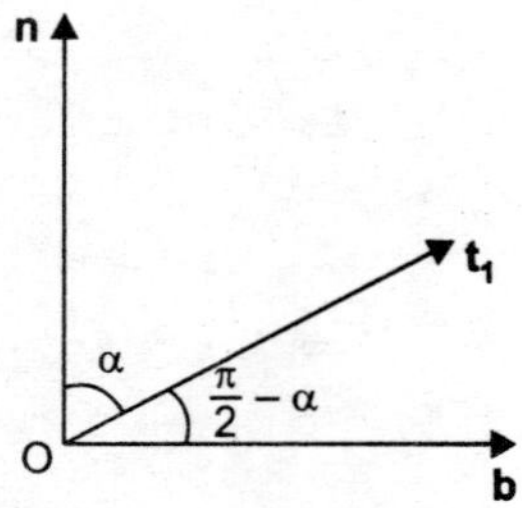

$$\therefore \quad \mathbf{t}_1 \cdot \mathbf{n} = 1 \,.\, 1 \,.\, \cos\alpha = \cos\alpha$$

$$\Rightarrow \quad (\rho\tau\mathbf{b} + \rho'\mathbf{n})\frac{ds}{ds_1} \cdot \mathbf{n} = \cos\alpha$$

$$\Rightarrow \quad 0 + \rho'\frac{ds}{ds_1} = \cos\alpha, \textit{ i.e., } \cos\alpha = \rho'\frac{ds}{ds_1}.$$

Also, angle between $\mathbf{t}_1$ and $\mathbf{b} = \dfrac{\pi}{2} - \alpha$

$$\therefore \quad \mathbf{t}_1 \cdot \mathbf{b} = 1 \,.\, 1 \,.\, \cos\left(\frac{\pi}{2} - \alpha\right) = \sin\alpha$$

$$\Rightarrow \quad (\rho\tau\mathbf{b} + \rho'\mathbf{n})\frac{ds}{ds_1} \cdot \mathbf{b} = \sin\alpha$$

$$\Rightarrow \quad \rho\tau\frac{ds}{ds_1} + 0 = \sin\alpha, \textit{ i.e., } \sin\alpha = \rho\tau\frac{ds}{ds_1}.$$

Dividing, we get $\quad \tan\alpha = \dfrac{\rho\,\tau}{\rho'} = \dfrac{\rho}{\rho'\sigma}$

$$\therefore \quad \alpha = \tan^{-1}\left(\frac{\rho}{\rho'\sigma}\right).$$

Property II. *If the original curve C has a constant curvature κ, then the curvature of the locus C_1 of centre of curvature is also constant and the torsion of C_1 varies inversely as that of C.*

Proof. Let P($\mathbf{r}$) be any point on the curve C given by $\mathbf{r} = \mathbf{r}(s)$. Let $\mathbf{c}$ be the position vector of the centre of curvature at the point P.

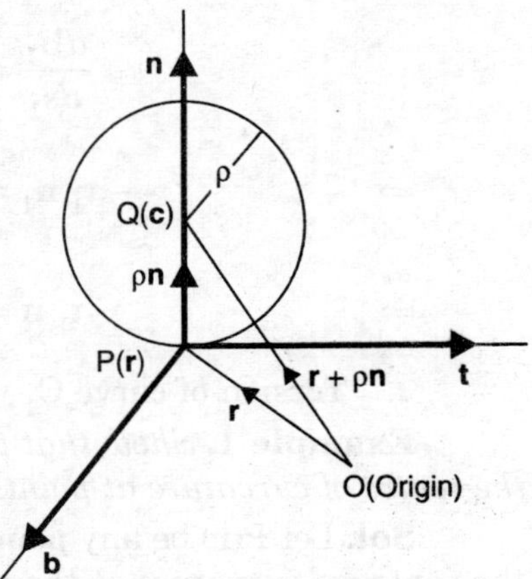

$$\therefore \quad \mathbf{c} = \mathbf{r} + \rho\mathbf{n} \qquad \text{...(1)}$$

We shall use suffix '1' with quantities corresponding to the curve C_1 of the locus of centre of curvature.

Differentiating (1) w.r.t. s_1, we get

$$\frac{d\mathbf{c}}{ds_1} = \frac{d}{ds}(\mathbf{r} + \rho\mathbf{n})\frac{ds}{ds_1}.$$

$$\Rightarrow \quad \mathbf{t}_1 = (\mathbf{r}' + \rho\mathbf{n}' + \rho'\mathbf{n})\frac{ds}{ds_1} \qquad \text{...(2)}$$

Since κ is constant, we have $\rho' = \left(\dfrac{1}{\kappa}\right)' = 0.$

$$\therefore \quad (2) \Rightarrow \quad \mathbf{t}_1 = (\mathbf{r}' + \rho(-\kappa\mathbf{t} + \tau\mathbf{b}))\frac{ds}{ds_1}$$

$$\Rightarrow \qquad \mathbf{t}_1 = (\mathbf{t} - \mathbf{t} + \rho\tau\mathbf{b})\frac{ds}{ds_1}$$

$$\Rightarrow \qquad \mathbf{t}_1 = \rho\tau\mathbf{b}\frac{ds}{ds_1} \qquad ...(3)$$

$$\Rightarrow \qquad \mathbf{t}_1 \cdot \mathbf{t}_1 = \left(\rho\tau\mathbf{b}\frac{ds}{ds_1}\right)\cdot\left(\rho\tau\mathbf{b}\frac{ds}{ds_1}\right)$$

$$\Rightarrow \qquad 1 = \rho^2\tau^2\left(\frac{ds}{ds_1}\right)^2 \Rightarrow \frac{ds}{ds_1} = \frac{1}{\rho\tau}$$

$$\therefore \quad (3) \Rightarrow \qquad \mathbf{t}_1 = (\rho\tau\mathbf{b})\frac{1}{\rho\tau} = \mathbf{b} \qquad ...(4)$$

Differentiating w.r.t. s_1, we get

$$\frac{d\mathbf{t}_1}{ds_1} = \frac{d\mathbf{b}}{ds}\frac{ds}{ds_1}.$$

$$\Rightarrow \qquad \mathbf{t}_1' = \mathbf{b}'\frac{1}{\rho\tau} \Rightarrow \rho\tau\mathbf{t}_1' = -\tau\mathbf{n} \Rightarrow \mathbf{t}_1' = -\frac{1}{\rho}\mathbf{n}$$

$$\Rightarrow \qquad \mathbf{t}_1' = -\kappa\mathbf{n}$$

$$\Rightarrow \qquad \kappa_1\mathbf{n}_1 = -\kappa\mathbf{n}$$

$\therefore$ Vectors $\mathbf{n}_1$ and $\mathbf{n}$ are parallel. Choosing the direction of $\mathbf{n}_1$ opposite to that of $\mathbf{n}$, we have $\mathbf{n}_1 = -\mathbf{n}$.

$$\therefore \qquad \kappa_1 = \kappa$$

$\therefore$ The curvature of the curve C_1 is also constant. ($\because$ κ is constant)

Also $\qquad \mathbf{b}_1 = \mathbf{t}_1 \times \mathbf{n}_1 = \mathbf{b} \times (-\mathbf{n}) = -\mathbf{b} \times \mathbf{n} = \mathbf{n} \times \mathbf{b} = \mathbf{t}$ (Using (4))

Differentiating w.r.t. s_1, we get

$$\frac{d\mathbf{b}_1}{ds_1} = \frac{d\mathbf{t}}{ds}\frac{ds}{ds_1}$$

$$\Rightarrow \qquad -\tau_1\mathbf{n}_1 = \mathbf{t}'\frac{ds}{ds_1} \Rightarrow -\tau_1\mathbf{n}_1 = (\kappa\mathbf{n})\frac{1}{\rho\tau}$$

$$\Rightarrow \qquad \tau_1\mathbf{n} = \frac{\kappa}{\rho\tau}\mathbf{n} \Rightarrow \tau_1 = \left(\frac{\kappa}{\rho}\right)\frac{1}{\tau} = \kappa^2 \cdot \frac{1}{\tau} \qquad (\because \mathbf{n}_1 = -\mathbf{n})$$

$\therefore$ Torsion of curve C_1 varies inversely as that of C.

Example 1. *Show that the principal normal to a curve is perpendicular to the locus of the centre of curvature at points, where curvature* κ *is constant.*

Sol. Let P(**r**) be any point on a curve C given by $\mathbf{r} = \mathbf{r}(s)$. Let **c** be the position vector of the centre of curvature at the point P.

$$\therefore \qquad \mathbf{c} = \mathbf{r} + \rho\mathbf{n} \qquad ...(1)$$

We shall use suffix '1' with quantities corresponding to the curve C_1 of the locus of centre of curvature.

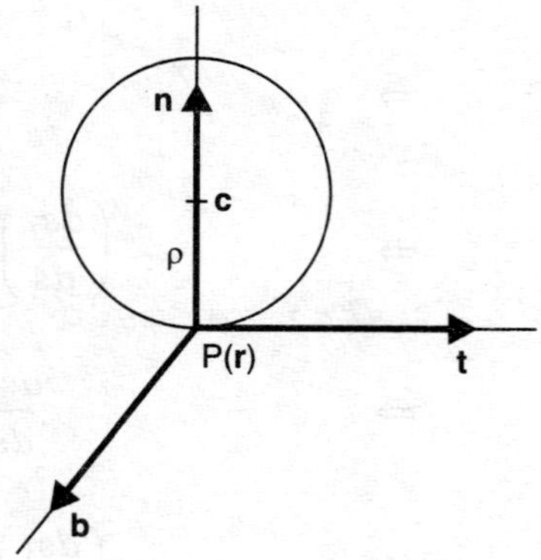

Differentiating (1) w.r.t. s_1, we get

$$\frac{d\mathbf{c}}{ds_1} = \frac{d}{ds}(\mathbf{r} + \rho\mathbf{n})\frac{ds}{ds_1}$$

$$\Rightarrow \quad \mathbf{t}_1 = (\mathbf{r}' + \rho\mathbf{n}' + \rho'\mathbf{n})\frac{ds}{ds_1}$$

$$\Rightarrow \quad \mathbf{t}_1 = (\mathbf{t} + \rho(-\kappa\mathbf{t} + \tau\mathbf{b}) + 0\mathbf{n})\frac{ds}{ds_1}$$

$$\left(\rho' = 0 \text{ because } \rho = \frac{1}{\kappa} \text{ is constant}\right)$$

$$\Rightarrow \quad \mathbf{t}_1 = \left(\rho\tau\frac{ds}{ds_1}\right)\mathbf{b} \qquad (\because \rho\kappa = 1)$$

$$\Rightarrow \quad \mathbf{n}.\,\mathbf{t}_1 = \left(\rho\tau\frac{ds}{ds_1}\right)\mathbf{n}\,.\,\mathbf{b} = 0$$

$\Rightarrow$ **n** is perpendicular to the tangent vector to the curve C_1 at the point with position vector **c**.

$\therefore$ Principal normal to the curve C is perpendicular to the curve C_1 *i.e.*, the locus of centre of curvature.

Example 2. *If s_1 is the arc length of the locus of centre of curvature, show that*

$$\frac{ds_1}{ds} = \frac{\sqrt{\kappa^2\tau^2 + \kappa'^2}}{\kappa^2} = \sqrt{\left(\frac{\rho}{\sigma}\right)^2 + \rho'^2}\,.$$

Sol. Let the given curve be $\mathbf{r} = \mathbf{r}(s)$. Let suffix '1' be used for quantities corresponding to the locus of centre of curvature.

Let $\mathbf{r}_1$ be the position vector of the centre of curvature corresponding to the point **r** on the curve $\mathbf{r} = \mathbf{r}(s)$.

$$\therefore \quad \mathbf{r}_1 = \mathbf{r} + \rho\mathbf{n}$$

$$\Rightarrow \quad \mathbf{r}_1 = \mathbf{r} + \frac{1}{\kappa}\mathbf{n}$$

Differentiating w.r.t. s_1, we get

$$\frac{d\mathbf{r}_1}{ds_1} = \frac{d}{ds}\left(\mathbf{r} + \frac{1}{\kappa}\mathbf{n}\right)\frac{ds}{ds_1}$$

$$\Rightarrow \quad \mathbf{t}_1 = \left(\mathbf{r}' + \frac{1}{\kappa}\mathbf{n}' + \left(-\frac{\kappa'}{\kappa^2}\right)\mathbf{n}\right)\frac{ds}{ds_1}$$

$$\Rightarrow \quad \mathbf{t}_1 = \left(\mathbf{t} + \frac{1}{\kappa}(-\kappa\mathbf{t} + \tau\mathbf{b}) - \frac{\kappa'}{\kappa^2}\mathbf{n}\right)\frac{ds}{ds_1}$$

$$\Rightarrow \quad \mathbf{t}_1 = \left(-\frac{\kappa'}{\kappa^2}\mathbf{n} + \frac{\tau}{\kappa}\mathbf{b}\right)\frac{ds}{ds_1}$$

$$\Rightarrow \quad \mathbf{t}_1\,.\,\mathbf{t}_1 = \left(-\frac{\kappa'}{\kappa^2}\mathbf{n} + \frac{\tau}{\kappa}\mathbf{b}\right).\left(-\frac{\kappa'}{\kappa^2}\mathbf{n} + \frac{\tau}{\kappa}\mathbf{b}\right)\left(\frac{ds}{ds_1}\right)^2$$

$$\Rightarrow \quad 1 = \left(\frac{\kappa'^2}{\kappa^4} + \frac{\tau^2}{\kappa^2}\right)\left(\frac{ds}{ds_1}\right)^2 = \left(\frac{\kappa'^2 + \kappa^2\tau^2}{\kappa^4}\right)\left(\frac{ds}{ds_1}\right)^2$$

$$\Rightarrow \quad \left(\frac{ds_1}{ds}\right)^2 = \frac{\kappa^2\tau^2 + \kappa'^2}{\kappa^4} \qquad \ldots(1)$$

$$\Rightarrow \quad \frac{ds_1}{ds} = \frac{\sqrt{\kappa^2\tau^2 + \kappa'^2}}{\kappa^2}$$

$$(1) \Rightarrow \quad \left(\frac{ds_1}{ds}\right)^2 = \frac{\tau^2}{\kappa^2} + \frac{\kappa'^2}{\kappa^4} = \frac{\rho^2}{\sigma^2} + \left(\left(\frac{1}{\kappa}\right)'\right)^2 = \frac{\rho^2}{\sigma^2} + \rho'^2$$

$$\therefore \quad \frac{ds_1}{ds} = \sqrt{\left(\frac{\rho}{\sigma}\right)^2 + \rho'^2}\,.$$

$\therefore$ The result holds.

9. OSCULATING SPHERE TO A CURVE

A sphere having at least 4-point contact with a given curve C at a point P on the curve is called the **osculating sphere** to the curve C at the point P.

The centre of the osculating sphere at point P is called the **centre of spherical curvature** of the curve C at the point P.

Remark. The radius of osculating sphere is also referred as the **radius of spherical curvature.**

10. EQUATION OF OSCULATING SPHERE

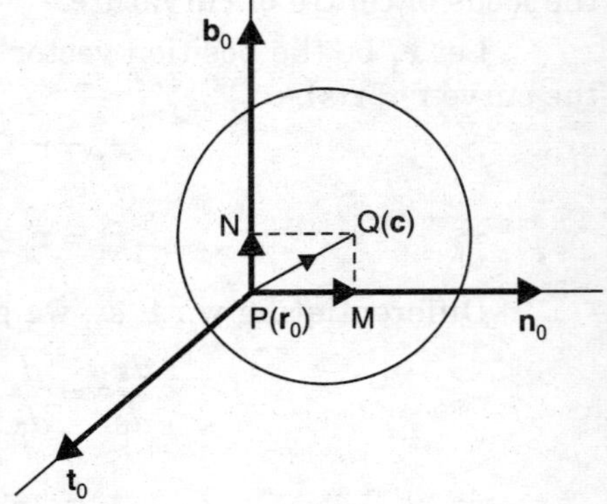

Let $\mathbf{r} = \mathbf{r}(s)$ be the equation of a curve C, where s is the parameter 'arc length'. Let P be any point on the curve C for the value s_0 of s. Let $\mathbf{r_0} = \mathbf{r}(s_0)$. Let curvature κ_0 (= $\kappa(s)$) and torsion τ_0 (= $\tau(s_0)$) be non-zero at P.

Let the equation of the osculating sphere be

$$|\,\mathbf{r} - \mathbf{c}\,|^2 = a^2$$

with centre at Q(**c**) and passing through P and having at least 4-point contact with the curve $\mathbf{r} = \mathbf{r}(s)$ at P.

$\therefore \quad |\,\mathbf{r}_0 - \mathbf{c}\,|^2 = a^2$

Let $\quad f(s) = |\,\mathbf{r}(s) - \mathbf{c}\,|^2 - a^2.$

$$\therefore \quad f(s) = (\mathbf{r}(s) - \mathbf{c}) \,.\, (\mathbf{r}(s) - \mathbf{c}) - a^2$$

$$f'(s) = (\mathbf{r}(s) - \mathbf{c}) \,.\, \mathbf{r}'(s)) + \mathbf{r}'(s) \,.\, (\mathbf{r}(s) - \mathbf{c}) - 0 = 2\,(\mathbf{r}(s) - \mathbf{c}) \,.\, \mathbf{t}(s)$$

$$f''(s) = 2(\mathbf{r}(s) - \mathbf{c}) \,.\, \mathbf{t}'(s) + 2\,(\mathbf{t}(s)) \,.\, \mathbf{t}(s)$$

$$= 2(\mathbf{r}(s) - \mathbf{c}) \,.\, \kappa(s)\,\mathbf{n}(s) + 2(1) = 2\kappa(s)\,(\mathbf{r}(s) - \mathbf{c}) \,.\, \mathbf{n}(s) + 2$$

$$f'''(s) = 2\kappa'(s)\,(\mathbf{r}(s) - \mathbf{c}) \,.\, \mathbf{n}(s) + 2\kappa(s)\,(\mathbf{t}(s) - 0) \,.\, \mathbf{n}(s) + 2\kappa(s)\,(\mathbf{r}(s) - \mathbf{c}) \,.\, \mathbf{n}'(s) + 0$$

$$= 2\kappa'(s)\,(\mathbf{r}(s) - \mathbf{c}) \,.\, \mathbf{n}(s) + 2\kappa(s)\,(0) + 2\kappa(s)\,(\mathbf{r}(s) - \mathbf{c}) \,.\, (-\,\kappa(s)\,\mathbf{t}(s) + \tau(s)\,\mathbf{b}(s))$$

$$= 2\kappa'(s)\,(\mathbf{r}(s) - \mathbf{c}) \,.\, \mathbf{n}(s) - 2\kappa^2(s)\,(\mathbf{r}(s) - \mathbf{c}) \,.\, \mathbf{t}(s) + 2\kappa(s)\,\tau(s)\,(\mathbf{r}(s) - \mathbf{c}) \,.\, \mathbf{b}(s)$$

Since the sphere has at least 4-point contact at P, we have

$$f(s_0) = f'(s_0) = f''(s_0) = f'''(s_0) = 0.$$

$f(s_0) = 0 \Rightarrow (\mathbf{r_0} - \mathbf{c}) . (\mathbf{r_0} - \mathbf{c}) - a^2 = 0 \Rightarrow |\mathbf{r_0} - \mathbf{c}|^2 - a^2 = 0$

$\Rightarrow a^2 - a^2 = 0$, which is true.

$f'(s_0) = 0 \Rightarrow 2(\mathbf{r_0} - \mathbf{c}) . \mathbf{t_0} = 0 \Rightarrow (\mathbf{c} - \mathbf{r_0}) . \mathbf{t_0} = 0$ ...(1)

$\Rightarrow$ $\mathbf{c} - \mathbf{r_0}$ lies in the normal plane at P

$\Rightarrow$ centre (Q) of the sphere is in the normal plane at P.

$f''(s_0) = 0 \Rightarrow 2\kappa_0(\mathbf{r_0} - \mathbf{c}) . \mathbf{n_0} + 2 = 0$

$$\Rightarrow (\mathbf{r_0} - \mathbf{c}) . \mathbf{n_0} = -\frac{1}{\kappa_0} = -\rho_0 \quad \text{...(2) (Using } \kappa_0 \neq 0)$$

$\Rightarrow (\mathbf{c} - \mathbf{r_0}) . \mathbf{n_0} = \rho_0$

$\Rightarrow$ Projection of $\mathbf{c} - \mathbf{r_0}$ on $\mathbf{n_0} = \rho_0$

$f'''(s_0) = 0 \Rightarrow 2\kappa_0' (\mathbf{r_0} - \mathbf{c}) . \mathbf{n_0} - 2\kappa_0^2 (\mathbf{r_0} - \mathbf{c}) . \mathbf{t_0} + 2\kappa_0\tau_0 (\mathbf{r_0} - \mathbf{c}) . \mathbf{b_0} = 0$

$$\Rightarrow 2\kappa_0' \left(-\frac{1}{\kappa_0}\right) - 2\kappa_0^2(0) + 2\kappa_0\tau_0(\mathbf{r_0} - \mathbf{c}) . \mathbf{b_0} = 0$$

(Using (1) and (2))

$$\Rightarrow -\frac{2\kappa_0'}{\kappa_0} + 2\kappa_0\tau_0 (\mathbf{r_0} - \mathbf{c}) . \mathbf{b_0} = 0$$

$$\Rightarrow (\mathbf{r_0} - \mathbf{c}) . \mathbf{b_0} = \frac{\kappa_0'}{\kappa_0^2\tau_0} \quad \text{(Using } \tau_0 \neq 0)$$

$$\Rightarrow (\mathbf{c} - \mathbf{r_0}) . \mathbf{b_0} = -\frac{\kappa_0'}{\kappa_0^2\tau_0} = \left(\frac{d}{ds}\left(\frac{1}{\kappa_0}\right)\right)\sigma_0 = \rho_0'\sigma_0$$

$\Rightarrow$ projection of $\mathbf{c} - \mathbf{r_0}$ on $\mathbf{b_0} = \rho_0'\sigma_0$

$\therefore$ The components of the vector $\mathbf{c} - \mathbf{r_0}$ along the vectors $\mathbf{t}_0$, $\mathbf{n}_0$ and $\mathbf{b}_0$ are 0, ρ_0 and $\rho_0'\,\sigma_0$ respectively.

$$\therefore \quad \mathbf{c} - \mathbf{r_0} = 0\mathbf{t}_0 + \rho_0\mathbf{n} + \rho_0'\sigma_0\mathbf{b}$$

$$\therefore \quad |\mathbf{c} - \mathbf{r_0}| = \sqrt{\rho_0^2 + \rho_0'^2\,\sigma_0^2}$$

and $\mathbf{c} = \mathbf{r_0} + \rho_0\,\mathbf{n} + \rho_0'\sigma_0\,\mathbf{b}_0$.

The centre Q(**c**) of the osculating sphere is the centre of spherical curvature of the curve C at P.

Radius of the osculating sphere at P = PQ = $|\mathbf{c} - \mathbf{r}_0| = \sqrt{\rho_0^2 + \rho_0'^2\,\sigma_0^2}$.

Also, the position vector of the centre of spherical curvature of the curve C at P($\mathbf{r_0}$) = P.V. of Q = $\mathbf{c} = \mathbf{r}_0 + \rho_0\mathbf{n_0} + \rho_0'\sigma_0\mathbf{b_0}$ and it lies on the normal plane of the given curve at P.

Remark 1. In terms of κ_0 and τ_0, we have

(*i*) radius of osculating sphere at $P(\mathbf{r_0}) = \sqrt{\left(\frac{1}{\kappa_0}\right)^2 + \left(\frac{\kappa_0'}{\kappa_0^2\,\tau_0}\right)^2}$ and

(*ii*) p.v. of centre of spherical curvature $= \mathbf{r}_0 + \frac{1}{\kappa_0}\,\mathbf{n_0} - \frac{\kappa_0'}{\kappa_0^2\,\tau_0}\,\mathbf{b_0}$.

Remark 2. If curvature of $\mathbf{r} = \mathbf{r}(s)$ at P is constant, then radius of osculating sphere at

$P(\mathbf{r}_0) = \sqrt{\rho_0^2 + 0.\sigma_0^2} = \rho_0$ and p.v. of centre of spherical curvature at

$P(\mathbf{r_0}) = \mathbf{r_0} + \rho_0\mathbf{n_0} + (0)\sigma_0\mathbf{b} = \mathbf{r_0} + \rho_0\mathbf{n_0}.$

$\therefore$ Centre of the osculating sphere coincides with the centre of osculating circle at points where curvature vanishes.

11. LOCUS OF CENTRE OF SPHERICAL CURVATURE

Let $\mathbf{r} = \mathbf{r}(s)$ be the equation of a curve C. For each point P with non-zero curvature and torsion on the curve C there exists an osculating sphere. Let C_1 denote the locus of the centre of spherical curvature *i.e.*, the centre of osculating sphere as the point P moves along the curve C. We shall prove some properties regarding the curve C_1, the locus of centre of spherical curvature.

Property I. *The tangent to the locus of centre of spherical curvature is parallel to the corresponding binormal to the original curve.*

Proof. Let $P(\mathbf{r})$ be any point on a curve C given by $\mathbf{r} = \mathbf{r}(s)$. Let $\kappa \neq 0$, $\tau \neq 0$ at P. Let $\mathbf{r}_1$ be the position vector of the centre of spherical curvature at the point P.

$$\therefore \qquad \mathbf{r}_1 = \mathbf{r} + \rho\mathbf{n} + \rho'\sigma\mathbf{b} \qquad ...(1)$$

We shall use suffix '1' with quantities corresponding to the curve C_1 of the locus of centre of spherical curvature.

Differentiating (1) w.r.t. s_1, we get

$$\frac{d\mathbf{r}_1}{ds_1} = \frac{d}{ds}(\mathbf{r} + \rho\mathbf{n} + \rho'\sigma\mathbf{b})\frac{ds}{ds_1}$$

$$\Rightarrow \qquad \mathbf{t}_1 = (\mathbf{t} + \rho\mathbf{n}' + \rho'\mathbf{n} + \rho''\sigma\mathbf{b} + \rho'\sigma'\mathbf{b} + \rho'\sigma\mathbf{b}')\frac{ds}{ds_1}$$

$$\Rightarrow \qquad \mathbf{t}_1 = (\mathbf{t} + \rho(-\kappa\mathbf{t} + \tau\mathbf{b}) + \rho'\mathbf{n} + \rho''\sigma\mathbf{b} + \rho'\sigma'\mathbf{b} - \rho'\sigma\tau\mathbf{n})\frac{ds}{ds_1}$$

$$\Rightarrow \qquad \mathbf{t}_1 = ((1 - \rho\kappa)\mathbf{t} + \rho'(1 - \sigma\tau)\mathbf{n} + (\rho\tau + \rho''\sigma + \rho'\sigma')\mathbf{b})\frac{ds}{ds_1}$$

$$\Rightarrow \qquad \mathbf{t}_1 = (\rho\tau + \rho''\sigma + \rho'\sigma')\frac{ds}{ds_1}\mathbf{b} \qquad (\because \kappa\rho = 1, \tau\sigma = 1)$$

$\therefore$ $\mathbf{t}_1$ is parallel to $\mathbf{b}$.

$\therefore$ The tangent to C_1 is parallel to the corresponding binormal to C.

Property II. *The principal normal to the locus of centre of spherical curvature is parallel to the corresponding principal normal to the original curve.*

Proof. Let $P(\mathbf{r})$ be any point on a curve C given by $\mathbf{r} = \mathbf{r}(s)$. Let $\kappa \neq 0$, $\tau \neq 0$ at P. Let $\mathbf{r}_1$ be the position vector of the centre of spherical curvature at the point P.

$$\therefore \qquad \mathbf{r}_1 = \mathbf{r} + \rho\mathbf{n} + \rho'\sigma\mathbf{b} \qquad ...(1)$$

We shall use suffix '1' with quantities corresponding to the curve C_1 of the locus of centre of spherical curvature.

Differentiating (1) w.r.t. s_1, we get

$$\frac{d\mathbf{r}_1}{ds_1} = \frac{d}{ds}(\mathbf{r} + \rho\mathbf{n} + \rho'\sigma\mathbf{b})\frac{ds}{ds_1}$$

$$\Rightarrow \qquad \mathbf{t}_1 = (\rho\tau + \rho''\sigma + \rho'\sigma')\frac{ds}{ds_1}\,\mathbf{b} \qquad ...(2)$$

(For detail see **property I**)

$$\Rightarrow \qquad |\mathbf{t}_1| = (\rho\tau + \rho''\sigma + \rho'\sigma')\frac{ds}{ds_1}\,|\mathbf{b}| \;\Rightarrow\; 1 = (\rho\tau + \rho''\sigma + \rho'\sigma')\frac{ds}{ds_1}\,.\,1$$

$$\therefore \qquad \frac{ds}{ds_1} = \frac{1}{\rho\tau + \rho''\sigma + \rho'\sigma'}$$

$$\therefore \;(2) \Rightarrow \qquad \mathbf{t}_1 = \mathbf{b} \qquad ...(3)$$

Differentiating (3) w.r.t. s_1, we get

$$\frac{d\mathbf{t}_1}{ds_1} = \frac{d\mathbf{b}}{ds}\frac{ds}{ds_1}$$

$$\Rightarrow \qquad \kappa_1\mathbf{n}_1 = -\tau\mathbf{n}\frac{ds}{ds_1}$$

$\therefore$ $\mathbf{n}_1$ is parallel to $\mathbf{n}$.

$\therefore$ The principal normal to C_1 is parallel to the corresponding principal normal to C.

Property III. *The binormal to the locus of centre of spherical curvature is parallel to the corresponding tangent to the original curve.*

Proof. Let P($\mathbf{r}$) be any point on a curve C given by $\mathbf{r} = \mathbf{r}(s)$. Let $\kappa \neq 0, \tau \neq 0$ at P. Let $\mathbf{r}_1$ be the position vector of the centre of spherical curvature at the point P.

$$\therefore \qquad \mathbf{r}_1 = \mathbf{r} + \rho\mathbf{n} + \rho'\sigma\mathbf{b} \qquad ...(1)$$

We shall use suffix '1' with quantities corresponding to the curve C_1 of the locus of centre of spherical curvature.

Differentiating (1) w.r.t. s_1, we get

$$\frac{d\mathbf{r}_1}{ds_1} = \frac{d}{ds}(\mathbf{r} + \rho\mathbf{n} + \rho'\sigma\mathbf{b})\frac{ds}{ds_1}$$

$$\Rightarrow \qquad \mathbf{t}_1 = (\rho\tau + \rho''\sigma + \rho'\sigma')\frac{ds}{ds_1}\,\mathbf{b} \qquad ...(2)$$

(For detail see **property I**)

$$\Rightarrow \qquad |\mathbf{t}_1| = (\rho\tau + \rho''\sigma + \rho'\sigma')\frac{ds}{ds_1}|\mathbf{b}|$$

$$\Rightarrow \qquad \frac{ds}{ds_1} = \frac{1}{\rho\tau + \rho''\sigma + \rho'\sigma'}$$

$$\therefore \;(2) \Rightarrow \qquad \mathbf{t}_1 = \mathbf{b} \qquad ...(3)$$

Differentiating (3) w.r.t. s_1, we get

$$\frac{d\mathbf{t}_1}{ds_1} = \frac{d\mathbf{b}}{ds}\frac{ds}{ds_1}$$

$$\Rightarrow \qquad \kappa_1\mathbf{n}_1 = -\tau\mathbf{n}\frac{ds}{ds_1} \qquad ...(4)$$

$\Rightarrow \quad \kappa_1 |\mathbf{n}_1| = |-1| \, \tau \dfrac{ds}{ds_1} |\mathbf{n}|$

$\Rightarrow \quad \kappa_1 . 1 = 1 . \tau \dfrac{ds}{ds_1} . 1 \quad \Rightarrow \quad \kappa_1 = \tau \dfrac{ds}{ds_1}$

$\therefore$ 4) $\Rightarrow \quad \mathbf{n}_1 = -\mathbf{n}$

$\Rightarrow \quad \mathbf{t}_1 \times \mathbf{n}_1 = \mathbf{b} \times (-\mathbf{n})$ (Using (3))

$\Rightarrow \quad \mathbf{b}_1 = \mathbf{t}$

$\therefore$ $\mathbf{b}_1$ is parallel to $\mathbf{t}$.

$\therefore$ The binormal to C_1 is parallel to the corresponding tangent to C.

Property IV. *The product of curvatures at the corresponding points on the locus of centre of spherical curvature and the original curve is equal to the product of their torsions.*

Proof. Let P(**r**) be any point on a curve C given by $\mathbf{r} = \mathbf{r}(s)$. Let $\kappa \neq 0$, $\tau \neq 0$ at P. Let $\mathbf{r}_1$ be the position vector of the centre of spherical curvature at the point P.

$$\therefore \quad \mathbf{r}_1 = \mathbf{r} + \rho\mathbf{n} + \rho'\sigma\,\mathbf{b} \qquad ...(1)$$

We shall use suffix '1' with quantities corresponding to the curve C_1 of locus of centre of spherical curvature.

Differentiating (1) w.r.t. s_1, we get

$$\frac{d\mathbf{r}_1}{ds_1} = \frac{d}{ds}(\mathbf{r} + \rho\mathbf{n} + \rho'\sigma\mathbf{b})\frac{ds}{ds_1}$$

$$\Rightarrow \quad \mathbf{t}_1 = (\rho\tau + \rho''\sigma + \rho'\sigma')\frac{ds}{ds_1}\mathbf{b} \qquad ...(2)$$

(For detail see **property I**)

$$\Rightarrow \quad |\,\mathbf{t}_1\,| = (\rho\tau + \rho''\sigma + \rho'\sigma')\frac{ds}{ds_1}|\,\mathbf{b}\,|$$

$$\Rightarrow \quad \frac{ds}{ds_1} = \frac{1}{\rho\tau + \rho''\sigma + \rho'\sigma'}$$

$$\therefore \ (2) \Rightarrow \quad \mathbf{t}_1 = \mathbf{b} \qquad ...(3)$$

Differentiating (3) w.r.t. s_1, we get

$$\frac{d\mathbf{t}_1}{ds_1} = \frac{d\mathbf{b}}{ds}\frac{ds}{ds_1}$$

$$\Rightarrow \quad \kappa_1\mathbf{n}_1 = -\tau\mathbf{n}\frac{ds}{ds_1} \qquad ...(4)$$

$$\Rightarrow \quad \kappa_1|\,\mathbf{n}_1\,| = |-1|\,\tau\frac{ds}{ds_1}|\,\mathbf{n}\,|$$

$$\Rightarrow \quad \kappa_1 . 1 = 1 . \tau\frac{ds}{ds_1} . 1 \quad \Rightarrow \quad \kappa_1 = \tau\frac{ds}{ds_1}$$

$\therefore$ (4) $\Rightarrow \quad \mathbf{n}_1 = -\mathbf{n}$

$\Rightarrow \quad \mathbf{t}_1 \times \mathbf{n}_1 = \mathbf{b} \times (-\mathbf{n})$ (Using (3)

$\Rightarrow \quad \mathbf{b}_1 = \mathbf{t}$

Differentiating w.r.t. s_1, we get

$$\frac{d\mathbf{b}_1}{ds_1} = \frac{d\mathbf{t}}{ds}\frac{ds}{ds_1}$$

$$\Rightarrow \quad -\tau_1\mathbf{n}_1 = (\kappa\mathbf{n})\frac{\kappa_1}{\tau} \qquad \left(\text{Using } \kappa_1 = \tau\frac{ds}{ds_1}\right)$$

$$\Rightarrow \quad -\tau_1(-\mathbf{n}) = \frac{\kappa\kappa_1}{\tau}\mathbf{n} \qquad (\text{Using } \mathbf{n}_1 = -\mathbf{n})$$

$$\Rightarrow \quad \tau\tau_1\mathbf{n} = \kappa\kappa_1\mathbf{n} \Rightarrow \kappa\kappa_1 = \tau\tau_1.$$

$\therefore$ The product of curvatures at the corresponding points is equal to the product of the torsions.

Property V. *If curvature κ of a curve C is constant, then the curvature κ_1 of the curve C_1 of the locus of centre of spherical curvature is also constant.*

Proof. Let P(**r**) be any point on a curve C given by $\mathbf{r} = \mathbf{r}(s)$. Let $\kappa \neq 0$, $\tau \neq 0$ at P. Let $\mathbf{r}_1$ be the position vector of the centre of spherical curvature at the point P.

$$\therefore \quad \mathbf{r}_1 = \mathbf{r} + \rho\mathbf{n} + \rho'\sigma\mathbf{b} \qquad ...(1)$$

We shall use suffix '1' with quantities corresponding to the curve C_1 of the locus of centre of spherical curvature.

Differentiating (1) w.r.t. s_1, we get

$$\frac{d\mathbf{r}_1}{ds_1} = \frac{d}{ds}(\mathbf{r} + \rho\mathbf{n} + \rho'\sigma\mathbf{b})\frac{ds}{ds_1}$$

$$\Rightarrow \quad \mathbf{t}_1 = (\rho\tau + \rho''\sigma + \rho'\sigma')\frac{ds}{ds_1}\mathbf{b} \qquad ...(2)$$

(For detail see **property I**)

$$\Rightarrow \quad |\,\mathbf{t}_1\,| = (\rho\tau + \rho''\sigma + \rho'\sigma')\frac{ds}{ds_1}\,|\,\mathbf{b}\,|$$

$$\Rightarrow \quad \frac{ds}{ds_1} = \frac{1}{\rho\tau + \rho''\sigma + \rho'\sigma'}$$

$$\therefore \ (2) \Rightarrow \quad \mathbf{t}_1 = \mathbf{b} \qquad ...(3)$$

Differentiating (3) w.r.t. s_1, we get

$$\frac{d\mathbf{t}_1}{ds_1} = \frac{d\mathbf{b}}{ds}\frac{ds}{ds_1}$$

$$\Rightarrow \quad \kappa_1\mathbf{n}_1 = -\tau\mathbf{n}\frac{ds}{ds_1}$$

$$\Rightarrow \quad \kappa_1\mathbf{n}_1 = -\tau\left(\frac{\kappa}{\tau}\right)\mathbf{n} \quad \left(\rho' = \left(\frac{1}{\kappa}\right)' = 0, \rho'' = 0 \Rightarrow \frac{ds}{ds_1} = \frac{1}{\rho\tau + 0\sigma + 0.\sigma'} = \frac{\kappa}{\tau}\right)$$

$$\Rightarrow \quad \kappa_1\mathbf{n}_1 = -\kappa\mathbf{n}$$

$$\Rightarrow \quad \kappa_1\,|\,\mathbf{n}_1\,| = |-1|\,\kappa\,|\,\mathbf{n}\,| \Rightarrow \kappa_1 = \kappa$$

$\Rightarrow$ κ_1 is constant because κ is constant.

$\therefore$ The curvature of the curve C_1 of the locus of centre of spherical curvature is also constant.

Example 3. *If R is the radius of osculating sphere to a curve* $\mathbf{r} = \mathbf{r}(s)$ *at point 's', then show that*

$$R = \left| \frac{\mathbf{t} \times \mathbf{t}''}{\kappa^2 \tau} \right|.$$

Sol. We have $\mathbf{t}' = \kappa \mathbf{n}$

and $\mathbf{t}'' = \kappa \mathbf{n}' + \kappa' \mathbf{n} = \kappa(-\kappa \mathbf{t} + \tau \mathbf{b}) + \kappa' \mathbf{n} = -\kappa^2 \mathbf{t} + \kappa' \mathbf{n} + \kappa\tau \mathbf{b}$

$$\therefore \quad \mathbf{t} \times \mathbf{t}'' = \mathbf{t} \times (-\kappa^2 \mathbf{t} + \kappa' \mathbf{n} + \kappa\tau \mathbf{b})$$

$$= -\kappa^2 (\mathbf{t} \times \mathbf{t}) + \kappa'(\mathbf{t} \times \mathbf{n}) + \kappa\tau(\mathbf{t} \times \mathbf{b})$$

$$= -\kappa^2(\mathbf{0}) + \kappa' \mathbf{b} + \kappa\tau(-\mathbf{n}) = -\kappa\tau \mathbf{n} + \kappa' \mathbf{b}$$

$$\therefore \quad \frac{\mathbf{t} \times \mathbf{t}''}{\kappa^2 \tau} = -\frac{\kappa\tau}{\kappa^2 \tau} \mathbf{n} + \frac{\kappa'}{\kappa^2 \tau} \mathbf{b} = -\rho \mathbf{n} - \left(\frac{1}{\kappa}\right)' \sigma \mathbf{b} = -\rho \mathbf{n} - \rho' \sigma \mathbf{b}$$

$$\Rightarrow \quad -\frac{\mathbf{t} \times \mathbf{t}''}{\kappa^2 \tau} = \rho \mathbf{n} + \rho' \sigma \mathbf{b}$$

$$\Rightarrow \quad (-1)^2 \frac{\mathbf{t} \times \mathbf{t}''}{\kappa^2 \tau} \cdot \frac{\mathbf{t} \times \mathbf{t}''}{\kappa^2 \tau} = (\rho \mathbf{n} + \rho' \sigma \mathbf{b}) \cdot (\rho \mathbf{n} + \rho' \sigma \mathbf{b})$$

$$\Rightarrow \quad \left| \frac{\mathbf{t} \times \mathbf{t}''}{\kappa^2 \tau} \right|^2 = \rho^2 \mathbf{n} \cdot \mathbf{n} + \rho'^2 \sigma^2 \mathbf{b} \cdot \mathbf{b} = \rho^2 + \rho'^2 \sigma^2$$

$$= R^2 \qquad (\because \ R^2 = \rho^2 + \rho'^2 \sigma^2)$$

$$\therefore \quad R = \left| \frac{\mathbf{t} \times \mathbf{t}''}{\kappa^2 \tau} \right|.$$

Example 4. *Show that the radius of osculating sphere of the circular helix*

$$x = a \cos \theta, \ y = a \sin \theta, \ z = a\theta \cot \alpha$$

is equal to the radius of curvature at each point on the helix.

Sol. Given helix is $x = a \cos \theta, y = a \sin \theta, z = a\, \theta \cot \alpha$.

Let $\mathbf{r}$ be the position vector of the point P(x, y, z) on the helix.

$$\therefore \quad \mathbf{r} = x\mathbf{i} + y\mathbf{j} + z\mathbf{k} = a \cos \theta \mathbf{i} + a \sin \theta \mathbf{j} + a\theta \cot \alpha \mathbf{k}$$

We know that $\kappa = \dfrac{|\dot{\mathbf{r}} \times \ddot{\mathbf{r}}|}{|\dot{\mathbf{r}}|^3}$ (A standard formula)

Now $\dot{\mathbf{r}} = -a \sin \theta \mathbf{i} + a \cos \theta \mathbf{j} + a \cot \alpha \mathbf{k}$

$$\ddot{\mathbf{r}} = -a \cos \theta \mathbf{i} - a \sin \theta \mathbf{j} + 0\mathbf{k}$$

$$\therefore \quad \dot{\mathbf{r}} \times \ddot{\mathbf{r}} = \begin{vmatrix} \mathbf{i} & \mathbf{j} & \mathbf{k} \\ -a \sin \theta & a \cos \theta & a \cot \alpha \\ -a \cos \theta & -a \sin \theta & 0 \end{vmatrix}$$

$$= -a^2 \begin{vmatrix} \mathbf{i} & \mathbf{j} & \mathbf{k} \\ -\sin \theta & \cos \theta & \cot \alpha \\ \cos \theta & \sin \theta & 0 \end{vmatrix}$$

$$= -a^2 [-\cot \alpha \sin \theta \mathbf{i} + \cot \alpha \cos \theta \mathbf{j} - \mathbf{k}]$$

$$\therefore \quad |\dot{\mathbf{r}} \times \ddot{\mathbf{r}}| = a^2\sqrt{\cot^2 \alpha \sin^2 \theta + \cot^2 \alpha \cos^2 \theta + 1}$$

$$= a^2\sqrt{\cot^2 \alpha + 1} = a^2 \operatorname{cosec} \alpha$$

Also
$$|\dot{\mathbf{r}}| = (a^2 \sin^2 \theta + a^2 \cos^2 \theta + a^2 \cot^2 \alpha)^{1/2}$$

$$= (a^2 + a^2 \cot^2 \alpha)^{1/2} = a \operatorname{cosec} \alpha$$

$$\therefore \quad \kappa = \frac{|\dot{\mathbf{r}} \times \ddot{\mathbf{r}}|}{|\dot{\mathbf{r}}|^3} = \frac{a^2 \operatorname{cosec} \alpha}{a^3 \operatorname{cosec}^3 \alpha} = \frac{\sin^2 \alpha}{a}$$

$\therefore$ κ is a constant quantity.

$$\therefore \quad \rho' = \left(\frac{1}{\kappa}\right)' = 0$$

Let R be the radius of osculating sphere at the point P on the curve.

$$\therefore \quad R = \sqrt{\rho^2 + \rho'^2\sigma^2} = \sqrt{\rho^2 + (0)^2\sigma^2} = \rho$$

Also the radius of curvature at P is ρ.

$\therefore$ The result holds.

Example 5. *Find the equation of the osculating sphere to the curve $x = 2t + 1$, $y = 3t^2 + 2$, $z = 4t^3 + 3$ at the point (1, 2, 3).*

Sol. The given curve is

$$x = 2t + 1, y = 3t^2 + 2, z = 4t^3 + 3.$$

The point P(1, 2, 3) corresponds to the value 0 of t.

Let $\mathbf{r}$ be the position vector of the point (x, y, z) on the given curve.

$$\therefore \quad \mathbf{r} = x\mathbf{i} + y\mathbf{j} + z\mathbf{k} = (2t + 1)\mathbf{i} + (3t^2 + 2)\mathbf{j} + (4t^3 + 3)\mathbf{k}$$

Let (α, β, γ) and R be the centre and the radius of the osculating sphere at P(1, 2, 3) respectively.

$\therefore$ The equation of the osculating sphere is $|\mathbf{r} - \mathbf{c}|^2 = R^2$, where $\mathbf{c} = \alpha\mathbf{i} + \beta\mathbf{j} + \gamma\mathbf{k}$.

This sphere passes through (1, 2, 3) and has at least 4-point contact with the given curve at P.

$$\therefore \quad |(\mathbf{i} + 2\mathbf{j} + 3\mathbf{k}) - (\alpha\mathbf{i} + \beta\mathbf{j} + \gamma\mathbf{k})|^2 = R^2$$

$$\Rightarrow \quad (1 - \alpha)^2 + (2 - \beta)^2 + (3 - \gamma)^2 = R^2 \qquad ...(1)$$

Let
$$f(t) = |(2t + 1)\mathbf{i} + (3t^2 + 2)\mathbf{j} + (4t^3 + 3)\mathbf{k} - (\alpha i + \beta j + \gamma k)|^2 - R^2$$

$$= (2t + 1 - \alpha)^2 + (3t^2 + 2 - \beta)^2 + (4t^3 + 3 - \gamma)^2 - R^2$$

$$\therefore \quad f'(t) = 2(2t + 1 - \alpha)2 + 2(3t^2 + 2 - \beta)6t + 2(4t^3 + 3 - \gamma)12t^2$$

$$f''(t) = 8 + 108t^2 + 12(2 - \beta) + 480t^4 + 48(3 - \gamma)t$$

$$f'''(t) = 216t + 1920t^3 + 48(3 - \gamma)$$

$$f^{iv}(t) = 216 + 5760t^2$$

Since the sphere has at least 4-point contact at $t = 0$, we have

$$f(0) = f'(0) = f''(0) = f'''(0) = 0 \text{ and } f^{iv}(0) \neq 0$$

$$f(0) = 0 \quad \Rightarrow \quad (1 - \alpha)^2 + (2 - \beta)^2 + (3 - \gamma)^2 - R^2 = 0, \text{ which is true.}$$

$$f'(0) = 0 \quad \Rightarrow \quad 4(1 - \alpha) = 0 \quad \Rightarrow \quad \alpha = 1 \qquad \text{(By using (1))}$$

$$f''(0) = 0 \quad \Rightarrow \quad 8 + 12(2 - \beta) = 0 \quad \Rightarrow \quad \beta = 8/3$$

$$f'''(0) = 0 \quad \Rightarrow \quad 48(3 - \gamma) = 0 \quad \Rightarrow \quad \gamma = 3$$

$$\therefore \quad \mathbf{c} = \alpha\mathbf{i} + \beta\mathbf{j} + \gamma\mathbf{k} = \mathbf{i} + \frac{8}{3}\mathbf{j} + 3\mathbf{k}$$

$$(1) \Rightarrow \quad R^2 = (1-1)^2 + \left(2 - \frac{8}{3}\right)^2 + (3-3)^2 = \frac{4}{9} \quad \Rightarrow \quad R = \frac{2}{3}$$

$\therefore$ Centre and radius of the osculating sphere at the point P are (1, 8/3, 3) and 2/3 respectively.

$\therefore$ The equation of the osculating sphere is

$$\left| (x\mathbf{i} + y\mathbf{j} + z\mathbf{k}) - \left(\mathbf{i} + \frac{8}{3}\mathbf{j} + 3\mathbf{k}\right) \right| = \left(\frac{2}{3}\right)^2.$$

$$\Rightarrow \quad (x-1)^2 + \left(y - \frac{8}{3}\right)^2 + (z-3)^2 = \frac{4}{9}$$

$$\Rightarrow \quad 3x^2 + 3y^2 + 3z^2 - 6x - 16y - 18z + 50 = 0.$$

Example 6. *If the radius of the osculating sphere of a curve is constant, prove that the curve lies on a sphere or has constant curvature.*

Sol. Let the given curve be $\mathbf{r} = \mathbf{r}(s)$.

Let R be the radius of the osculating sphere at each point on the curve $\mathbf{r} = \mathbf{r}(s)$. Let P($\mathbf{r}$) be any point on this curve.

$$\therefore \quad R = \sqrt{\rho^2 + \rho'^2 \sigma^2}$$

$$\Rightarrow \quad R^2 = \rho^2 + \rho'^2\sigma^2 \qquad ...(1)$$

Differentiating (1) w.r.t. s, we get

$$0 = 2\rho\rho' + \rho'^2(2\sigma\sigma') + (2\rho'\rho'')\sigma^2$$

$$\Rightarrow \quad 0 = 2\rho'(\rho + \rho'\sigma\sigma' + \rho''\sigma^2)$$

$\Rightarrow$ Either $\quad \rho' = 0 \quad ...(2) \quad$ or $\quad \rho + \rho'\sigma\sigma' + \rho''\sigma^2 = 0 \quad ...(3)$

(2) $\Rightarrow$ ρ is constant $\Rightarrow$ κ is constant.

Let (3) hold. Let $\mathbf{r}_1$ be the position vector of the centre of spherical curvature at the point P($\mathbf{r}$).

$$\therefore \quad \mathbf{r}_1 = \mathbf{r} + \rho\mathbf{n} + \rho'\sigma\mathbf{b}$$

Differentiating w.r.t. s, we get

$$\frac{d\mathbf{r}_1}{ds} = \mathbf{r}' + (\rho\mathbf{n}' + \rho'\mathbf{n}) + (\rho''\sigma\mathbf{b} + \rho'\sigma'\mathbf{b} + \rho'\sigma\mathbf{b}')$$

$$= \mathbf{t} + \rho(-\kappa\mathbf{t} + \tau\mathbf{b}) + \rho'\mathbf{n} + \rho''\sigma\mathbf{b} + \rho'\sigma'\mathbf{b} + \rho'\sigma(-\tau\mathbf{n})$$

$$= (1 - \rho\kappa)\mathbf{t} + (\rho' - \rho'\sigma\tau)\mathbf{n} + (\rho\tau + \rho''\sigma + \rho'\sigma')\mathbf{b}$$

$$= 0\mathbf{t} + 0\mathbf{n} + (\rho\tau + \rho''\sigma + \rho'\sigma')\mathbf{b} \qquad (\because \rho\kappa = 1, \sigma\tau = 1)$$

$$= \left(\frac{\rho}{\sigma} + \rho''\sigma + \rho'\sigma'\right)\mathbf{b}$$

$$= \frac{1}{\sigma}(\rho + \rho''\sigma^2 + \rho'\sigma'\sigma)\mathbf{b}$$

$$= \frac{1}{\sigma}(0)\mathbf{b} = \mathbf{0} \qquad \text{(Using (3))}$$

$$\Rightarrow \quad \frac{d\mathbf{r}_1}{ds} = \mathbf{0} \quad \Rightarrow \quad \mathbf{r}_1 \text{ is constant.}$$

∴ The centre of the osculating sphere is independent of the point P(**r**) on the curve **r** = **r**(s).

∴ The centre of osculating sphere is same at every point on the curve **r** = **r**(s).

Also, the radius of the osculating sphere is same for each point on the curve **r** = **r**(s).

∴ The osculating sphere is a fixed sphere for each point on the curve **r** = **r**(s).

∴ The curve **r** = **r**(s) lies itself on this sphere.

Example 7. *Show that the necessary and sufficient condition for a curve* **r** = **r**(s) *to lie on a sphere is* $s\frac{\rho}{\sigma} + \frac{d}{ds}(\rho'\sigma) = 0$ *at every point on the curve.*

Sol. Necessity. Let the curve **r** = **r**(s) lie on a sphere.

∴ This sphere is the osculating sphere to the curve **r** = **r**(s) at every point of the curve.

∴ The radius R of the osculating sphere at each point is constant.

We have $R^2 = \rho^2 + \rho'^2\sigma^2$

Differentiating w.r.t. s, we get

$$0 = 2\rho\rho' + \rho'^2(2\sigma\sigma') + (2\rho'\rho'')\sigma^2$$

$$\Rightarrow \quad 0 = 2\rho'\sigma\left(\frac{\rho}{\sigma} + \rho'\sigma' + \rho''\sigma\right)$$

$$\Rightarrow \quad \frac{\rho}{\sigma} + \rho'\sigma' + \rho''\sigma = 0 \qquad (\text{Assuming } \rho' \neq 0, \sigma \neq 0)$$

$$\Rightarrow \quad \frac{\rho}{\sigma} + \frac{d}{ds}(\rho'\sigma) = 0.$$

Sufficiency. Let $\frac{\rho}{\sigma} + \frac{d}{ds}(\rho'\sigma) = 0.$

$$\Rightarrow \quad \frac{\rho}{\sigma} + \rho'\sigma' + \rho''\sigma = 0$$

$$\Rightarrow \quad 2\rho'\sigma\left(\frac{\rho}{\sigma} + \rho'\sigma' + \rho''\sigma\right) = 0$$

$$\Rightarrow \quad 2\rho\rho' + \rho'^2(2\sigma\sigma') + (2\rho'\rho'')\sigma^2 = 0$$

$$\Rightarrow \quad \frac{d}{ds}(\rho^2 + \rho'^2\sigma^2) = 0$$

$$\Rightarrow \quad \rho^2 + \rho'^2\sigma^2 = \lambda, \text{ a constant.}$$

$$\Rightarrow \quad R^2 = \lambda \quad \textit{i.e.}, \text{ R is constant.}$$

∴ Radius of osculating sphere is independent of the point on the curve **r** = **r**(s).

Let $\mathbf{r}_1$ be the position vector of the centre of osculating sphere at the point s.

$\therefore \qquad \mathbf{r}_1 = \mathbf{r} + \rho\mathbf{n} + \rho'\sigma\mathbf{b}$

Differentiating w.r.t. s, we get

$$\frac{d\mathbf{r}_1}{ds} = \mathbf{r}' + (\rho\mathbf{n}' + \rho'\mathbf{n}) + (\rho''\sigma\mathbf{b} + \rho'\sigma'\mathbf{b} + \rho'\sigma\mathbf{b}')$$
$$= \mathbf{t} + \rho(-\kappa\mathbf{t} + \tau\mathbf{b}) + \rho'\mathbf{n} + \rho''\sigma\mathbf{b} + \rho'\sigma'\mathbf{b} + \rho'\sigma(-\tau\mathbf{n})$$
$$= (1 - \rho\kappa)\,\mathbf{t} + (\rho' - \rho'\sigma\tau)\,\mathbf{n} + (\rho\tau + \rho''\sigma + \rho'\sigma')\,\mathbf{b}$$
$$= 0\mathbf{t} + 0\mathbf{n} + \left(\frac{\rho}{\sigma} + \frac{d}{ds}(\rho'\sigma)\right)\mathbf{b} = 0\,\mathbf{b} = \mathbf{0}$$

$$\Rightarrow \qquad \frac{d\mathbf{r}_1}{ds} = 0 \;\Rightarrow\; \mathbf{r}_1 \text{ is constant.}$$

The centre of the osculating sphere is independent of the point on the curve $\mathbf{r} = \mathbf{r}(s)$.

$\therefore$ The centre of the osculating sphere is same at every point on the curve $\mathbf{r} = \mathbf{r}(s)$.

Also, the radius of the osculating sphere is same for each point on the curve $\mathbf{r} = \mathbf{r}(s)$.

$\therefore$ The osculating sphere is a fixed sphere for each point on the curve $\mathbf{r} = \mathbf{r}(s)$.

$\therefore$ The curve $\mathbf{r} = \mathbf{r}(s)$ lies itself on this sphere.

WORKING RULES FOR SOLVING PROBLEMS

Rule I. *A curve $\mathbf{r} = x(t)\mathbf{i} + y(t)\mathbf{j} + z(t)\mathbf{k}$ of sufficiently high class is said to have n-***point contact** *with a surface $F(x, y, z) = 0$ at the point t_0 if the function $f(t) = F(x(t), y(t), z(t))$ satisfies:*

$$f(t_0) = \dot{f}(t_0) = \ddot{f}(t_0) = \ldots = f^{(n-1)}(t_0) = 0 \text{ and } f^{(n)}(t_0) \neq 0.$$

Rule II. *A curve $\mathbf{r} = x(t)\mathbf{i} + y(t)\mathbf{j} + z(t)\mathbf{k}$ of sufficiently high class is said to have n-***point contact** *with a curve $F(x, y, z) = 0$, $G(x, y, z) = 0$ at the point t_0 if the functions $f(t) = F(x(t), y(t), z(t))$ and $g(t) = G(x(t), y(t), z(t))$ satisfy:*

$$f(t_0) = \dot{f}(t_0) = \ddot{f}(t_0) = \ldots = f^{(n-1)}(t_0) = 0$$
$$g(t_0) = \dot{g}(t_0) = \ddot{g}(t_0) = \ldots = g^{(n-1)}(t_0) = 0$$

and either $f^{(n)}(t_0) \neq 0$ or $g^{(n)}(t_0) \neq 0$.

Rule III. *A circle having at least 3-point contact with a given curve C at a point P on the curve is called the* **osculating circle** *to the curve C at the point P. The centre of the osculating circle at the point P is called the* **centre of curvature** *of the curve C at the point P.*

If $\mathbf{r}_1$ be the position vector of the centre of curvature at point $\mathbf{r}$ then,

$\mathbf{r}_1 = \mathbf{r} + \rho\mathbf{n}$ and radius of osculating circle $= \rho$.

Rule IV. *A sphere having at least 4-point contact with a given curve C at a point P on the curve is called the* **osculating sphere** *to the curve C at the point P. The centre of the osculating sphere at the point P is called the* **centre of spherical curvature** *of the curve C at the point P.*

If $\mathbf{r}_1$ be the position vector of the centre of spherical curvature at point $\mathbf{r}$, then

$\mathbf{r}_1 = \mathbf{r} + \rho\mathbf{n} + \rho'\sigma\mathbf{b}$ *and radius of osculating sphere* $= \sqrt{\rho^2 + \rho'^2\sigma^2}$.

EXERCISE 2.4

1. Show that the tangent to the locus of the centre of curvature lies in the normal plane of the original curve and makes an angle $\tan^{-1} \dfrac{\rho}{\sigma\rho'}$ with the principal normal of the original curve.
2. If C is a curve of constant curvature κ, show that the locus C_1 of its centre of curvature is also a curve of constant curvature κ_1 such that $\kappa_1 = \kappa$ and its torsion τ_1 is given by the relation $\tau_1 = \dfrac{\kappa^2}{\tau}$.
3. For a curve of constant curvature, show that the centre of spherical curvature coincides with the centre of circular curvature.
4. If R is the radius of the osculating sphere to a curve $\mathbf{r} = \mathbf{r}(s)$ at point s, then show that:
$$R^2 = \rho^4\sigma^2 \,|\mathbf{r}'''|^3 - \sigma^2.$$
5. For the curve $\mathbf{r} = x(s)\mathbf{i} + y(s)\mathbf{j} + z(s)\mathbf{k}$, show that:
$$x'''^2 + y'''^2 + z'''^2 = \frac{1}{\rho^2\sigma^2} + \frac{1+\rho'^2}{\rho^4}.$$
6. For the curve $\mathbf{r} = x(s)\mathbf{i} + y(s)\mathbf{j} + z(s)\mathbf{k}$, show that:
$$\rho^4(x'''^2 + y'''^2 + z'''^2) = 1 + \frac{R^2}{\sigma^2},$$
where R is the radius of spherical curvature at the point (x, y, z).
7. Show that the radius of spherical curvature of a circular helix is equal to the radius of its circular curvature.

Hint

4. $\mathbf{r}''' = (\mathbf{r}'')' = (\mathbf{t}')' = (\kappa\mathbf{n})' = \kappa\mathbf{n}' + \kappa'\mathbf{n} = \kappa(-\kappa\mathbf{t} + \tau\mathbf{b}) + \kappa'\mathbf{n} = -\kappa^2\mathbf{t} + \kappa'\mathbf{n} + \kappa\tau\,\mathbf{b}$

$\therefore \quad |\mathbf{r}'''|^2 = (-\kappa^2)^2 + (\kappa')^2 + (\kappa\tau)^2$

$$= \frac{1}{\rho^4} + \left(-\frac{\rho'}{\rho^2}\right)^2 + \frac{1}{\rho^2\sigma^2} = \frac{\sigma^2 + \rho'^2\sigma^2 + \rho^2}{\rho^4\sigma^2} = \frac{\sigma^2 + R^2}{\rho^4\sigma^2}.$$

3

Surfaces in Space

1. INTRODUCTION

A sphere, a portion of a cylinder are examples of surfaces in space. Till now we have been discussing the differential geometry of curves in space. In the present chapter , we shall learn few quantities regarding surfaces in space.

2. SURFACE IN SPACE

We know that a curve in space is the locus of a point whose coordinates x, y, z are functions of a single parameter. On the same lines, we shall define a surface in space as the locus of a point whose coordinates are functions of two independent parameters.

A **surface in space** is defined as the locus of a point whose position vector relative to a fixed origin may be expressed as a function of two independent parameters.

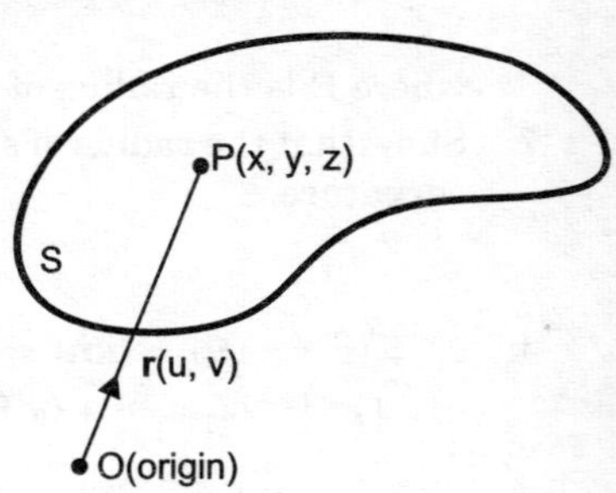

Thus, a surface S in space may be represented by a vector function

$$\mathbf{r}(u, v) = x(u, v)\mathbf{i} + y(u, v)\mathbf{j} + z(u, v)\mathbf{k},$$

where u and v are independent parameters such that $(u, v) \in$ R, where R is some region in the uv-plane.

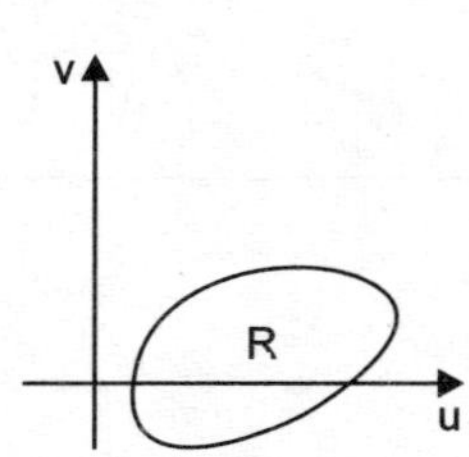

Here $\mathbf{r}(u, v)$ is the position vector of the point P on the surface S and $(x(u, v), y(u, v), z(u, v))$ are the cartesian coordinates of the point P. The above representation of a surface is called a **parametric representation of a surface** and is due to *Gauss*. Also, (u, v) are called the **curvilinear coordinates** of the point P. In particular cases, one or even two of the functions $x(u, v)$, $y(u, v)$, $z(u, v)$ may involve only one parameter.

Let $\mathbf{r}(u, v) = x(u, v)\mathbf{i} + y(u, v)\mathbf{j} + z(u, v)\mathbf{k}$ be the parametric representation of a surface S. This surface is also represented by the equations:

$$x = x(u, v), y = y(u, v), z = z(u, v). \qquad ...(1)$$

If u and v are eliminated from the equation (1), we obtain a relation between the coordinates x, y and z which may be written as

$$F(x, y, z) = 0.$$

This form of equation of a surface is considered to be the oldest form of the equation of a surface. The equation $F(x, y, z) = 0$ is also called the **implicit equation of a surface**. In particular, if the equation $F(x, y, z) = 0$ of a surface is of the form $z = f(x, y)$, then it is called the **Monge's form of representation of the surface**. In this case, the position vector **r** of the point (x, y, z) on the surface is given by $\mathbf{r} = x\mathbf{i} + y\mathbf{j} + f(x, y)\mathbf{k}$ and it is regarded as the parametric representation of the surface with parameters x and y.

The parametric representation of a surface may not be unique. For example, the surface $z = x^2 - y^2$ is represented by $x = u, y = v, z = u^2 - v^2$, where u and v are parameters.

This surface can also be represented by parametric equations $x = u + v$, $y = u - v$, $z = 4uv$, where u and v are parameters.

3. TANGENT LINE AND TANGENT PLANE

Any tangent line to any curve drawn on a surface is called a **tangent line** to the surface under consideration.

Let P be any point on a surface S. The plane containing all tangent lines of curves on the surface S and through P is called the **tangent plane** of surface S at the point P.

Equation of tangent plane.

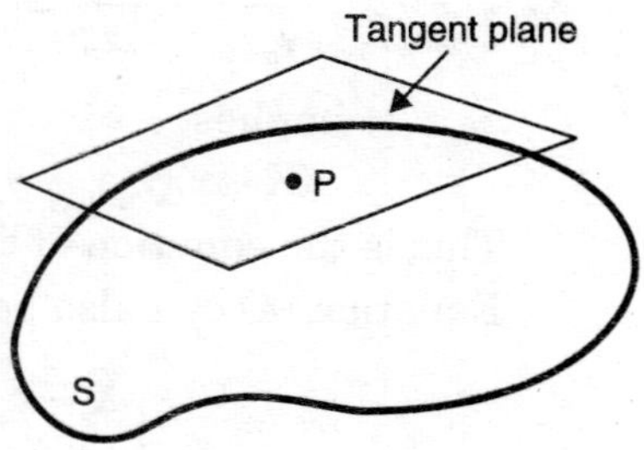

Let $P(x, y, z)$ be any point on a surface S given by $x = x(u, v)$, $y = y(u, v)$, $z = z(u, v)$, where u and v are independent parameters. Let the equation of the surface S after eliminating u and v be $F(x, y, z) = 0$. For any arbitrary but fixed curve drawn on the surface S and passing through P, let s be the arc length measured from a fixed point to the current point (x, y, z). Since the function $F(x, y, z)$ has the same value at all points on the surface, it would also have the same constant value along the curve as s varies.

$$\therefore \quad \frac{d}{ds}(F(x, y, z)) = 0$$

$$\Rightarrow \quad \frac{\partial F}{\partial x}\frac{dx}{ds} + \frac{\partial F}{\partial y}\frac{dy}{ds} + \frac{\partial F}{\partial z}\frac{dz}{ds} = 0$$

$$\Rightarrow \quad F_x x' + F_y y' + F_z z' = 0$$

$$\Rightarrow \quad (F_x\,\mathbf{i} + F_y\,\mathbf{j} + F_z\,\mathbf{k}) \,.\, (x'\mathbf{i} + y'\mathbf{j} + z'\mathbf{k}) = 0 \qquad ...(1)$$

$x'\mathbf{i} + y'\mathbf{j} + z'\mathbf{k}$ is the unit tangent vector to the curve at the point $P(x, y, z)$.

$\therefore$ (1) implies that the unit tangent vector at $P(x, y, z)$ is perpendicular to the vector $F_x\mathbf{i} + F_y\mathbf{j} + F_z\mathbf{k}$.

$\therefore$ By definition, all tangent lines to the surface S at the point $P(x, y, z)$ are perpendicular to the vector $F_x\mathbf{i} + F_y\mathbf{j} + F_z\mathbf{k}$.

$\therefore$ All tangent lines to the surface S at the point $P(x, y, z)$ lie in the plane through $P(x, y, z)$ and perpendicular to the vector $F_x\mathbf{i} + F_y\mathbf{j} + F_z\mathbf{k}$.

$\therefore$ By definition, the tangent plane to the surface $F(x, y, z) = 0$ at the point $P(x, y, z)$ passes through the point $P(x, y, z)$ and is perpendicular to the vector $F_x\mathbf{i} + F_y\mathbf{j} + F_z\mathbf{k}$. Let (X, Y, Z) be a current point on the tangent plane at $P(x, y, z)$.

$\therefore$ The equation of the tangent plane is

$$[(X\mathbf{i} + Y\mathbf{j} + Z\mathbf{k}) - (x\mathbf{i} + y\mathbf{j} + z\mathbf{k})] \cdot (F_x\,\mathbf{i} + F_y\,\mathbf{j} + F_z\,\mathbf{k}) = 0$$

$$\Rightarrow \quad (X - x)F_x + (Y - y)F_y + (Z - z)F_z = 0. \qquad ...(1)$$

This is the cartesian equation of the tangent plane to the surface S given by $F(x, y, z) = 0$ at the point (x, y, z).

We have $F(x, y, z) = 0$ and x, y, z are functions of u and v.

$$\therefore \quad \frac{\partial F}{\partial u} = 0 \quad \text{and} \quad \frac{\partial F}{\partial v} = 0$$

$$\Rightarrow \quad \frac{\partial F}{\partial x}\frac{\partial x}{\partial u} + \frac{\partial F}{\partial y}\frac{\partial y}{\partial u} + \frac{\partial F}{\partial z}\frac{\partial z}{\partial u} = 0$$

and $$\frac{\partial F}{\partial x}\frac{\partial x}{\partial v} + \frac{\partial F}{\partial y}\frac{\partial y}{\partial v} + \frac{\partial F}{\partial z}\frac{\partial z}{\partial v} = 0$$

$$\Rightarrow \quad F_x x_u + F_y y_u + F_z z_u = 0 \qquad ...(2)$$

and $$F_x x_v + F_y y_v + F_z z_v = 0 \qquad ...(3)$$

Solving (2) and (3), we get

$$\frac{F_x}{y_u z_v - y_v z_u} = \frac{F_y}{z_u x_v - z_v x_u} = \frac{F_z}{x_u y_v - x_v y_u}.$$

$\therefore$ (1) implies

$$(X - x)(y_u z_v - y_v z_u) + (Y - y)(z_u x_v - z_v x_u) + (Z - z)(x_u y_v - x_v y_u) = 0 \qquad ...(4)$$

This is the equation of the tangent plane at (x, y, z) in terms of the parameters u and v. Equation (4) can also be written as

$$\begin{vmatrix} X - x & Y - y & Z - z \\ x_u & y_u & z_u \\ x_v & y_v & z_v \end{vmatrix} = 0 \qquad ...(5)$$

If $\mathbf{r} = x(u, v)\mathbf{i} + y(u, v)\mathbf{j} + z(u, v)\mathbf{k}$, then we denote $\frac{\partial \mathbf{r}}{\partial u}$ by $\mathbf{r}_1$ and $\frac{\partial \mathbf{r}}{\partial v}$ by $\mathbf{r}_2$.

$$\therefore \quad \mathbf{r}_1 = \frac{\partial \mathbf{r}}{\partial u} = x_u\mathbf{i} + y_u\mathbf{j} + z_u\mathbf{k}$$

and $$\mathbf{r}_2 = \frac{\partial \mathbf{r}}{\partial v} = x_v\mathbf{i} + y_v\mathbf{j} + z_v\mathbf{k}$$

$\therefore$ Equation (5) can also be written as

$[\mathbf{R} - \mathbf{r} \;\; \mathbf{r}_1 \;\; \mathbf{r}_2] = 0$, where $\mathbf{R} = X\mathbf{i} + Y\mathbf{j} + Z\mathbf{k}$ and $\mathbf{r} = x\mathbf{i} + y\mathbf{j} + z\mathbf{k}$.

4. NORMAL LINE

The line perpendicular to the tangent plane to a surface at the point of contact is called the **normal line** to the surface at that point.

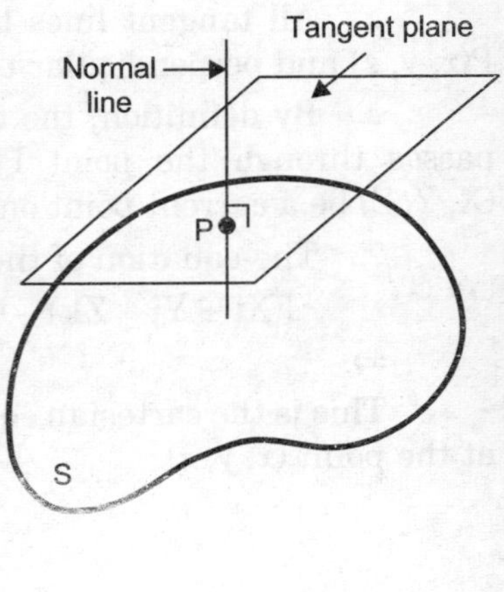

Equation of normal line

Let $P(x, y, z)$ be any point on a surface S given by $x = x(u, v)$, $y = y(u, v)$, $z = z(u, v)$, where u and v are independent parameters. Let the equation of the surface S after eliminating u and v be $F(x, y, z) = 0$.

The equation of the tangent plane to the surface S at the point P is $(X - x)F_x + (Y - y)F_y + (Z - z)F_z = 0$.

∴ The tangent plane at P is perpendicular to the line with d.r.'s F_x, F_y, F_z.

∴ F_x, F_y, F_z are d.r.'s of the normal line to the surface S at the point P.

∴ The equations of the normal line to the surface $F(x, y, z) = 0$ at the point $P(x, y, z)$ are

$$\frac{X - x}{F_x} = \frac{Y - y}{F_y} = \frac{Z - z}{F_z},$$

where (X, Y, Z) are the coordinates of a current point on the normal line.

Also,
$$\frac{F_x}{y_u z_v - y_v z_u} = \frac{F_y}{z_u x_v - z_v x_u} = \frac{F_z}{x_u y_v - x_v y_u}$$

∴ In terms of parameters u and v, the equations of the normal line at $P(x, y, z)$ are

$$\frac{X - x}{y_u z_v - y_v z_u} = \frac{Y - y}{z_u x_v - z_v x_u} = \frac{Z - z}{x_u y_v - x_v y_u}.$$

These equations implies

$$X - x = \lambda(y_u z_v - y_v z_u),$$
$$Y - y = \lambda(z_u x_v - z_v x_u)$$

and
$$Z - z = \lambda(x_u y_v - x_v y_u), \text{ where } \lambda \text{ is a parameter.}$$

∴ $\mathbf{r} = x(u, v)\mathbf{i} + y(u, v)\mathbf{j} + z(u, v)\mathbf{k}$, then the equation of the normal line at the point $P(x, y, z)$ is

$$\mathbf{R} - \mathbf{r} = \lambda(\mathbf{r}_1 \times \mathbf{r}_2) \text{ } i.e., \text{ } \mathbf{R} = \mathbf{r} + \lambda(\mathbf{r}_1 \times \mathbf{r}_2),$$

where **R** is the position vector of a current point (X, Y, Z) on the normal line to the given surface at the point $P(x, y, z)$.

Example 1. *Find a unit normal vector to the surface $2xz^2 - 3xy - 4x = 7$ at the point (1, –1, 2) on the surface.*

Sol. Given surface is $2xz^2 - 3xy - 4x - 7 = 0$.

Let
$$F(x, y, z) = 2xz^2 - 3xy - 4x - 7$$

∴
$$F_x = 2z^2 - 3y - 4, \; F_y = -3x, \; F_z = 4xz$$

∴ At the point (1, – 1, 2),

$$F_x = 2(2)^2 - 3(-1) - 4 = 7, \; F_y = -3(1) = -3, \; F_z = 4(1)(2) = 8$$

∴ D.r.'s of normal line at (1, – 1, 2) are 7, – 3, 8.

Now
$$\sqrt{(7)^2 + (-3)^2 + (8)^2} = \sqrt{122}$$

∴ D.c.'s of normal line at (1, – 1, 2) are

$$\frac{7}{\sqrt{122}}, -\frac{3}{\sqrt{122}}, \frac{8}{\sqrt{122}}.$$

∴ Unit normal vector to the given surface at (1, – 1, 2)

$$= \frac{7}{\sqrt{122}}\mathbf{i} - \frac{3}{\sqrt{122}}\mathbf{j} + \frac{8}{\sqrt{122}}\mathbf{k}.$$

Example 2. *Find the equation of the tangent plane and the normal line to the surface $xyz = 4$ at the point (1, 2, 2).*

Sol. Given surface is $xyz - 4 = 0$.

Let
$$F(x, y, z) = xyz - 4$$

$\therefore \quad F_x = yz, F_y = xz, F_z = xy$

$\therefore$ At the point (1, 2, 2),

$$F_x = 2(2) = 4, F_y = 1(2) = 2, F_z = 1(2) = 2$$

$\therefore$ The equation of the tangent plane at (1, 2, 2) is

$$(x-1)F_x + (y-2)F_y + (z-2)F_z = 0$$

or
$$(x-1)4 + (y-2)2 + (z-2)2 = 0$$

or
$$2x + y + z = 6.$$

The equations of the normal line at (1, 2, 2) are

$$\frac{x-1}{F_x} = \frac{y-2}{F_y} = \frac{z-2}{F_z}$$

or
$$\frac{x-1}{4} = \frac{y-2}{2} = \frac{z-2}{2} \quad \text{or} \quad \frac{x-1}{2} = \frac{y-2}{1} = \frac{z-2}{1}.$$

Example 3. *Find the equation of the tangent plane and the normal line to the surface given by* $\mathbf{r} = u\mathbf{i} + v\mathbf{j} + (u^2 - v^2)\mathbf{k}$ *at the point corresponding to* $u = 1, v = 1$.

Sol. The given surface is

$$\mathbf{r} = u\mathbf{i} + v\mathbf{j} + (u^2 - v^2)\mathbf{k}.$$

Let (x, y, z) be the point on the surface with position vector $\mathbf{r}$.

$$\therefore \quad x = u, \; y = v, \; z = u^2 - v^2$$

$$u = 1, v = 1 \;\Rightarrow\; x = 1, y = 1, z = (1)^2 - (1)^2 = 0$$

$\therefore$ The point corresponding to $u = 1, v = 1$ is (1, 1, 0).

We have $\quad x_u = 1, x_v = 0, y_u = 0, y_v = 1, z_u = 2u, z_v = -2v$

$\therefore$ At (1, 1, 0), $\quad x_u = 1, x_v = 0, y_u = 0, y_v = 1, z_u = 2(1) = 2, z_v = -2(1) = -2.$

The equation of the tangent plane at (1, 1, 0) is

$$\begin{vmatrix} x-1 & y-1 & z-0 \\ x_u & y_u & z_u \\ x_v & y_v & z_v \end{vmatrix} = 0$$

or
$$\begin{vmatrix} x-1 & y-1 & z \\ 1 & 0 & 2 \\ 0 & 1 & -2 \end{vmatrix} = 0$$

or
$$-2(x-1) + 2(y-1) + z = 0 \quad \text{or} \quad 2x - 2y - z = 0. \qquad ...(1)$$

(1) $\Rightarrow$ D.r.'s of normal to the tangent plane at (1, 1, 0) are 2, – 2, – 1.

$\therefore$ The equations of the normal line at (1, 1, 0) are

$$\frac{x-1}{2} = \frac{y-1}{-2} = \frac{z-0}{-1} \quad \text{or} \quad \frac{x-1}{-2} = \frac{y-1}{2} = \frac{z}{1}.$$

Example 4. *Show that the tangent plane to the surface* $xyz = a^3$ *at any point and the coordinate planes bound a tetrahedron of constant volume.*

Sol. Given surface is $\quad xyz = a^3$.

Let $\quad F(x, y, z) = xyz - a^3$

$$\therefore \quad F_x = yz, F_y = xz, F_z = xy$$

Let P(α, β, γ) be any point on the given surface.

$$\therefore \quad \alpha\beta\gamma = a^3$$

At the point P, $F_x = \beta\gamma,\ F_y = \alpha\gamma,\ F_z = \alpha\beta$

∴ The equations of the tangent plane at P(α, β, γ) is

$$(x-\alpha)F_x + (y-\beta)F_y + (z-\gamma)F_z = 0$$

or

$$(x-\alpha)\beta\gamma + (y-\beta)\alpha\gamma + (z-\gamma)\alpha\beta = 0$$

$$\Rightarrow \quad \beta\gamma\, x + \alpha\gamma\, y + \alpha\beta z = 3\alpha\beta\gamma$$

$$\Rightarrow \quad \frac{x}{3\alpha} + \frac{y}{3\beta} + \frac{z}{3\gamma} = 1.$$

This plane makes intercepts 3α, 3β, 3γ on the coordinate axes.

∴ Volume of the tetrahedron formed by the tangent plane and the coordinate axes

$$= \frac{1}{6}(3\alpha)(3\beta)(3\gamma) = \frac{9}{2}\alpha\beta\gamma = \frac{9}{2}a^3, \text{ which is constant.}$$

Example 5. *The normal line at any point P of the ellipsoid*

$$\frac{x^2}{a^2} + \frac{y^2}{b^2} + \frac{z^2}{c^2} = 1$$

meets the coordinate planes in G_1, G_2, G_3. Show that the ratio $PG_1 : PG_2 : PG_3$ is constant.

Sol. Given surface is

$$\frac{x^2}{a^2} + \frac{y^2}{b^2} + \frac{z^2}{c^2} - 1 = 0.$$

Let

$$F(x, y, z) = \frac{x^2}{a^2} + \frac{y^2}{b^2} + \frac{z^2}{c^2} - 1$$

$$\therefore \quad F_x = \frac{2x}{a^2},\ F_y = \frac{2y}{b^2},\ F_z = \frac{2z}{c^2}$$

Let P(α, β, γ) be any point on the given surface.

∴ At the point P,

$$F_x = \frac{2\alpha}{a^2},\ F_y = \frac{2\beta}{b^2},\ F_z = \frac{2\gamma}{c^2}$$

∴ The equations of the normal line at P(α, β, γ) are

$$\frac{x-\alpha}{F_x} = \frac{y-\beta}{F_y} = \frac{z-\gamma}{F_z}$$

or

$$\frac{x-\alpha}{\frac{2\alpha}{a^2}} = \frac{y-\beta}{\frac{2\beta}{b^2}} = \frac{z-\gamma}{\frac{2\gamma}{c^2}} \quad \text{or} \quad \frac{x-\alpha}{\frac{\alpha}{a^2}} = \frac{y-\beta}{\frac{\beta}{b^2}} = \frac{z-\gamma}{\frac{\gamma}{c^2}} \qquad ...(1)$$

(1) meets the yz-plane, where $x = 0$. Putting $x = 0$ in (1), we get

$$\frac{-\alpha}{\frac{\alpha}{a^2}} = \frac{y-\beta}{\frac{\beta}{b^2}} = \frac{z-\gamma}{\frac{\gamma}{c^2}}$$

$$\Rightarrow \qquad y = \beta + \frac{a^2}{b^2}\beta,\ z = \gamma + \frac{a^2}{c^2}\gamma$$

$\therefore$ Coordinates of G_1 are $\left(0, \beta + \frac{a^2}{b^2}\beta, \gamma + \frac{a^2}{c^2}\gamma\right)$

$$\therefore \qquad PG_1 = \left[(\alpha - 0)^2 + \left(\beta - \beta - \frac{a^2}{b^2}\beta\right)^2 + \left(\gamma - \gamma - \frac{a^2}{c^2}\gamma\right)^2\right]^{1/2}$$

$$= \left[\alpha^2 + \frac{a^4}{b^4}\beta^2 + \frac{a^4}{c^4}\gamma^2\right]^{1/2} = a^2\left[\frac{\alpha^2}{a^4} + \frac{\beta^2}{b^4} + \frac{\gamma^2}{c^4}\right]^{1/2}$$

Similarly, $PG_2 = b^2\left[\frac{\alpha^2}{a^4} + \frac{\beta^2}{b^4} + \frac{\gamma^2}{c^4}\right]^{1/2}$ and $PG_3 = c^2\left[\frac{\alpha^2}{a^4} + \frac{\beta^2}{b^4} + \frac{\gamma^2}{c^4}\right]^{1/2}$.

$\therefore$ $PG_1 : PG_2 : PG_3 = a^2 : b^2 : c^2$, which is a constant ratio.

Example 6. *Prove that at points common to the surface $a(yz + zx + xy) = xyz$ and a sphere whose centre is at origin, the tangent plane to the surface makes intercepts on the axes whose sum is constant.*

Sol. Given surface is

$$a(yz + zx + xy) = xyz.$$

Let $F(x, y, z) = a(yz + zx + xy) - xyz$

$\therefore$ $F_x = a(z + y) - yz,\ F_y = a(z + x) - xz,\ F_z = a(y + x) - xy$

Let $P(\alpha, \beta, \gamma)$ be a point on the given surface and also on the sphere $x^2 + y^2 + z^2 = r^2$.

$\therefore$ At $P(\alpha, \beta, \gamma)$,

$$F_x = a(\gamma + \beta) - \beta\gamma, \quad F_y = a(\gamma + \alpha) - \alpha\gamma, \quad F_z = a(\beta + \alpha) - \alpha\beta$$

$\therefore$ The equation of the tangent plane at (α, β, γ) is

$$(x - \alpha)F_x + (y - \beta)F_y + (z - \gamma)F_z = 0$$

or $(x - \alpha)[a(\beta + \gamma) - \beta\gamma] + (y - \beta)[a(\alpha + \gamma) - \alpha\gamma] + (z - \gamma)[a(\alpha + \beta) - \alpha\beta] = 0$

$$\Rightarrow \qquad [a(\beta + \gamma) - \beta\gamma]x + [a(\alpha + \gamma) - \alpha\gamma]y + [a(\alpha + \beta) - \alpha\beta]z = 2a(\alpha\beta + \beta\gamma + \gamma\alpha) - 3\alpha\beta\gamma$$

$$\Rightarrow \qquad [a(\alpha\beta + \alpha\gamma) - \alpha\beta\gamma]\frac{x}{\alpha} + [a(\alpha\beta + \gamma\beta) - \alpha\beta\gamma]\frac{y}{\beta} + [a(\alpha\gamma + \beta\gamma) - \alpha\beta\gamma]\frac{z}{\gamma}$$

$$= 2a(\alpha\beta + \beta\gamma + \gamma\alpha) - 3\alpha\beta\gamma \qquad \text{...(1)}$$

P(α, β, γ) is on the given surface.

$$\therefore \quad a(\beta\gamma + \gamma\alpha + \alpha\beta) = \alpha\beta\gamma$$

$$\Rightarrow \quad a(\alpha\beta + \alpha\gamma) - \alpha\beta\gamma = -a\beta\gamma,$$

$$a(\alpha\beta + \gamma\beta) - \alpha\beta\gamma = -a\alpha\gamma$$

and

$$a(\alpha\gamma + \beta\gamma) - \alpha\beta\gamma = -a\beta\alpha$$

$$\therefore \quad (1) \Rightarrow (-a\beta\gamma)\frac{x}{\alpha} + (-a\alpha\gamma)\frac{y}{\beta} + (-a\beta\alpha)\frac{z}{\gamma}$$

$$= 2\alpha\beta\gamma - 3\alpha\beta\gamma = -\alpha\beta\gamma$$

Dividing by $-\alpha\beta\gamma$, we get

$$\frac{ax}{\alpha^2} + \frac{ay}{\beta^2} + \frac{az}{\gamma^2} = 1$$

$$\Rightarrow \quad \frac{x}{\alpha^2/a} + \frac{y}{\beta^2/a} + \frac{z}{\gamma^2/a} = 1.$$

∴ The intercepts made by the tangent plane with the coordinate axes are

$$\frac{\alpha^2}{a}, \frac{\beta^2}{a}, \frac{\gamma^2}{a}.$$

∴ Sum of intercepts $= \dfrac{\alpha^2}{a} + \dfrac{\beta^2}{a} + \dfrac{\gamma^2}{a} = \dfrac{\alpha^2 + \beta^2 + \gamma^2}{a}$

$= \dfrac{r^2}{a}$, a constant. $\quad (\because \ (\alpha, \beta, \gamma)$ is on $x^2 + y^2 + z^2 = r^2)$

WORKING RULES FOR SOLVING PROBLEMS

Rule I. *The equation of the tangent plane to the surface F(x, y, z) = 0 at the point (x, y, z) is*

$$(X - x)F_x + (Y - y)F_y + (Z - z)F_z = 0,$$

where (X, Y, Z) are the coordinates of a current point on the tangent plane.

Rule II. *The equations of the normal line to the surface F(x, y, z) = 0 at the point (x, y, z) are*

$$\frac{X - x}{F_x} = \frac{Y - y}{F_y} = \frac{Z - z}{F_z},$$

where (X, Y, Z) are the coordinates of a current point on the normal line.

Rule III. *The equation of the tangent plane to the surface* **r** = **r**(*u*, *v*) *at the point with position vector* **r** *is*

$[\mathbf{R} - \mathbf{r}\ \mathbf{r}_1\ \mathbf{r}_2] = 0$, *where* **R** *is the position vector of a current point (X, Y, Z) on the tangent plane.*

Rule IV. *The equation of the normal line to the surface* **r** = **r**(*u*, *v*) *at the point with position vector* **r** *is*

$\mathbf{R} = \mathbf{r} + \lambda(\mathbf{r}_1 \times \mathbf{r}_2)$, *where* **R** *is the position vector of a current point (X, Y, Z) on the normal line.*

EXERCISE 3.1

1. Find a unit normal vector to the following surfaces at the point (x, y, z) of the surface:
   (*i*) $4x - 4y + 7z = -3$ (*ii*) $y^2 + z^2 = a^2$
   (*iii*) $x^2 + y^2 + z^2 = 49$ (*iv*) $z = \sqrt{x^2 + y^2}$.
2. Find the equations of the tangent plane and the normal line to the surface $z = xy$ at the point (2, 3, 6).
3. Find the equations of the tangent plane and the normal line to the surface $2xz^2 - 3x^2y - 4x - 7 = 0$ at the point (1, – 1, 2).
4. Find the equations of the tangent plane and the normal line to the surface $z = x^2 + y^2$ at the point (1, – 1, 2).
5. Show that the sum of the squares of the intercepts on the coordinate axes made by a tangent plane to the surface $x^{2/3} + y^{2/3} + z^{2/3} = a^{2/3}$ is always constant.

Answers

1. (*i*) $\frac{1}{9}(4\mathbf{i} - 4\mathbf{j} + 7\mathbf{k})$ (*ii*) $\frac{1}{a}(y\mathbf{j} + z\mathbf{k})$
   (*iii*) $\frac{1}{7}(x\mathbf{i} + y\mathbf{j} + z\mathbf{k})$ (*iv*) $\frac{1}{\sqrt{2}}\left(\frac{x}{\sqrt{x^2+y^2}}\mathbf{i} + \frac{y}{\sqrt{x^2+y^2}}\mathbf{j} - \mathbf{k}\right)$
2. $3x + 2y - z = 6$; $\frac{x-2}{3} = \frac{y-3}{2} = \frac{z-6}{-1}$
3. $10x - 3y + 8z = 29$; $\frac{x-1}{10} = \frac{y+1}{-3} = \frac{z-2}{8}$
4. $2x - 2y - z = 2$; $\frac{x-1}{-2} = \frac{y+1}{2} = \frac{z-2}{1}$.

---

5. ENVELOPE

If a is a constant, then an equation of the form

$$F(x, y, z, a) = 0 \qquad ...(1)$$

represents a surface and if the value of a is changed, the surface is changed.

Thus the equation (1) represents a one-parameter **family of surfaces** with a as a parameter. A particular value of the parameter a determines a particular member of the family of surfaces (1).

The surfaces corresponding to the values a and $a + \delta a$ of the parameter are $F(x, y, z, a) = 0$ and $F(x, y, z, a + \delta a) = 0$. The curve of intersection of these surfaces is given by the equations

$$F(x, y, z, a) = 0,\ F(x, y, z, a + \delta a) = 0.$$

The equations of this curve of intersection can also be written as

$$F(x, y, z, a) = 0, \quad \frac{F(x, y, z, a + \delta a) - F(x, y, z, a)}{\delta a} = 0.$$

Now as $\delta a \to 0$, the curve of intersection of the surfaces $F(x, y, z, a) = 0$ and $F(x, y, z, a + \delta a) = 0$ becomes the curve of intersection of consecutive members of the family and its defining equations becomes

$$F(x, y, z, a) = 0, \quad \lim_{\delta a \to 0} \frac{F(x, y, z, a + \delta a) - F(x, y, z, a)}{\delta a} = 0$$

or equivalently $F(x, y, z, a) = 0$, $\frac{\partial}{\partial a}(F(x, y, z, a)) = 0$. This curve is called the **characteristic** of the surface for the value a of the parameter. As the value of the parameter varies, we get a family of characteristics. The locus of the family of characteristics of the family of surfaces $F(x, y, z, a) = 0$ is called the **envelope** of the family of surfaces $F(x, y, z, a) = 0$. The envelope of the family of surfaces $F(x, y, z, a) = 0$ is a surface and its equation is obtained by eliminating a from the equations $F(x, y, z, a) = 0$ and $\frac{\partial}{\partial a}(F(x, y, z, a)) = 0$. For the sake of simplicity, we write these equations as

$$F(a) = 0 \quad \text{and} \quad \frac{\partial}{\partial a}(F(a)) = 0.$$

Theorem. *Prove that the envelope of a one-parameter family of surfaces touches each member of the family of surfaces at all points of its characteristic.*

Proof. Let $F(x, y, z, a) = 0$ be a one-parameter family of surfaces, where a is a parameter. The equation of the envelope of this family is obtained by eliminating a from the equations $F(x, y, z, a) = 0$ and $\frac{\partial}{\partial a}(F(x, y, z, a)) = 0$. By definition, two surfaces touch each other at a common point when they have the same tangent plane at that common point.

$\therefore$ Two surfaces touch each other at a common point when they have the same normal line at the common point. For any particular value of the parameter a, the corresponding characteristic lies on the surface for that value of a and also on the envelope of the family of surfaces. Thus all points of a characteristic are common to the corresponding surface and the envelope of the family.

At point (x, y, z), the normal to the surface $F(x, y, z, a)$ has d.r.'s F_x, F_y, F_z.

We know that the envelope of the family $F(x, y, z, a) = 0$ is obtained by eliminating a from the equations

$$F(x, y, z, a) = 0 \quad \text{and} \quad \frac{\partial}{\partial a}(F(x, y, z, a)) = 0.$$

If a is regarded as a function of x, y and z given by the equation $\frac{\partial}{\partial a}(F(x, y, z, a)) = 0$, then the equation of the envelope is represented by the single equation $F(x, y, z, a) = 0$.

$\therefore$ At point $(x, y, z,)$, the normal to the envelope $F(x, y, z, a) = 0$ has d.r.'s $F_x + F_a \frac{\partial a}{\partial x}$, $F_y + F_a \frac{\partial a}{\partial y}$, $F_z + F_a \frac{\partial a}{\partial z}$, because a is a function of x, y and z.

Since $\frac{\partial F}{\partial a} = 0$ *i.e.*, $F_a = 0$, the d.r.'s of the normal to the envelope at (x, y, z) reduces to F_x, F_y, F_z. Thus at all common points, the surface and the envelope have the same normal.

$\therefore$ The envelope touches each member of the family of surfaces at all points of its characteristic.

Example 1. *Find the envelope of the family of paraboloids $x^2 + y^2 = 4a(z - a)$, where a is a parameter.*

Sol. Given family of surfaces is

$$x^2 + y^2 = 4a(z - a).$$

Let $$F(a) = x^2 + y^2 - 4a(z - a)$$

$$\therefore \quad \frac{\partial}{\partial a}(F(a)) = -4z + 8a$$

$\therefore$ The equation of the envelope is obtained by eliminating a from the equations

$$x^2 + y^2 - 4a(z - a) = 0 \qquad ...(1)$$

and $$-4z + 8a = 0 \qquad ...(2)$$

(2) $\Rightarrow$ $$a = \frac{z}{2}$$

Putting the value of a in (1), we get

$$x^2 + y^2 - 4\left(\frac{z}{2}\right)\left(z - \frac{z}{2}\right) = 0$$

or $$x^2 + y^2 - z^2 = 0 \quad \text{or} \quad x^2 + y^2 = z^2.$$

$\therefore$ The envelope of the given family of paraboloids is the surface $x^2 + y^2 = z^2$.

Example 2. *Spheres of constant radius b have their centres on the fixed circle $x^2 + y^2 = a^2$, $z = 0$. Prove that the envelope of this family of spheres is the surface*

$$(x^2 + y^2 + z^2 + a^2 - b^2)^2 = 4a^2(x^2 + y^2).$$

Sol. Fixed circle is $x^2 + y^2 = a^2$, $z = 0$.

Any point on this circle is $(a \cos t, a \sin t, 0)$, where t is a parameter. Equation of the sphere with centre at $(a \cos t, a \sin t, 0)$ and radius b is

$$(x - a \cos t)^2 + (y - a \sin t)^2 + z^2 = b^2 \qquad ...(1)$$

This equation represents a family of spheres with parameter t.

Let $$F(t) = (x - a \cos t)^2 + (y - a \sin t)^2 + z^2 - b^2$$

$$\therefore \quad \frac{\partial}{\partial t}(F(t)) = 2a \sin t\,(x - a \cos t) - 2a \cos t\,(y - a \sin t)$$

The equation of the envelope of the family of spheres (1) is obtained by eliminating t from the equations

$$(x - a \cos t)^2 + (y - a \sin t)^2 + z^2 - b^2 = 0 \qquad ...(2)$$

and $$2a \sin t\,(x - a \cos t) - 2a \cos t\,(y - a \sin t) = 0 \qquad ...(3)$$

(3) $\Rightarrow$ $$(a \sin t)x - (a \cos t)y = 0$$

$\Rightarrow$ $$\tan t = \frac{y}{x}$$

$$\therefore \quad \sin t = \frac{y}{\sqrt{x^2 + y^2}} \quad \text{and} \quad \cos t = \frac{x}{\sqrt{x^2 + y^2}}$$

(2) $\Rightarrow$ $x^2 + y^2 + z^2 - 2ax \cos t - 2ay \sin t + a^2 - b^2 = 0$

Putting the values of $\sin t$ and $\cos t$, we get

$$x^2 + y^2 + z^2 - 2ax\left(\frac{x}{\sqrt{x^2 + y^2}}\right) - 2ay\left(\frac{y}{\sqrt{x^2 + y^2}}\right) + a^2 - b^2 = 0$$

$$\Rightarrow \quad x^2 + y^2 + z^2 + a^2 - b^2 = 2a\left(\frac{x^2 + y^2}{\sqrt{x^2 + y^2}}\right) = 2a\sqrt{x^2 + y^2}.$$

Squaring, we get $(x^2 + y^2 + z^2 + a^2 - b^2)^2 = 4a^2(x^2 + y^2)$.

This represents the envelope of the given family of spheres.

Example 3. *Find the envelope of the family of planes $3a^2x - 3ay + z = a^3$, where a is a parameter.*

Sol. Given family of planes is $3a^2x - 3ay + z = a^3$.

Let $$F(a) = 3a^2x - 3ay + z - a^3$$

$$\therefore \quad \frac{\partial}{\partial a}(F(a)) = 6ax - 3y - 3a^2$$

The equation of the envelope is obtained by eliminating a from the equations

$$3a^2x - 3ay + z - a^3 = 0 \quad \text{...(1)}$$

and $$6ax - 3y - 3a^2 = 0 \quad \text{...(2)}$$

$$(1) \Rightarrow \quad a^3 - 3xa^2 + 3ya - z = 0 \quad \text{...(3)}$$

$$(2) \Rightarrow \quad a^2 - 2xa + y = 0 \quad \text{...(4)}$$

$$(4) \times a \Rightarrow \quad a^3 - 2xa^2 + ya = 0 \quad \text{...(5)}$$

$$(3) - (5) \Rightarrow \quad -xa^2 + 2ya - z = 0 \quad \text{...(6)}$$

Solving (4) and (6), we get

$$\frac{a^2}{2xz - 2y^2} = \frac{a}{-xy + z} = \frac{1}{2y - 2x^2}$$

$$\Rightarrow \quad a^2 = \frac{xz - y^2}{y - x^2} \quad \text{and} \quad a = \frac{z - xy}{2y - 2x^2}$$

Eliminating a, we get

$$\frac{xz - y^2}{y - x^2} = \left(\frac{z - xy}{2y - 2x^2}\right)^2$$

$$\Rightarrow \quad 4(y - x^2)(xz - y^2) = (z - xy)^2.$$

This represents the envelope of the given family of planes.

Example 4. *Find the envelope of the family of planes $\frac{x}{a+u} + \frac{y}{b+u} + \frac{z}{c+u} = 1$, where u is a parameter.*

Sol. Given family of planes is $\frac{x}{a+u} + \frac{y}{b+u} + \frac{z}{c+u} = 1$ ...(1)

$$\Rightarrow \quad (b + u)(c + u)x + (c + u)(a + u)y + (a + u)(b + u)z - (a + u)(b + u)(c + u) = 0$$

$\Rightarrow \quad u^3 + lu^2 + mu + n = 0$, where

$$l = a + b + c - (x + y + z)$$

$$m = ab + bc + ca - [(b + c)x + (c + a)y + (a + b)z]$$

and $$n = abc - (bcx + cay + abz)$$ (Note this step carefully)

Let $$F(u) = u^3 + lu^2 + mu + n$$

$\therefore$ $$\frac{\partial}{\partial u}(F(u)) = 3u^2 + 2lu + m$$

The equation of the envelope of the given family is obtained by eliminating u from the equations

$$F(u) = 0 \quad \text{and} \quad \frac{\partial}{\partial u}(F(u)) = 0.$$

$\therefore$ $$u^3 + lu^2 + mu + n = 0 \qquad \text{...(2)}$$

$$3u^2 + 2lu + m = 0 \qquad \text{...(3)}$$

Multiplying (2) by 3 and (3) by u, we get

$$3u^3 + 3lu^2 + 3mu + 3n = 0 \qquad \text{...(4)}$$

$$3u^3 + 2lu^2 + mu = 0 \qquad \text{...(5)}$$

(4) – (5) $\Rightarrow$ $$lu^2 + 2mu + 3n = 0 \qquad \text{...(6)}$$

Solving (3) and (6), we get

$$\frac{u^2}{6ln - 2m^2} = \frac{u}{ml - 9n} = \frac{1}{6m - 2l^2}$$

$\Rightarrow$ $$u^2 = \frac{6ln - 2m^2}{6m - 2l^2} = \frac{3ln - m^2}{3m - l^2} \quad \text{and} \quad u = \frac{ml - 9n}{6m - 2l^2}$$

Eliminating u, we get

$\therefore$ $$\left(\frac{ml - 9n}{6m - 2l^2}\right)^2 = \frac{3ln - m^2}{3m - l^2}$$

$\Rightarrow$ $$(ml - 9n)^2 = 4(3ln - m^2)(3m - l^2).$$

This is the equation of the required equation of the envelope, where l, m and n are as given above.

6. IMPORTANT RESULTS

I. The envelope of the two-parameter family of surfaces $F(x, y, z, a, b) = 0$, **where** a, b **are parameters is found by eliminating** a **and** b **from the equations:**

$$F = 0, \frac{\partial f}{\partial a} = 0, \frac{\partial f}{\partial b} = 0.$$

II. The envelope of the two-parameter family of surfaces $F(x, y, z, a, b) = 0$, **where** a, b **are parameters connected by the equation** $\phi(a, b) = 0$ **is found by eliminating** a **and** b **from the equations:**

$$F = 0, \phi = 0, \frac{\partial F/\partial a}{\partial \phi/\partial a} = \frac{\partial F/\partial b}{\partial \phi/\partial b}.$$

III. The envelope of the three-parameter family of surface $F(x, y, z, a, b, c) = 0$, **where** a, b, c **are parameters connected by the equation** $\phi(a, b, c) = 0$ **and** F **and** ϕ **are homogeneous with respect to** a, b **and** c **is found by eliminating** a, b **and** c **from the equations:**

$$F = 0, \phi = 0, \frac{\partial F/\partial a}{\partial \phi/\partial a} = \frac{\partial F/\partial b}{\partial \phi/\partial b} = \frac{\partial F/\partial c}{\partial \phi/\partial c}.$$

IV. The envelope of the four-parameter family of surface F(x, y, z, a, b, c, d) = 0, **where** a, b, c, d **are parameters connected by the equation** $\phi(a, b, c, d) = 0$ **and** F **and** ϕ **are homogeneous with respect to** a, b, c **and** d **is found by eliminating** a, b, c **and** d **from the equations:**

$$F = 0, \phi = 0, \frac{\partial F/\partial a}{\partial\phi/\partial a} = \frac{\partial F/\partial b}{\partial\phi/\partial b} = \frac{\partial F/\partial c}{\partial\phi/\partial c} = \frac{\partial F/\partial d}{\partial\phi/\partial d}.$$

Example 5. *Find the envelope of the family of planes* $\frac{x}{a}\cos\theta\sin\phi + \frac{y}{b}\sin\theta\sin\phi + \frac{z}{c}\cos\phi = 1$, *where* θ *and* ϕ *are arbitrary parameters.*

Sol. Given two-parameter family of planes is

$$\frac{x}{a}\cos\theta\sin\phi + \frac{y}{b}\sin\theta\sin\phi + \frac{z}{c}\cos\phi = 1.$$

Let $$F(\theta, \phi) = \frac{x}{a}\cos\theta\sin\phi + \frac{y}{b}\sin\theta\sin\phi + \frac{z}{c}\cos\phi - 1$$

$\therefore$ $$\frac{\partial}{\partial\theta}(F(\theta, \phi)) = -\frac{x}{a}\sin\theta\sin\phi + \frac{y}{b}\cos\theta\sin\phi$$

and $$\frac{\partial}{\partial\phi}(F(\theta, \phi)) = \frac{x}{a}\cos\theta\cos\phi + \frac{y}{b}\sin\theta\cos\phi - \frac{z}{c}\sin\phi$$

The equation of the envelope is obtained by eliminating θ and ϕ from the equations:

$$F(\theta, \phi) = 0, \frac{\partial}{\partial\theta}(F(\theta, \phi)) = 0, \frac{\partial}{\partial\phi}(F(\theta, \phi)) = 0. \quad \text{(c.f. **Result I**)}$$

$$\Rightarrow \quad \frac{x}{a}\cos\theta\sin\phi + \frac{y}{b}\sin\theta\sin\phi + \frac{z}{c}\cos\phi - 1 = 0 \quad ...(1)$$

$$-\frac{x}{a}\sin\theta\sin\phi + \frac{y}{b}\cos\theta\sin\phi = 0 \quad ...(2)$$

$$\frac{x}{a}\cos\theta\cos\phi + \frac{y}{b}\sin\theta\cos\phi - \frac{z}{c}\sin\phi = 0 \quad ...(3)$$

(3) $$\Rightarrow \quad \left(\frac{x}{a}\cos\theta + \frac{y}{b}\sin\theta\right)\cos\phi = \frac{z}{c}\sin\phi$$

$$\Rightarrow \quad \frac{x}{a}\cos\theta + \frac{y}{b}\sin\theta = \frac{z}{c}\tan\phi \quad ...(4)$$

(1) $$\Rightarrow \quad \left(\frac{x}{a}\cos\theta + \frac{y}{b}\sin\theta\right)\sin\phi = 1 - \frac{z}{c}\cos\phi$$

$$\Rightarrow \quad \left(\frac{z}{c}\tan\phi\right)\sin\phi = 1 - \frac{z}{c}\cos\phi \quad \text{(Using (4))}$$

$$\Rightarrow \quad \frac{z}{c}\left(\frac{\sin^2\phi + \cos^2\phi}{\cos\phi}\right) = 1 \Rightarrow \frac{z}{c} = \cos\phi$$

$\therefore$ (4) $$\Rightarrow \quad \frac{x}{a}\cos\theta + \frac{y}{b}\sin\theta = \frac{z}{c}\sqrt{\frac{1}{\cos^2\phi} - 1} = \frac{z}{c}\sqrt{\frac{c^2}{z^2} - 1}$$

$$\therefore \quad \frac{x}{a}\cos\theta + \frac{y}{b}\sin\theta = \frac{\sqrt{c^2 - z^2}}{c} \qquad ...(5)$$

$$(2) \Rightarrow \quad \frac{x}{a}\sin\theta - \frac{y}{b}\cos\theta = 0 \qquad ...(6)$$

(By cancelling sin ϕ)

Squaring and adding (5) and (6), we get

$$\frac{x^2}{a^2} + \frac{y^2}{b^2} = \frac{c^2 - z^2}{c^2}$$

$$\Rightarrow \quad \frac{x^2}{a^2} + \frac{y^2}{b^2} + \frac{z^2}{c^2} = 1.$$

This is the equation of the envelope of the given family of planes.

Example 6. *Find the envelope of the family of planes whose members form a tetrahedron of volume $d^3/6$ with the coordinate planes.*

Sol. Let the family of planes be $\frac{x}{a} + \frac{y}{b} + \frac{z}{c} = 1$, where a, b, c are parameters.

The intercepts of this plane on the coordinate axes are a, b, c.

$\therefore$ Volume of tetrahedron $= \frac{1}{6}abc$ (This is a standard formula)

$\therefore$ By the given condition,

$$\frac{1}{6}abc = \frac{d^3}{6}$$

$$\therefore \quad abc = d^3$$

Let $\quad F(a, b, c) = \frac{x}{a} + \frac{y}{b} + \frac{z}{c} - 1$ and $\phi(a, b, c) = abc - d^3$

The equation of the envelope is obtained by eliminating a, b and c from the equations:

$$F = 0, \phi = 0, \frac{\partial F/\partial a}{\partial \phi/\partial a} = \frac{\partial F/\partial b}{\partial \phi/\partial b} = \frac{\partial F/\partial c}{\partial \phi/\partial c}. \qquad \text{(c.f. **Result III**)}$$

$$\therefore \quad \frac{x}{a} + \frac{y}{b} + \frac{z}{c} - 1 = 0 \qquad ...(1)$$

$$abc - d^3 = 0 \qquad ...(2)$$

$$\text{and} \quad \frac{-x/a^2}{bc} = \frac{-y/b^2}{ac} = \frac{-z/c^2}{ab} \qquad ...(3)$$

$$(3) \Rightarrow \quad \frac{x}{a^2bc} = \frac{y}{ab^2c} = \frac{z}{abc^2}$$

$$\Rightarrow \quad \frac{x}{a} = \frac{y}{b} = \frac{z}{c} = \lambda, \text{ say}$$

$$\therefore \ \bullet(1) \Rightarrow \lambda + \lambda + \lambda - 1 = 0 \Rightarrow \lambda = \frac{1}{3}$$

$$\therefore \quad \frac{x}{a} = \frac{y}{b} = \frac{z}{c} = \frac{1}{3} \quad \text{or} \quad a = 3x, b = 3y, c = 3z$$

Putting the values of a, b and c in (2), we get

$$(3x)(3y)(3z) - d^3 = 0 \quad \text{or} \quad 27xyz = d^3.$$

This is the equation of the envelope of the given family of planes.

Example 7. *Find the envelope of the family of planes $lx + my + nz = p$, where l, m, n and p are parameters connected by the equation $a^2l^2 + b^2m^2 + c^2n^2 = p^2$.*

Sol. Given family of planes is

$$lx + my + nz = p.$$

The parameters l, m, n and p are connected by the equation $a^2l^2 + b^2m^2 + c^2n^2 = p^2$.

Let $\quad F(l, m, n, p) = lx + my + nz - p$

and $\quad \phi(l, m, n, p) = a^2l^2 + b^2m^2 + c^2n^2 - p^2$

The equation of the envelope is obtained by eliminating l, m, n and p from the equations:

$$F = 0, \phi = 0, \frac{\partial F/\partial l}{\partial \phi/\partial l} = \frac{\partial F/\partial m}{\partial \phi/\partial m} = \frac{\partial F/\partial n}{\partial \phi/\partial n} = \frac{\partial F/\partial p}{\partial \phi/\partial p}. \quad \text{(c.f. **Result IV**)}$$

$$\therefore \quad lx + my + nz - p = 0 \qquad \text{...(1)}$$

$$a^2l^2 + b^2m^2 + c^2n^2 - p^2 = 0 \qquad \text{...(2)}$$

and

$$\frac{x}{2a^2l} = \frac{y}{2b^2m} = \frac{z}{2c^2n} = \frac{-1}{-2p} = \lambda, \text{ say}$$

$$\therefore \quad l = \frac{x}{2\lambda a^2}, m = \frac{y}{2\lambda b^2}, n = \frac{z}{2\lambda c^2}, p = \frac{1}{2\lambda}$$

Putting the values l, m, n and p in (1), we get

$$\left(\frac{x}{2\lambda a^2}\right)x + \left(\frac{y}{2\lambda b^2}\right)y + \left(\frac{z}{2\lambda c^2}\right)z - \frac{1}{2\lambda} = 0$$

$$\Rightarrow \quad \frac{x^2}{a^2} + \frac{y^2}{b^2} + \frac{z^2}{c^2} = 1.$$

This is the equation of the envelope of the given family of planes.

WORKING RULES FOR SOLVING PROBLEMS

Rule I. *For the one-parameter family of surfaces $F(x, y, z, a) = 0$, the curve represented by the equations $F(x, y, z, a) = 0$, $\frac{\partial}{\partial a}(F(x, y, z, a)) = 0$ is called the* **characteristic** *of the surface for the value a of the parameter.*

Rule II. *The locus of the family of characteristics of the one-parameter family of surfaces $F(x, y, z, a) = 0$ is called the* **envelope** *of the family $F(x, y, z, a) = 0$. The envelope of the family $F(x, y, z, a) = 0$ is a surface and its equation is obtained by eliminating a from the equations $F(x, y, z, a) = 0$ and $\frac{\partial}{\partial a}(F(x, y, z, a)) = 0$.*

Rule III. *The envelope of the two-parameter family of surfaces $F(x, y, z, a, b) = 0$ is obtained by eliminating a and b from the equations:*

$$F = 0, \frac{\partial F}{\partial a} = 0, \frac{\partial F}{\partial b} = 0.$$

Rule IV. *The envelope of the two-parameter family of surfaces $F(x, y, z, a, b) = 0$ with the condition $\phi(a, b) = 0$ is obtained by eliminating a and b from the equations:*

$$F = 0,\ \phi = 0,\ \frac{\partial F/\partial a}{\partial \phi/\partial a} = \frac{\partial F/\partial b}{\partial \phi/\partial b}.$$

Rule V. *The envelope of the three-parameter family of surfaces $F(x, y, z, a, b, c) = 0$ with the condition $\phi(a, b, c) = 0$ and F and ϕ are homogeneous w.r.t. a, b and c is obtained by eliminating a, b and c from the equations:*

$$F = 0,\ \phi = 0,\ \frac{\partial F/\partial a}{\partial \phi/\partial a} = \frac{\partial F/\partial b}{\partial \phi/\partial b} = \frac{\partial F/\partial c}{\partial \phi/\partial c}.$$

Rule VI. *The envelope of the four-parameter family of surfaces $F(x, y, z, a, b, c, d) = 0$ with the condition $\phi(a, b, c, d) = 0$ and F and ϕ are homogeneous w.r.t. a, b, c, and d is obtained by eliminating a, b, c,* and *d from the equations:*

$$F = 0,\ \phi = 0,\ \frac{\partial F/\partial a}{\partial \phi/\partial a} = \frac{\partial F/\partial b}{\partial \phi/\partial b} = \frac{\partial F/\partial c}{\partial \phi/\partial c} = \frac{\partial F/\partial d}{\partial \phi/\partial d}.$$

EXERCISE 3.2

1. Find the envelope of the family of paraboloids $y^2 + z^2 = 4a(x - a)$, where a is a parameter.
2. Find the envelope of the family of spheres $(x - a \cos \theta)^2 + (y - a \sin \theta)^2 + z^2 = b^2$, where θ is a parameter.
3. Find the envelope of the family of planes $3xt^2 + y - 3zt = t^3$, where t is a parameter.
4. Show that the envelope of the family of planes $x \cos \theta \sin \phi + y \sin \theta \sin \phi + z \cos \phi = \lambda$, where θ and ϕ are parameters, is the sphere $x^2 + y^2 + z^2 = \lambda^2$.
5. Find the envelope of the family of planes whose members form a tetrahedron of volume of 10 units with the coordinate planes.
6. Find the envelope of the family of planes $lx + my + nz = 0$, where l, m amd n are parameters connected by the equation $al^2 + bm^2 + cn^2 = 0$.
7. Find the envelope of the family of planes $\frac{x}{a} + \frac{y}{b} + \frac{z}{c} = 1$, where a, b, c are parameters, and $\frac{1}{a^2} + \frac{1}{b^2} + \frac{1}{c^2} = \frac{1}{k^2}$.

Answers

1. $y^2 + z^2 = x^2$
2. $(x^2 + y^2 + z^2 + a^2 - b^2)^2 = 4a^2(x^2 + y^2)$
3. $4(z - x^2)(xy - z^2) = (y - xz)^2$
5. $9xyz = 20$
6. $\frac{x^2}{a} + \frac{y^2}{b} + \frac{z^2}{c} = 0$
7. $x^2 + y^2 + z^2 = k^2$.

7. EDGE OF REGRESSION

Let $F(x, y, z, a) = 0$ be a one-parameter family of surfaces, where a is a parameter. Let $F(x, y, z, \alpha) = 0$ be a member of this family for the value α of a. We know that the curve in which the surface $F(x, y, z, \alpha) = 0$ is met by a consecutive surface of $F(x, y, z, \alpha) = 0$ and belonging to the family $F(x, y, z, a) = 0$ is the *characteristic* of the surface $F(x, y, z, \alpha) = 0$. Thus for each member of the family of surfaces $F(x, y, z, a) = 0$, we get one characteristic. The locus of the points of intersection of consecutive characteristics of the family of surfaces $F(x, y, z, a) = 0$ is called the **edge of regression** of the family of surfaces $F(x, y, z, a) = 0$.

Since each characteristic corresponding to the members of the family $F(x, y, z, a) = 0$ lies on the envelope of the family $F(x, y, z, a) = 0$, the curve *edge of regression* also lies on the envelope of $F(x, y, z, a) = 0$ and is also called as the **edge of regression of the envelope** of the family of surfaces $F(x, y, z, a) = 0$.

8. EQUATION OF EDGE OF REGRESSION

Let $F(x, y, z, a) = 0$ be a one-parameter family of surfaces. By definition, the edge of regression of this family is a curve in space and it is the locus of the points of intersection of consecutive characteristics of the family of surfaces $F(x, y, z, a) = 0$.

The characteristic with parameter value a is the curve of intersections of the surfaces

$$F(a) = 0 \quad \text{and} \quad \frac{\partial}{\partial a}(F(a)) = 0. \qquad ...(1)$$

The consecutive characteristic with parameter value $a + \delta a$ is the curve of intersection of the surfaces

$$F(a + \delta a) = 0 \quad \text{and} \quad \frac{\partial}{\partial a}(F(a + \delta a)) = 0. \qquad ...(2)$$

Using **Taylor's expansion**, equations (2) can be written as

$$F(a) + \delta a \frac{\partial}{\partial a}(F(a)) = 0 \quad \text{and} \quad \frac{\partial}{\partial a}(F(a)) + \delta a \frac{\partial^2}{\partial a^2}(F(a)) = 0 \qquad ...(3)$$

Since for the point of intersection of consecutive characteristics, all four equations in (1) and (3) must be satisfied and for this to happen it is sufficient to have

$$F(a) = 0, \frac{\partial}{\partial a}(F(a)) = 0 \quad \text{and} \quad \frac{\partial^2}{\partial a^2}(F(a)) = 0.$$

$\therefore$ The equation of the curve edge of regression for the family of surfaces $F(x, y, z, a) = 0$ is obtained by eliminating a from the equations:

$$F(a) = 0, \frac{\partial}{\partial a}(F(a)) = 0 \quad \text{and} \quad \frac{\partial^2}{\partial a^2}(F(a)) = 0.$$

Theorem. *Prove that the characteristic of each surface of a one-parameter family of surfaces touches the edge of regression of the envelope of that family of surfaces.*

Proof. Let $F(x, y, z, a) = 0$ be a one-parameter family of surfaces, where a is a parameter. The characteristic with parameter value a is the curve of intersection of the surfaces

$$F(a) = 0 \quad \text{and} \quad \frac{\partial}{\partial a}(F(a)) = 0.$$

The tangent to the characteristic at the point (x, y, z) is therefore perpendicular to the normal to both surfaces $F(a) = 0$ and $\frac{\partial}{\partial a}(F(a)) = 0$ at this point. It is therefore perpendicular to the lines with d.r.'s

$$F_x, F_y, F_z \quad \text{and} \quad \left(\frac{\partial F}{\partial a}\right)_x, \left(\frac{\partial F}{\partial a}\right)_y, \left(\frac{\partial F}{\partial a}\right)_z.$$

We know that the edge of regression of the envelope of the family $F(x, y, z, a) = 0$ is obtained by eliminating a from the equations:

$$F(a) = 0, \quad \frac{\partial}{\partial a}(F(a)) = 0 \quad \text{and} \quad \frac{\partial^2}{\partial a^2}(F(a)) = 0.$$

If a is regarded as a function of x, y, z given by the equation $\frac{\partial^2}{\partial a^2}(F(a)) = 0$, then the equation of the edge of regression is represented by the equations:

$$F(a) = 0 \quad \text{and} \quad \frac{\partial}{\partial a}(F(a)) = 0.$$

$\because$ **At point (x, y, z), the tangent to the edge of regression, being perpendicular to the normals to both surfaces $F(a) = 0$ and $\frac{\partial}{\partial a}(F(a)) = 0$, is perpendicular to the lines with d.r.'s**

$$F_x + F_a \frac{\partial a}{\partial x}, F_y + F_a \frac{\partial a}{\partial y}, F_z + F_a \frac{\partial a}{\partial z}$$

and
$$\left(\frac{\partial F}{\partial a}\right)_x + F_{aa} \frac{\partial a}{\partial x}, \left(\frac{\partial F}{\partial a}\right)_y + F_{aa} \frac{\partial a}{\partial y}, \left(\frac{\partial F}{\partial a}\right)_z + F_{aa} \frac{\partial a}{\partial z},$$

because a is a function of x, y, and z.

Since
$$\frac{\partial F}{\partial a} = 0, \frac{\partial^2 F}{\partial a^2} = 0 \quad i.e., \quad F_a = 0, F_{aa} = 0,$$

the d.r.'s of the tangent to the edge of regression at (x, y, z) is perpendicular to the lines with d.r.'s

$$F_x, F_y, F_z \text{ and } \left(\frac{\partial F}{\partial a}\right)_x, \left(\frac{\partial F}{\partial a}\right)_y, \left(\frac{\partial F}{\partial a}\right)_z.$$

Thus at all common points, the characteristic and the edge of regression have the same tangent.

$\therefore$ **The characteristic of each surface of the family of surfaces touches the edge of regression of the family of surfaces.**

WORKING RULES FOR SOLVING PROBLEMS

Rule I. *The locus of the point of intersection of consecutive characteristics of a one parameter family of surfaces is called its* **edge of regression.**

Rule II. *The equation of the curve 'edge of regression' for the family of surfaces $F(x, y, z, a) = 0$ is obtained by eliminating a from the equations:*

$$F(x, y, z, a) = 0, \quad \frac{\partial}{\partial a}(F(x, y, z, a)) = 0 \quad \text{and} \quad \frac{\partial^2}{\partial a^2}(F(x, y, z, a)) = 0.$$

Example 1. *Show that the edge of regression of the envelope of the family of planes $3xt^2 - 3yt + z = t^3$, where t is a parameter, is the curve of intersection of the surfaces $y^2 = xz$ and $xy = z$*

Sol. Given family of planes is

$$3xt^2 - 3yt + z = t^3.$$

Let
$$F(t) = 3xt^2 - 3yt + z - t^3$$

$\therefore$
$$\frac{\partial}{\partial t}(F(t)) = 6xt - 3y - 3t^2$$

and
$$\frac{\partial^2}{\partial t^2}(F(t)) = 6x - 6t.$$

The equation of the edge of regression of the envelope of the given family is obtained by eliminating t from the equations:

$$F(t) = 0, \quad \frac{\partial}{\partial t}(F(t)) = 0 \quad \text{and} \quad \frac{\partial^2}{\partial t^2}(F(t)) = 0.$$

$$\therefore \quad 3xt^2 - 3yt + z - t^3 = 0 \qquad ...(1)$$

$$6xt - 3y - 3t^2 = 0 \qquad ...(2)$$

$$6x - 6t = 0 \qquad ...(3)$$

$$(1) \Rightarrow \quad t^3 - 3xt^2 + 3yt - z = 0 \qquad ...(4)$$

$$(2) \Rightarrow \quad t^2 - 2xt + y = 0 \qquad ...(5)$$

$$(3) \Rightarrow \quad t = x$$

$$\therefore \ (5) \Rightarrow \quad x^2 - 2x^2 + y = 0 \quad \Rightarrow \quad y = x^2 \Rightarrow y = t^2$$

$$(4) \Rightarrow \quad xy - 3xy + 3yx - z = 0 \quad \Rightarrow \quad z = xy$$

Also

$$y = x^2 \quad \Rightarrow \quad y^2 = x^2y = x(xy) = xz$$

$\therefore$ The edge of regression is the curve of intersection of surfaces $z = xy$ and $y^2 = xz$.

Example 2. *Find the edge of regression of the envelope of the family of planes* $\frac{x}{a+u} + \frac{y}{b+u} + \frac{z}{c+u} = 1$*, where u is a parameter.*

Sol. Given family of planes is

$$\frac{x}{a+u} + \frac{y}{b+u} + \frac{z}{c+u} = 1. \qquad ...(1)$$

Let

$$F(u) = \frac{x}{a+u} + \frac{y}{b+u} + \frac{z}{c+u} - 1$$

$$\therefore \quad \frac{\partial}{\partial u}(F(u)) = -\frac{x}{(a+u)^2} - \frac{y}{(b+u)^2} - \frac{z}{(c+u)^2}$$

and

$$\frac{\partial^2}{\partial u^2}(F(u)) = \frac{2x}{(a+u)^3} + \frac{2y}{(b+u)^3} + \frac{2z}{(c+u)^3}$$

The equation of the edge of regression of the envelope of given family is obtained by eliminating a from the equations:

$$F(u) = 0, \quad \frac{\partial}{\partial u}(F(u)) = 0 \quad \text{and} \quad \frac{\partial^2}{\partial u^2}(F(u)) = 0\,.$$

$$\Rightarrow \quad \frac{x}{a+u} + \frac{y}{b+u} + \frac{z}{c+u} - 1 = 0 \qquad ...(2)$$

$$\frac{x}{(a+u)^2} + \frac{y}{(b+u)^2} + \frac{z}{(c+u)^2} = 0 \qquad ...(3)$$

$$\frac{x}{(a+u)^3} + \frac{y}{(b+u)^3} + \frac{z}{(c+u)^3} = 0 \qquad ...(4)$$

Multiplying (3) by $\frac{1}{c+u}$ and subtracting (4) from it, we get

$$\frac{x}{(a+u)^2}\left[\frac{1}{c+u} - \frac{1}{a+u}\right] + \frac{y}{(b+u)^2}\left[\frac{1}{c+u} - \frac{1}{b+u}\right] = 0$$

$$\Rightarrow \quad \frac{(a-c)x}{(a+u)^3(c+u)} + \frac{(b-c)y}{(b+u)^3(c+u)} = 0$$

$$\Rightarrow \quad \frac{(c-a)x}{(a+u)^3} = \frac{(b-c)y}{(b+u)^3}$$

$$\Rightarrow \quad \frac{x}{(a+u)^3(b-c)} = \frac{y}{(b+u)^3(c-a)}$$

$\therefore$ By symmetry, we have

$$\frac{x}{(a+u)^3(b-c)} = \frac{y}{(b+u)^3(c-a)} = \frac{z}{(c+u)^3(a-b)}.$$

$$\Rightarrow \quad \frac{\frac{x}{a+u}}{(a+u)^2(b-c)} = \frac{\frac{y}{b+u}}{(b+u)^2(c-a)} = \frac{\frac{z}{c+u}}{(c+u)^2(a-b)}$$

Each term of this equality

$$= \frac{\frac{x}{a+u} + \frac{y}{b+u} + \frac{z}{c+u}}{(a+u)^2(b-c)+(b+u)^2(c-a)+(c+u)^2(a-b)}$$

$$= \frac{1}{a^2(b-c)+b^2(c-a)+c^2(a-b)} \qquad \text{(Using (2))}$$

$$= \frac{1}{(b-c)(c-a)(a-b)}$$

$$\therefore \quad \frac{\frac{x}{a+u}}{(a+u)^2(b-c)} = \frac{1}{(b-c)(c-a)(a-b)}$$

$$\Rightarrow \quad x = \frac{(a+u)^3}{(c-a)(a-b)}$$

Similarly, $y = \dfrac{(b+u)^3}{(b-c)(a-b)}$, $z = \dfrac{(c+u)^3}{(b-c)(c-a)}$

$\therefore$ The equation of the edge of regression of the given family of surfaces is

$x = \dfrac{(a+u)^3}{(c-a)(a-b)}$, $y = \dfrac{(b+u)^3}{(b-c)(a-b)}$, $z = \dfrac{(c+u)^3}{(b-c)(c-a)}$, where u is a parameter.

9. FUNDAMENTAL MAGNITUDES OF THE FIRST ORDER

Let $x = x(u, v)$, $y = y(u, v)$, $z = z(u, v)$ be the equations of a surface, where u and v are independent parameters. Let $\mathbf{r}$ be the position vector of the point (x, y, z) on the surface.

$\therefore$ $\mathbf{r}$ is a function of the parameters u and v.

We know that the first order partial derivatives $\dfrac{\partial \mathbf{r}}{\partial u}$ and $\dfrac{\partial \mathbf{r}}{\partial v}$ of $\mathbf{r}$ are denoted by $\mathbf{r}_1$ and $\mathbf{r}_2$ respectively.

$$\therefore \quad \mathbf{r}_1 = \frac{\partial \mathbf{r}}{\partial u} \quad \text{and} \quad \mathbf{r}_2 = \frac{\partial \mathbf{r}}{\partial v}$$

The quantities $\mathbf{r}_1 \cdot \mathbf{r}_1$, $\mathbf{r}_1 \cdot \mathbf{r}_2$ and $\mathbf{r}_2 \cdot \mathbf{r}_2$ are called the **fundamental magnitudes of the first order** and are denoted by E, F and G respectively.

$\therefore$ $$E = \mathbf{r}_1 \cdot \mathbf{r}_1,\ F = \mathbf{r}_1 \cdot \mathbf{r}_2 \text{ and } G = \mathbf{r}_2 \cdot \mathbf{r}_2.$$

These quantities are of the greatest importance in differential geometry. The magnitudes E and G are always positive because

$$E = |\mathbf{r}_1|^2 > 0 \text{ and } F = |\mathbf{r}_2|^2 > 0.$$

Theorem. *Prove that $EG - F^2 \geq 0$.*

Proof. Let $\mathbf{r} = \mathbf{r}(u, v)$ be the equation of a surface, where u and v are independent parameters.

We have $$E = \mathbf{r}_1 \cdot \mathbf{r}_1,\ F = \mathbf{r}_1 \cdot \mathbf{r}_2 \text{ and } G = \mathbf{r}_2 \cdot \mathbf{r}_2.$$

$$\therefore \quad EG - F^2 = (\mathbf{r}_1 \cdot \mathbf{r}_1)(\mathbf{r}_2 \cdot \mathbf{r}_2) - (\mathbf{r}_1 \cdot \mathbf{r}_2)^2$$

$$= |\mathbf{r}_1|^2 |\mathbf{r}_2|^2 - |\mathbf{r}_1|^2 |\mathbf{r}_2|^2 \cos^2\theta,$$

where θ is the angle between $\mathbf{r}_1$ and $\mathbf{r}_2$

$$= |\mathbf{r}_1|^2 |\mathbf{r}_2|^2 (1 - \cos^2\theta) = |\mathbf{r}_1|^2 |\mathbf{r}_2|^2 \sin^2\theta$$

$$= |\mathbf{r}_1 \times \mathbf{r}_2|^2 \geq 0.$$

$$\therefore \quad EG - F^2 \geq 0.$$

Remark. The quantity $|\mathbf{r}_1 \times \mathbf{r}_2|$ is denoted by H.

$$\therefore \quad H^2 = EG - F^2 \quad \text{or} \quad H = \sqrt{EG - F^2}.$$

Example 3. *Find the fundamental magnitudes of the first order for the surface $z = f(x, y)$.*

Sol. Given surface is $z = f(x, y)$. Let $u = x, v = y$.

$\therefore$ Given surface is $x = u, y = v, z = f(u, v)$.

Let $\mathbf{r}$ be the position vector of the point (x, y, z) on the surface.

$$\therefore \quad \mathbf{r} = u\mathbf{i} + v\mathbf{j} + f(u, v)\mathbf{k}$$

$$\therefore \quad \mathbf{r}_1 = \frac{\partial \mathbf{r}}{\partial u} = \mathbf{i} + f_u\mathbf{k} \text{ and } \mathbf{r}_2 = \frac{\partial \mathbf{r}}{\partial v} = \mathbf{j} + f_v\mathbf{k}$$

$$\therefore \quad E = \mathbf{r}_1 \cdot \mathbf{r}_1 = (\mathbf{i} + f_u\mathbf{k}) \cdot (\mathbf{i} + f_u\mathbf{k})$$

$$= 1(1) + 0(0) + f_u(f_u) = 1 + f_u^2$$

$$F = \mathbf{r}_1 \cdot \mathbf{r}_2 = (\mathbf{i} + f_u\mathbf{k}) \cdot (\mathbf{j} + f_v\mathbf{k})$$

$$= 1(0) + 0(1) + f_u f_v = f_u f_v$$

$$G = \mathbf{r}_2 \cdot \mathbf{r}_2 = (\mathbf{j} + f_v\mathbf{k}) \cdot (\mathbf{j} + f_v\mathbf{k}) = 0(0) + 1(1) + f_v(f_v) = 1 + f_v^2$$

$$\therefore \quad E = 1 + f_u^2,\ F = f_u f_v \text{ and } G = 1 + f_v^2.$$

Example 4. *Calculate the values of E, F and G for the surface $x = u \sin\alpha \cos v$, $y = u \sin\alpha \sin v$, $z = u \cos\alpha + av$, where u and v are parameters.*

Sol. Given surface is

$$x = u \sin\alpha \cos v,\ y = u \sin\alpha \sin v,\ z = u \cos\alpha + av.$$

Let $\mathbf{r}$ be the position vector of the point (x, y, z) on the given surface.

$$\therefore \quad \mathbf{r} = u \sin\alpha \cos v\mathbf{i} + u \sin\alpha \sin v\mathbf{j} + (u \cos\alpha + av)\mathbf{k}$$

$$\therefore \quad \mathbf{r}_1 = \frac{\partial \mathbf{r}}{\partial u} = \sin\alpha \cos v\mathbf{i} + \sin\alpha \sin v\mathbf{j} + \cos\alpha\mathbf{k}$$

and $$\mathbf{r}_2 = \frac{\partial \mathbf{r}}{\partial v} = -u \sin\alpha \sin v\mathbf{i} + u \sin\alpha \cos v\mathbf{j} + a\mathbf{k}$$

$$\therefore \quad E = \mathbf{r}_1 \cdot \mathbf{r}_1 = (\sin\alpha \cos v\mathbf{i} + \sin\alpha \sin v\mathbf{j} + \cos\alpha\mathbf{k}) \cdot (\sin\alpha \cos v\mathbf{i} + \sin\alpha \sin v\mathbf{j} + \cos\alpha\mathbf{k})$$

$$= \sin^2 \alpha \cos^2 v + \sin^2 \alpha \sin^2 v + \cos^2 \alpha$$

$$= \sin^2 \alpha(\cos^2 v + \sin^2 v) + \cos^2 \alpha = 1$$

$$F = \mathbf{r}_1 . \mathbf{r}_2$$

$$= (\sin \alpha \cos v\mathbf{i} + \sin \alpha \sin v\mathbf{j} + \cos \alpha\mathbf{k}) . (-u \sin \alpha \sin v\, \mathbf{i} + u \sin \alpha \cos v\mathbf{j} + a\mathbf{k})$$

$$= -u \sin^2 \alpha \sin v \cos v + u \sin^2 \alpha \sin v \cos v + a \cos \alpha$$

$$= a \cos \alpha$$

$$G = \mathbf{r}_2 \cdot \mathbf{r}_2 = (-u \sin \alpha \sin v\mathbf{i} + u \sin \alpha \cos v\mathbf{j} + a\mathbf{k}) . (-u \sin \alpha \sin v\mathbf{i} + u \sin \alpha \cos v\mathbf{j} + a\mathbf{k})$$

$$= u^2 \sin^2 \alpha \sin^2 v + u^2 \sin^2 \alpha \cos^2 v + a^2$$

$$= u^2 \sin^2 \alpha (\sin^2 v + \cos^2 v) + a^2$$

$$= u^2 \sin^2 \alpha + a^2$$

$$\therefore \quad E = 1, F = a \cos \alpha \quad \text{and} \quad G = u^2 \sin^2 \alpha + a^2.$$

10. PARAMETRIC CURVES

We know that a surface is the locus of a point whose position vector **r** is a function of two independent parameters u and v. The cartesian coordinates x, y, z of the point with position vector **r** are then known functions of u and v.

Let $\mathbf{r} = \mathbf{r}(u, v)$ be the equation of a surface. For each value of pair (u, v), we get a point on the surface. Any relation of the form $\phi(u, v) = 0$ represents a curve on the surface $\mathbf{r} = \mathbf{r}(u, v)$, because **r** in this case becomes a function of only one parameter. In particular, if either u or v is kept constant, the locus of $\mathbf{r} = \mathbf{r}(u, v)$ would be a curve in space.

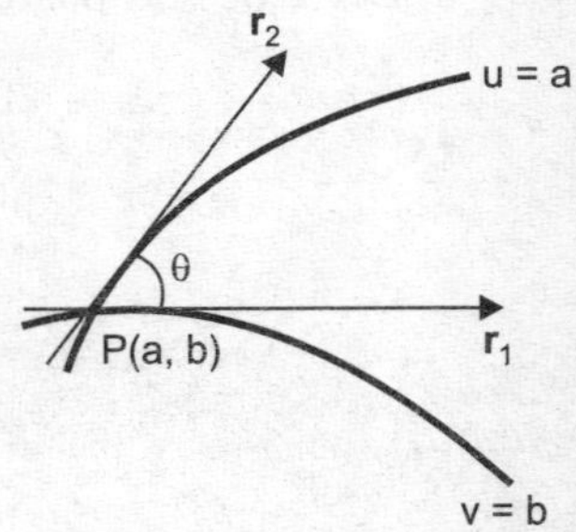

The curves on the surface $\mathbf{r} = \mathbf{r}(u, v)$ obtained by keeping one of the parameters constant are called **parametric curves** on the surface $\mathbf{r} = \mathbf{r}(u, v)$. Thus if u is kept constant, say a, then as v varies, the locus of the point $\mathbf{r} = \mathbf{r}(a, v)$ is a parametric curve on the surface $\mathbf{r} = \mathbf{r}(u, v)$ and is called the parametric curve $u = a$.

Similarly, if v is kept constant, say b, then as u varies, the locus of the point $\mathbf{r} = \mathbf{r}(u, b)$ is a parametric curve on the surface $\mathbf{r} = \mathbf{r}(u, v)$ and is called the parametric curve $v = b$.

Let P be the point on the surface $\mathbf{r} = \mathbf{r}(u, v)$ for the values $u = a$ and $v = b$ of the parameters. Through P passes exactly two parametric curves namely parametric curve $u = a$ and parametric curve $v = b$. The vector $\mathbf{r}_1$, *i.e.*, $\frac{\partial \mathbf{r}}{\partial u}$ is tangential to the parametric curve $v = b$ at the point P(**r**) because its direction is that of the displacement $\delta\mathbf{r}$ due to a small variation δu in the parameter u along the parametric curve $v = b$.

Similarly $\mathbf{r}_2$ *i.e.*, $\frac{\partial \mathbf{r}}{\partial v}$ is tangential to the parametric curve $u = a$ at the point P(**r**).

If $\mathbf{r}_1 . \mathbf{r}_2 = 0$ at the point P, then the parametric curves through the point P are called **orthogonal parametric curves.**

If the condition of orthogonality is satisfied at every point on the surface *i.e.*, for all possible values of u and v, then the given surface is said to have **orthogonal parametric curves.**

11. ANGLE BETWEEN TWO PARAMETRIC CURVES

Let $\mathbf{r} = \mathbf{r}(u, v)$ be the equation of a surface, where u and v are independent parameters. Let P be the point on the surface with position vector $\mathbf{r}(a, b)$.

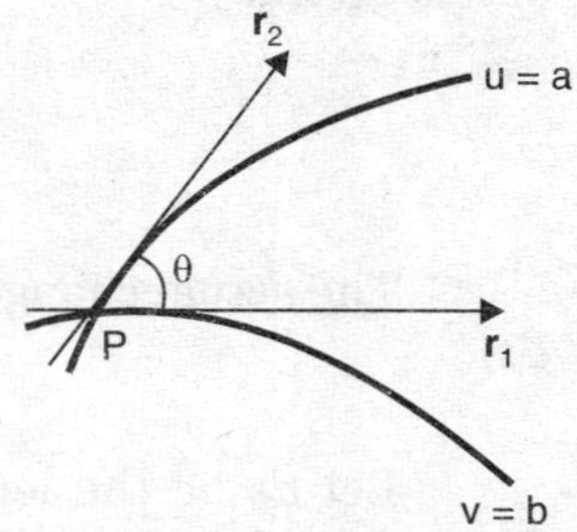

∴ The parametric curves, $u = a$ and $v = b$ intersect at the point P.

Let θ be the angle between the parametric curves $u = a$ and $v = b$ at the point P.

∴ The angle between the vectors $\mathbf{r}_1$ and $\mathbf{r}_2$ is also θ because $\mathbf{r}_1$ and $\mathbf{r}_2$ are tangent to the parametric curves $v = b$ and $u = a$ respectively.

$$\therefore \qquad \cos\theta = \frac{\mathbf{r}_1 . \mathbf{r}_2}{|\mathbf{r}_1||\mathbf{r}_2|} = \frac{F}{\sqrt{E}\sqrt{G}} = \frac{F}{\sqrt{EG}}$$

and
$$\sin\theta = \frac{|\mathbf{r}_1 \times \mathbf{r}_2|}{|\mathbf{r}_1||\mathbf{r}_2|} = \frac{H}{\sqrt{E}\sqrt{G}} = \frac{H}{\sqrt{EG}}$$

Dividing, we get
$$\tan\theta = \frac{H}{F}.$$

The equation $\tan\theta = \frac{H}{F}$ implies that the parametric curves $u = a$ and $v = b$ are orthogonal if and only if $F = 0$ at the point P, where $u = a$, $v = b$.

Also, the necessary and sufficient condition for parametric curves at each point of the surface to be orthogonal is that $F = 0$, identically.

In case $F = 0$ identically zero then we also say that the parametric curves at different points of the surface form an **orthogonal system**.

Example 5. *Show that for the surface $x = u\cos v$, $y = u\sin v$, $z = f(u)$, the parametric curves at each point on the surface are orthogonal.*

Sol. The given surface is $x = u\cos v, y = u\sin v, z = f(u)$.

Let $\mathbf{r}$ be the position vector of the point (x, y, z) on the given surface.

$$\therefore \qquad \mathbf{r} = u\cos v\,\mathbf{i} + u\sin v\,\mathbf{j} + f(u)\mathbf{k}$$

$$\therefore \qquad \mathbf{r}_1 = \frac{\partial \mathbf{r}}{\partial u} = \cos v\,\mathbf{i} + \sin v\,\mathbf{j} + f'(u)\mathbf{k}$$

and
$$\mathbf{r}_2 = \frac{\partial \mathbf{r}}{\partial v} = -u\sin v\,\mathbf{i} + u\cos v\,\mathbf{j} + 0\mathbf{k}$$

$$\begin{aligned} \therefore \qquad F &= \mathbf{r}_1 \cdot \mathbf{r}_2 \\ &= (\cos v\,\mathbf{i} + \sin v\,\mathbf{j} + f'(u)\mathbf{k}) . (-u\sin v\,\mathbf{i} + u\cos v\,\mathbf{j} + 0\mathbf{k}) \\ &= -u\cos v\sin v + u\sin v\cos v + (f'(u))(0) \\ &= 0. \end{aligned}$$

Since F is identically zero, the parametric curves at each point on the surface are orthogonal.

12. LENGTH OF ARC JOINING TWO POINTS ON A SURFACE

Let $\mathbf{r} = \mathbf{r}(u, v)$ be a surface, where u and v are independent parameters. Let P and Q be two neighbouring points on the surface with position vectors $\mathbf{r}$ and $\mathbf{r} + \delta\mathbf{r}$ corresponding to parameter values u, v and $u + \delta u$, $v + \delta v$ respectively.

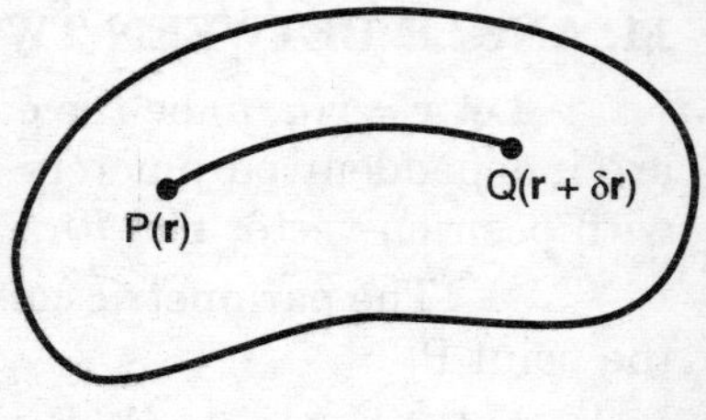

Since **r** is a function of u and v, the differential $d\mathbf{r}$ of **r** is given by

$$d\mathbf{r} = \frac{\partial \mathbf{r}}{\partial u}\,\delta u + \frac{\partial \mathbf{r}}{\partial v}\,\delta v$$

$$\Rightarrow \quad d\mathbf{r} = \mathbf{r}_1\,\delta u + \mathbf{r}_2\,\delta v$$

The actual change $\delta\mathbf{r}$ in **r** is approximately equal to $d\mathbf{r}$.

$$\therefore \quad \delta\mathbf{r} = \mathbf{r}_1\,\delta u + \mathbf{r}_2\,\delta v$$

Let ds be the length of the arc joining P and Q along any curve through P and Q. Since the points P and Q are neighbouring points, we have $ds = |\,\delta\mathbf{r}\,|$.

$$\therefore \quad (ds)^2 = |\,\delta\mathbf{r}\,|^2 = |\,d\mathbf{r}\,|\,.\,|\,d\mathbf{r}\,|$$

$$= (\mathbf{r}_1\,\delta u + \mathbf{r}_2\,\delta v)\,.\,(\mathbf{r}_1\,\delta u + \mathbf{r}_2\,\delta v)$$

$$= \mathbf{r}_1 .\, \mathbf{r}_1\,(\delta u)^2 + 2\,\mathbf{r}_1 \,.\, \mathbf{r}_2\,\delta u\,\delta v + \mathbf{r}_2 \,.\, \mathbf{r}_2\,(\delta v)^2$$

$$\therefore \quad (ds)^2 = \mathrm{E}(\delta u)^2 + 2\mathrm{F}\,\delta u\,\delta v + \mathrm{G}(\delta v)^2.$$

Since u and v being independent variables, for the function $\mathbf{r} = \mathbf{r}(u, v)$, we have

differential of $u = du = \delta u$

and differential of $v = dv = \delta v$

$$\therefore \quad (ds)^2 = \mathrm{E}(du)^2 + 2\mathrm{F}\,du\,dv + \mathrm{G}(dv)^2$$

or

$$ds = \sqrt{\mathrm{E}(du)^2 + 2\mathrm{F}\,du\,dv + \mathrm{G}(dv)^2}\,.$$

Corollary. (*i*) On the parametric curve u = constant, we have $du = \delta u = 0$.

$$\therefore \quad ds = \sqrt{\mathrm{E}(0)^2 + 2\mathrm{F}(0)dv + \mathrm{G}(dv)^2} = \sqrt{\mathrm{G}}\,dv.$$

(*ii*) On the parametric curve v = constant, we have $dv = \delta v = 0$.

$$\therefore \quad ds = \sqrt{\mathrm{E}(du)^2 + 2\mathrm{F}(du)(0) + \mathrm{G}(0)^2} = \sqrt{\mathrm{E}}\,du\,.$$

The quantity $\mathrm{E}(du)^2 + 2\mathrm{F}du\,dv + \mathrm{G}(dv)^2$ is called the **metric** or the **first fundamental form** of the surface $\mathbf{r} = \mathbf{r}(u, v)$.

Example 6. *For the surface $x = u \cos v$, $y = u \sin v$, $z = f(u)$, show that $(ds)^2 = (1 + (f'(u))^2)(du)^2 + u^2(dv)^2$, where ds is the length of arc between any two neighbouring points on the surface.*

Sol. Given surface is

$$x = u \cos v,\; y = u \sin v,\; z = f(u).$$

Let **r** be the position vector of the point (x, y, z) on the surface.

$$\therefore \quad \mathbf{r} = u \cos v\,\mathbf{i} + u \sin v\,\mathbf{j} + f(u)\mathbf{k}$$

$$\therefore \quad \mathbf{r}_1 = \frac{\partial \mathbf{r}}{\partial u} = \cos v\,\mathbf{i} + \sin v\,\mathbf{j} + f'(u)\mathbf{k}$$

and

$$\mathbf{r}_2 = \frac{\partial \mathbf{r}}{\partial v} = -\,u \sin v\,\mathbf{i} + u \cos v\,\mathbf{j} + 0\mathbf{k}$$

$$\therefore \quad \mathrm{E} = \mathbf{r}_1 \,.\, \mathbf{r}_1 = (\cos v\,\mathbf{i} + \sin v\,\mathbf{j} + f'(u)\mathbf{k})\,.\,(\cos v\,\mathbf{i} + \sin v\,\mathbf{j} + f'(u)\mathbf{k})$$

$$= \cos^2 u + \sin^2 v + (f'(u))^2 = 1 + (f'(u))^2$$

$$F = \mathbf{r}_1 \cdot \mathbf{r}_2 = (\cos v\mathbf{i} + \sin v\mathbf{j} + f'(u)\mathbf{k}) \cdot (-u \sin v\, \mathbf{i} + u \cos v\mathbf{j} + 0\mathbf{k})$$
$$= -u \cos v \sin v + u \sin v \cos v + 0 = 0$$
$$G = \mathbf{r}_2 \cdot \mathbf{r}_2 = (-u \sin v\mathbf{i} + u \cos v\mathbf{j}) \cdot (-u \sin v\mathbf{i} + u \cos v\mathbf{j})$$
$$= u^2 \sin^2 v + u^2 \cos^2 v = u^2$$
$$\therefore \quad (ds)^2 = E(du)^2 + 2F du dv + G(dv)^2$$
$$= (1 + (f'(u))^2)(du)^2 + 2(0)\, du\, dv + u^2(dv)^2$$
$$\therefore \quad (ds)^2 = (1 + (f'(u))^2)^2 (du)^2 + u^2 (dv)^2.$$

Example 7. *For the polar coordinates on the plane $z = 0$, show that $(ds)^2 = (dr)^2 + (r\, d\theta)^2$.*

Sol. Let $(r, \theta, 0)$ be the polar coordinates of a point P on the plane $z = 0$.

$\therefore$ The cartesian coordinates of the point P are $(r \cos \theta, r \sin \theta, 0)$.

Let $\mathbf{r}$ be the position vector of the point P on the surface $z = 0$.

$$\therefore \quad \mathbf{r} = r \cos \theta\mathbf{i} + r \sin \theta\mathbf{j} + 0\mathbf{k}$$
$$\therefore \quad \mathbf{r}_1 = \frac{\partial \mathbf{r}}{\partial r} = \cos \theta\mathbf{i} + \sin \theta\mathbf{j}$$

and
$$\mathbf{r}_2 = \frac{\partial \mathbf{r}}{\partial \theta} = -r \sin \theta\mathbf{i} + r \cos \theta\mathbf{j}$$
$$\therefore \quad E = \mathbf{r}_1 \cdot \mathbf{r}_1 = (\cos \theta\mathbf{i} + \sin \theta\mathbf{j}) \cdot (\cos \theta\mathbf{i} + \sin \theta\mathbf{j})$$
$$= \cos^2 \theta + \sin^2 \theta = 1$$
$$F = \mathbf{r}_1 \cdot \mathbf{r}_2 = (\cos \theta\mathbf{i} + \sin \theta\mathbf{j}) \cdot (-r \sin \theta\mathbf{i} + r \cos \theta\mathbf{j})$$
$$= -r \cos \theta \sin \theta + r \sin \theta \cos \theta = 0$$
$$G = \mathbf{r}_2 \cdot \mathbf{r}_2 = (-r \sin \theta\mathbf{i} + r \cos \theta\mathbf{j}) \cdot (-r \sin \theta\mathbf{i} + r \cos \theta\mathbf{j})$$
$$= r^2 \sin^2 \theta + r^2 \cos^2 \theta = r^2$$
$$\therefore \quad (ds)^2 = E(dr)^2 + 2F\, dr\, d\theta + G(d\theta)^2$$
$$= 1.\, (dr)^2 + 2(0)\, dr\, d\theta + r^2 (d\theta)^2$$
$$= (dr)^2 + (r\, d\theta)^2.$$

13. AREA ELEMENT ON A SURFACE

Let $\mathbf{r} = \mathbf{r}(u, v)$ be a surface, where u and v are independent parameters. Consider the figure PQRS on the surfaces with vertices with curvilinear coordinates $P(u, v)$, $Q(u + \delta u, v)$, $R(u + \delta u, v + \delta v)$ and $S(u, v + \delta v)$. Let these vertices be joined by corresponding parametric curves. Since δu and δv are small changes in u and v respectively, the figure PQRS is approximately a parallelogram.

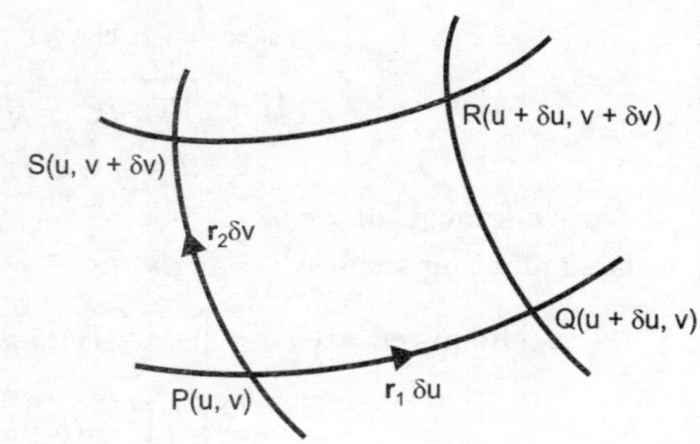

Now
$$\mathbf{PQ} = \text{P.V. of Q} - \text{P.V. of P}$$
$$= \mathbf{r}(u + \delta u, v) - \mathbf{r}(u, v)$$
$$= \left(\mathbf{r}(u,v) + \delta u \frac{\partial \mathbf{r}}{\partial u}\right) - \mathbf{r}(u, v) \qquad \text{(By neglecting other small terms)}$$
$$= \delta u \frac{\partial \mathbf{r}}{\partial u} = \mathbf{r}_1\, \delta u$$

Also $\quad$ $\mathbf{PS}$ = P.V. of S – P.V. of P = $\mathbf{r}(u, v + \delta v) - \mathbf{r}(u, v)$

$$= \left(\mathbf{r}(u,v) + \delta v \frac{\partial \mathbf{r}}{\partial v}\right) - \mathbf{r}(u, v) = \delta v \frac{\partial \mathbf{r}}{\partial v} = \mathbf{r}_2 \,\delta v.$$

Let dA represent the area of the parallelogram PQRS.

$$\therefore \quad dA = |\,\mathbf{PQ} \times \mathbf{PS}\,| = |\,\mathbf{r}_1\,\delta u \times \mathbf{r}_2\,\delta v\,|$$

$$= |\,\mathbf{r}_1 \times \mathbf{r}_2\,|\;\delta u\,\delta v = H\,\delta u\,\delta v = H\,du\,dv$$

($\because$ For independent variables u, v, we have $du = \delta u$ and $dv = \delta v$)

$\therefore$ Area element on the surface = H $du\ dv$.

Example 8. *Calculate the values of E, F, G and H and the area corresponding to the domain $0 \le u \le 2\pi$, $0 \le v \le 2\pi$ for the anchor ring:*

$$x = (b + a \cos u) \cos v, \quad y = (b + a \cos u) \sin v, \quad z = a \sin u.$$

Sol. Given surface is

$$x = (b + a \cos u) \cos v, \quad y = (b + a \cos u) \sin v, \quad z = a \sin u.$$

Let $\mathbf{r}$ be the position vector of the point (x, y, z) on the surface.

$$\therefore \quad \mathbf{r} = (b + a \cos u) \cos v\,\mathbf{i} + (b + a \cos u) \sin v\,\mathbf{j} + a \sin u\,\mathbf{k}$$

$$\therefore \quad \mathbf{r}_1 = \frac{\partial \mathbf{r}}{\partial u} = -\, a \sin u \cos v\,\mathbf{i} - a \sin u \sin v\,\mathbf{j} + a \cos u\,\mathbf{k}$$

$$\mathbf{r}_2 = \frac{\partial \mathbf{r}}{\partial v} = -\,(b + a \cos u) \sin v\,\mathbf{i} + (b + a \cos u) \cos v\,\mathbf{j} + 0\mathbf{k}$$

$$\therefore \quad E = \mathbf{r}_1 \cdot \mathbf{r}_1 = a^2 \sin^2 u \cos^2 v + a^2 \sin^2 u \sin^2 v + a^2 \cos^2 u$$

$$= a^2 \sin^2 u\,(\cos^2 v + \sin^2 v) + a^2 \cos^2 u = a^2 (\sin^2 u + \cos^2 u) = a^2$$

$$F = \mathbf{r}_1 \cdot \mathbf{r}_2 = a \sin u \cos v\,(b + a \cos u) \sin v$$

$$- a \sin u \sin v\,(b + a \cos u) \cos v + (a \cos u)(0) = 0$$

$$G = \mathbf{r}_2 \cdot \mathbf{r}_2 = (b + a \cos u)^2 \sin^2 v + (b + a \cos u)^2 \cos^2 v$$

$$= (b + a \cos u)^2$$

$$H = \sqrt{EG - F^2} = \sqrt{a^2(b + a \cos u)^2 - (0)^2}$$

$$= a(b + a \cos u).$$

We know that:

area element on surface $= H\,du\,dv$

$$\therefore \text{ Required area } = \int_0^{2\pi}\int_0^{2\pi} H\,du\,dv$$

$$= \int_0^{2\pi}\int_0^{2\pi} a(b + a \cos u)\,du\,dv$$

$$= \int_0^{2\pi} \left(a(b + a \cos u) \int_0^{2\pi} dv \right) du$$

$$= \int_0^{2\pi} a(b + a \cos u)\,2\pi\,du = 2a\pi(bu + a \sin u)\Big|_0^{2\pi}$$

$$= 2a\pi\,(2\pi b + 0) - 2a\pi(0 + 0) = 4\pi^2 ab.$$

WORKING RULES FOR SOLVING PROBLEMS

Rule I. *For the surface* $\mathbf{r} = \mathbf{r}(u, v)$, *we have*

$$\mathbf{r}_1 = \frac{\partial \mathbf{r}}{\partial u} \quad \text{and} \quad \mathbf{r}_2 = \frac{\partial \mathbf{r}}{\partial v}.$$

Rule II. *The quantities* $E = \mathbf{r}_1 . \mathbf{r}_1$, $F = \mathbf{r}_1 . \mathbf{r}_2$ *and* $G = \mathbf{r}_2 . \mathbf{r}_2$ *are called the* **fundamental magnitudes of the first order.** *E, F are always positive.*

Rule III. *We define* $H = | \mathbf{r}_1 \times \mathbf{r}_2 |$.

In terms of E, F and G, we have $H = \sqrt{EG - F^2}$.

Rule IV. *The curves on the surface* $\mathbf{r} = \mathbf{r}(u, v)$ *obtained by keeping one of the parameters constant are called* **parametric curves** *on the surface* $\mathbf{r} = \mathbf{r}(u, v)$.

Rule V. *Angle between parametric curves of a surface* $\mathbf{r} = \mathbf{r}(u, v)$ *at the point P(a, b) is given by* $\cos^{-1} \frac{F}{\sqrt{EG}} \left(\text{or } \sin^{-1} \frac{H}{\sqrt{EG}} \right)$.

EXERCISE 3.3

1. Find the values of E, F, G and H for the paraboloid $x = u, y = v, z = u^2 - v^2$, where u and v are independent parameters.
2. Calculate the fundamental magnitudes of the first order for the surface $\mathbf{r} = (u \cos v, u \sin v, f(u))$, where u and v are independent parameters.
3. Calculate the fundamental magnitudes of the first order for the surface $x = a(u + v), y = b(u - v), z = uv$, where u and v are independent parameters.
4. Calculate the fundamental magnitudes of the first order for the conicoid: $x = u \cos v, y = u \sin v, z = f(v)$ where u and v are independent parameters.
5. For the cartesian coordinates on the plane $z = 0$, show that $(ds)^2 = (dx)^2 + (dy)^2$.
6. For the sphere $\mathbf{r} = a \cos u \cos v\mathbf{i} + a \cos u \sin v\mathbf{j} + a \sin u\mathbf{k}$, find the value of the metric.

Answers

1. $E = 1 + 4u^2$, $F = -4uv$, $G = 1 + 4v^2$, $H = \sqrt{1 + 4u^2 + 4v^2}$
2. $E = 1 + (f'(u))^2$, $F = 0$, $G = u^2$
3. $E = a^2 + b^2 + v^2$, $F = a^2 - b^2 + uv$, $G = a^2 + b^2 + u^2$
4. $E = 1$, $F = 0$, $G = u^2 + f'^2$
6. $(adu)^2 + (a \cos u\, dv)^2$.

SECTION B

CALCULUS OF VARIATIONS

4

Variational Problems with Fixed Boundaries

1. INTRODUCTION

By a function we mean a correspondence between the elements of sets A and B such that to each element of A there correspond exactly one element of the set B. We have also studied the methods of finding maximum and minimum values of real functions of one or more variables. In calculus of variations, we shall deal with functionals and their maximum and minimum values. A functional associates a unique real number to each function belonging to a certain class of functions.

2. FUNCTIONAL

Let A be a class of functions. A correspondence between the functions of class A and the set of real numbers such that to each function belonging to A, there correspond exactly one real number is called a **functional**. In other words, a functional is a correspondence which assigns a unique real number to each function belonging to some class. Thus a functional is a function, where the independent variable takes functions as values.

A functional is denoted by capital letter I (or J). If $y(x)$ represents a function in the class of functions of a functional J, then we write $J = J[y(x)]$.

The class of functions $y(x)$ on which the functional $J[y(x)]$ is defined is called the **domain** of the functional J.

Illustrations. (*i*) Let $A = C_1[0, 4]$ be the class of functions $y(x)$ having continuous derivatives on $[0, 4]$.

Let $$J[y(x)] = y'(3).$$

$\therefore$ $J[y(x)]$ is a functional defined in the indicated class of functions and to each function of this class there exists a unique number, the value of the derivative of the function at '3'.

Here $$J[x^2] = (x^2)'\Big|_{x=3} = 2x\Big|_{x=3} = 2(3) = 6$$

$$J[x^4 - x + 2] = (x^4 - x + 2)'\Big|_{x=3} = (4x^3 - 1)\Big|_{x=3} = 4(3)^3 - 1 = 107$$

$$J[\log(x + 2)] = [\log(x + 2)]'\Big|_{x=3} = \frac{1}{x+2}\Big|_{x=3} = \frac{1}{5}.$$

(*ii*) Let $A = C[0, 2]$ be the class of functions $y(x)$ continuous on $[0, 2]$.

Let $$J[y(x)] = \frac{1}{3}\int_0^2 y(x)\, dx.$$

$\therefore$ J[$y(x)$] is a functional defined on C[0, 2] and to each function $y(x)$ of this class there exists a unique number given by $\frac{1}{3}\int_0^2 y(x)\,dx$.

Here
$$J[5] = \frac{1}{3}\int_0^2 5\,dx = \frac{1}{3}(5x)\Big|_0^2 = \frac{5}{3}(2-0) = \frac{10}{3}$$
$$J[x^3 + x] = \frac{1}{3}\int_0^2 (x^3 + x)dx = \left(\frac{1}{3}\left(\frac{x^4}{4} + \frac{x^2}{2}\right)\right)\Bigg|_0^2 = 2$$
$$J(\sin x) = \frac{1}{3}\int_0^2 \sin x\,dx = \frac{1}{3}(-\cos x)\Big|_0^2 = \frac{1}{3}(1 - \cos 2).$$

(*iii*) Let A = $C_1[a, b]$ be the class of functions $y(x)$ having a continuous derivative $y'(x)$ on the interval $[a, b]$. Then
$$J[y(x)] = \int_a^b \sqrt{1 + y'^2(x)}\,dx$$
is a functional defined on $C_1[a, b]$. This functional geometrically describes the arc length of the curve $y = y(x)$ with ends at the points $(a, y(a))$ and $(b, y(b))$.

(*iv*) Let A = C[– 1, 1] be the class of functions $y(x)$ continuous on the interval [– 1, 1].

Let
$$J[y(x)] = \int_{-1}^1 F(x, y(x))\,dx, \text{ where } F(x, y(x)) = \frac{x}{1 + (y(x))^2}.$$

Here J[$y(x)$] is a functional defined on C [– 1, 1].

For $y(x) = 4x$, we have
$$J[y(x)] = \int_{-1}^1 \frac{x}{1 + (4x)^2}\,dx = \int_{-1}^1 \frac{x}{1 + 16x^2}\,dx = 0.$$
($\because$ The integrand is an odd function)

3. CLOSENESS OF FUNCTIONS

Let $y_1(x)$ and $y_2(x)$ be any two functions.

$y_1(x)$ and $y_2(x)$ are said to be **close in the sense of zero-order proximity** if $|\, y_1(x) - y_2(x) \,|$ is small for values of x for which $y_1(x)$ and $y_2(x)$ are defined. Geometrically, this means that the curves of $y = y_1(x)$ and $y = y_2(x)$ are close as to the ordinates.

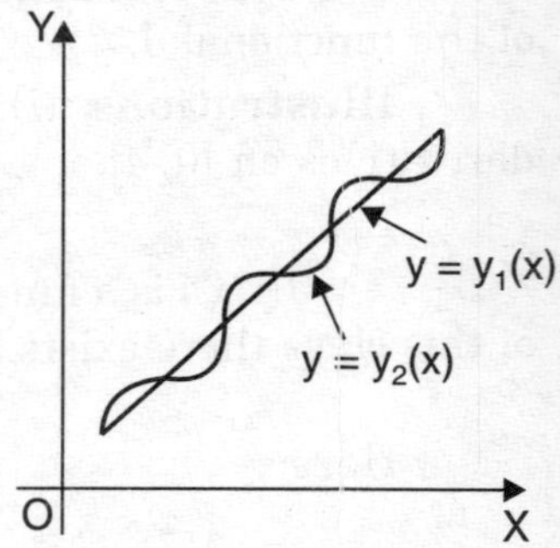

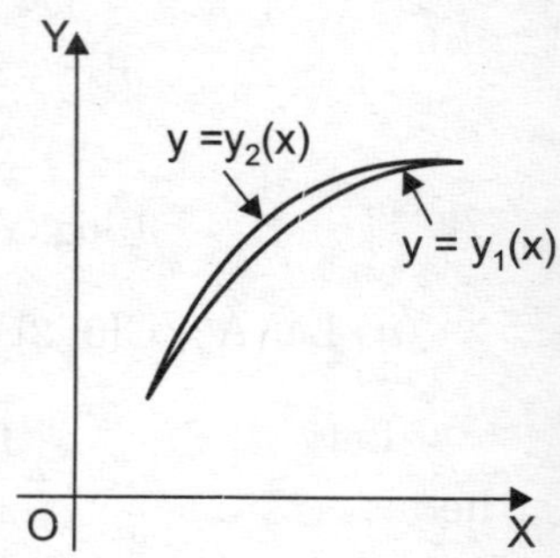

The functions $y_1(x)$ and $y_2(x)$ are said to be **close in the sense of first-order proximity** if $|\, y_1(x) - y_2(x) \,|$ and $|\, y_1'(x) - y_2'(x) \,|$ are small for values of x for which these functions are defined. Geometrically, this means that the curves of $y = y_1(x)$ and $y = y_2(x)$ are close both as to ordinates and as to directions of the tangents at the appropriate points.

The functions $y_1(x)$ and $y_2(x)$ are said to be **close in the sense of nth order proximity** if $|y_1(x)-y_2(x)|$, $|y_1'(x)-y_2'(x)|$, ..., $|y_1^{(n)}(x)-y_2^{(n)}(x)|$ are small for values of x for which these functions are defined.

From the above definitions it is clear that if two functions *i.e.,* curves are close in the sense of nth order proximity then they are also close in the sense of proximity of any smaller order.

4. MAXIMUM AND MINIMUM VALUES OF A FUNCTIONAL

A functional $J[y(x)]$ is said to have a **maximum value** on $y = y_0(x)$ if the value of the functional on any curve close to $y = y_0(x)$ is not greater than $J[y_0(x)]$ *i.e.,* $J[y(x)] \le J[y_0(x)]$ for all curves $y(x)$ close to $y_0(x)$.

A functional $J[y(x)]$ is said to have a **minimum value** on $y = y_0(x)$ if the values of the functional on any curve close to $y = y_0(x)$ is not less than $J[y_0(x)]$ *i.e.,* $J[y(x)] \ge J[y_0(x)]$ for all curves $y(x)$ close to $y_0(x)$.

The maximum and minimum values of a functional are called its **extremum values.**

If a functional $J[y(x)]$ attains a maximum (resp. minimum) on $y = y_0(x)$ with respect to all curves $y = y(x)$ in the sense of zero order proximity *i.e.,* $|y(x) - y_0(x)|$ is small, then the maximum (resp. minimum) is said to be **strong.**

Similarly, if a functional $J[y(x)]$ attains a maximum (resp. minimum) on $y = y_0(x)$ with respect to all curves $y = y(x)$ in the sense of first-order proximity *i.e.,* $|y(x) - y_0(x)|$ and $|y'(x) - y'_0(x)|$ are both small, then the maximum (resp. minimum) is said to be **weak.**

Let a functional $J[y(x)]$ attains a strong maximum on $y = y_0(x)$.

$\therefore$ $J[y(x)] \le J[y_0(x)]$ for all functions $y(x)$ such that $|y(x) - y_0(x)|$ is small.

$\therefore$ $J[y(x)] \le J[y_0(x)]$ is also true for functions $y(x)$ for which $|y(x) - y_0(x)|$ is small and at the same time $|y'(x) - y_0'(x)|$ is also small.

$\therefore$ $y_0(x)$ is a weak maximum of the functional $J[y(x)]$.

Similarly a strong minimum of a functional is also a weak minimum of that functional.

Remark. If a strong maximum (resp. minimum) of a functional $J[y(x)]$ is attained on the curve $y = y_0(x)$, then a weak maximum (resp. minimum) is also attained on the same curve. In other words, every strong extremum value is at the same time a weak extremum value. This does not hold conversely.

5. EULER'S EQUATION FOR FUNCTIONALS CONTAINING FIRST ORDER DERIVATIVE AND ONE INDEPENDENT VARIABLE

Theorem. *Let* $J[y(x)] = \int_{x_1}^{x_2} F(x, y(x), y'(x))\, dx$ *be a functional defined on the set of functions* $y(x)$ *which admits of continuous first order derivative and satisfy the boundary conditions :* $y(x_1) = y_1$ *and* $y(x_2) = y_2$, *where* y_1 *and* y_2 *are prescribed at the fixed boundary points* x_1 *and* x_2. *Also, F is differentiable three times with respect to all its arguments.*

If the functional J[y(x)] has an extremum on a function y(x) then y(x) satisfy the differential equation $\dfrac{\partial f}{\partial y} - \dfrac{d}{dx}\left(\dfrac{\partial f}{\partial y'}\right) = 0.$

Proof. We have $\qquad J[y(x)] = \int_{x_1}^{x_2} F(x, y(x)\, y'(x))\, dx.$

Let $y = y(x)$ be a function which extremise $J[y(x)]$ and whose curve is passing through the points $A(x_1, y_1)$ and $B(x_2, y_2)$ *i.e.,* $y(x_1) = y_1$ and $y(x_2) = y_2$.

Let $\eta(x)$ be any arbitrary but fixed function such that $\eta''(x)$ is continuous and $\eta(x_1) = \eta(x_2) = 0$.

Let $\quad \bar{y}(x) = y(x) + \alpha\eta(x)$, where α is a small parameter.

$\therefore$ $\bar{y}(x)$ admits of continuous first order derivative.

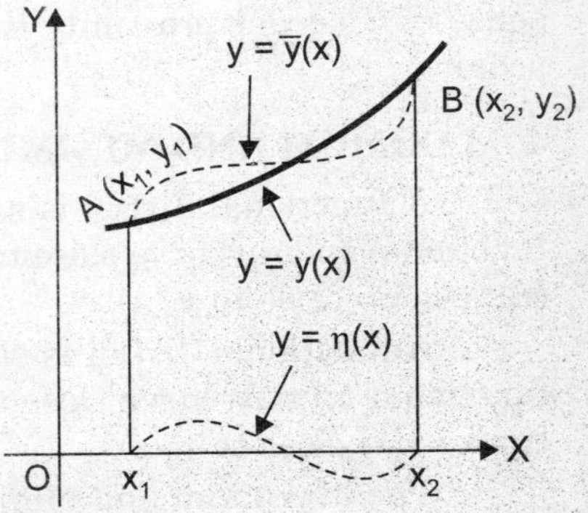

Now $\quad \bar{y}(x_1) = y(x_1) + \alpha\eta(x_1) = y_1 + \alpha(0) = y_1$

and $\quad \bar{y}(x_2) = y(x_2) + \alpha\eta(x_2) = y_2 + \alpha(0) = y_2$

$\therefore$ $\bar{y}(x)$ also satisfy the boundary conditions.

The functional J is defined at $\bar{y}(x)$.

$$\therefore \quad J[\bar{y}(x)] = \int_{x_1}^{x_2} F(x, \bar{y}(x), \bar{y}'(x))\, dx$$

$$= \int_{x_1}^{x_2} F(x, y(x) + \alpha\eta(x), y'(x) + \alpha\eta'(x))\, dx$$

On the curves of the family $y(x) + \alpha\eta(x)$, the right hand side is a function of α, say $\psi(\alpha)$.

$$\therefore \quad \psi(\alpha) = \int_{x_1}^{x_2} F(x, \bar{y}(x), \bar{y}'(x))\, dx = \int_{x_1}^{x_2} F(x, y(x) + \alpha\eta(x), y'(x) + \alpha\eta'(x))\, dx$$

Since the extremising curve $y = y(x)$ corresponds to $\alpha = 0$, it follows that $\psi(\alpha)$ is extremised for $\alpha = 0$.

$$\therefore \quad \frac{d\psi}{d\alpha} = 0 \text{ at } \alpha = 0$$

Differentiating $\psi(\alpha)$ under the integral sign, we have

$$\frac{d\psi}{d\alpha} = \int_{x_1}^{x_2} \frac{\partial}{\partial \alpha} F(x, \bar{y}(x), \bar{y}'(x))\, dx$$

$$= \int_{x_1}^{x_2} \left(\frac{\partial F}{\partial x}\frac{\partial x}{\partial \alpha} + \frac{\partial F}{\partial \bar{y}}\frac{\partial \bar{y}}{\partial \alpha} + \frac{\partial F}{\partial \bar{y}'}\frac{\partial \bar{y}'}{\partial \alpha} \right) dx$$

$$= \int_{x_1}^{x_2} \left(\frac{\partial F}{\partial x}.0 + \frac{\partial F}{\partial \bar{y}}\eta(x) + \frac{\partial F}{\partial \bar{y}'}\eta'(x) \right) dx$$

$$\therefore \quad \frac{d\psi}{d\alpha} = \int_{x_1}^{x_2} \left(\frac{\partial F}{\partial \bar{y}}\eta(x) + \frac{\partial F}{\partial \bar{y}'}\eta'(x) \right) dx \qquad \text{...(1)}$$

Now $\quad \dfrac{d\psi}{d\alpha} = 0 \quad$ at $\quad \alpha = 0$.

$\therefore$ Putting $\alpha = 0$ in (1), we get

$$\int_{x_1}^{x_2} \left(\frac{\partial F}{\partial y}\eta(x) + \frac{\partial F}{\partial y'}\eta'(x) \right) dx = 0$$

$$\Rightarrow \quad \int_{x_1}^{x_2} \frac{\partial F}{\partial y}\eta(x)\, dx + \int_{x_1}^{x_2} \frac{\partial F}{\partial y'}\eta'(x)\, dx = 0$$

$$\Rightarrow \quad \int_{x_1}^{x_2} \frac{\partial F}{\partial y}\eta(x)\, dx + \left[\frac{\partial F}{\partial y'}.\eta(x) \right]_{x_1}^{x_2} - \int_{x_1}^{x_2} \frac{d}{dx}\left(\frac{\partial F}{\partial y'} \right).\eta(x)\, dx = 0$$

$$\Rightarrow \quad \int_{x_1}^{x_2} \frac{\partial F}{\partial y}\eta(x)\, dx - \int_{x_1}^{x_2} \frac{d}{dx}\left(\frac{\partial F}{\partial y'} \right)\eta(x)\, dx = 0 \qquad (\because \eta(x_1) = \eta(x_2) = 0)$$

$$\Rightarrow \qquad \int_{x_1}^{x_2} \left[\frac{\partial F}{\partial y} - \frac{d}{dx}\left(\frac{\partial F}{\partial y'}\right)\right] \eta(x)\, dx = 0$$

This holds for any arbitrary continuous function $\eta(x)$.

$\therefore$ We have $$\frac{\partial F}{\partial y} - \frac{d}{dx}\left(\frac{\partial F}{\partial y'}\right) = 0.$$

$\therefore$ The functions $y(x)$ satisfies the equation $\mathbf{\dfrac{\partial F}{\partial y} - \dfrac{d}{dx}\left(\dfrac{\partial F}{\partial y'}\right) = 0.}$

This equation is known as the **Euler's equation.**

Remarks 1. If $y(x)$ is a function extremising the functional $\int_{x_1}^{x_2} F(x, y(x), y'(x))dx$, then $y(x)$ satisfies the Euler's equation. But if a function satisfies the Euler's equation then it may or may not extremise the functional. The solutions of Euler's equation satisfying the boundary conditions are called **stationary functions.** The value of the functional at a stationary function is called a **stationary value** of the functional.

2. In the above theorem, we require the extremum $y(x)$ to be *weak.* Since any strong extremum is at the same time a weak extremum, it follow that any condition necessary for a weak extremum is also necessary for a strong extremum.

6. EXTREMALS

The solutions of Euler's equation which are unrestricted by the boundary conditions are called **extremals.**

The Euler's equation for the functional $\int_{x_1}^{x_2} F(x, y, y')\, dx$ is

$$\frac{\partial F}{\partial y} - \frac{d}{dx}\left(\frac{\partial F}{\partial y'}\right) = 0 \qquad \text{...(1)}$$

$$(1) \Rightarrow \quad \frac{\partial F}{\partial y} - \left[\frac{\partial}{\partial x}\left(\frac{\partial F}{\partial y'}\right) + \frac{\partial}{\partial y}\left(\frac{\partial F}{\partial y'}\right)\frac{dy}{dx} + \frac{\partial}{\partial y'}\left(\frac{\partial F}{\partial y'}\right)\frac{dy'}{dx}\right] = 0$$

$$\Rightarrow \quad F_y - \left[F_{y'x} + F_{y'y}\frac{dy}{dx} + F_{y'y'}\frac{d^2y}{dx^2}\right] = 0$$

$$\Rightarrow \quad \mathbf{F_{y'y'}\frac{d^2y}{dx^2} + F_{y'y}\frac{dy}{dx} + (F_{y'x} - F_y) = 0.}$$

$\therefore$ The Euler's equation is a second order differential equation unless, $F_{y'y'} = 0$.

$\therefore$ In general, the extremals constitute a two-parameter family of curves. The extremals which satisfies the given boundary conditions are the stationary functions of the functional $\int_{x_1}^{x_2} F(x, y, y')\, dx$.

7. EXTREMALS IN PARTICULAR CASES

The Euler's equation for the functional $\int_{x_1}^{x_2} F(x, y, y')\, dx$ is $F_y - \dfrac{d}{dx}(F_{y'}) = 0$.

Case I. If x is missing from the function 'F' *i.e.,* the functional is of the form $\int_{x_1}^{x_2} F(y, y')\, dx$, then $F_x = 0$.

We have

$$\frac{d}{dx}\left(F - F_{y'}\frac{dy}{dx}\right) = \frac{d}{dx}(F) - \frac{d}{dx}\left(F_{y'}\frac{dy}{dx}\right)$$

$$= \left(F_x + F_y\frac{dy}{dx} + F_{y'}\frac{dy'}{dx}\right) - \left(F_{y'}\frac{d^2y}{dx^2} + \frac{dy}{dx}\frac{d}{dx}(F_{y'})\right)$$

$$= 0 + F_y\frac{dy}{dx} - \frac{dy}{dx}\frac{d}{dx}(F_{y'}) = \frac{dy}{dx}\left(F_y - \frac{d}{dx}(F_{y'})\right) = \frac{dy}{dx}\,.\,0 = 0.$$

$$\therefore \quad \frac{d}{dx}\left(F - F_{y'}\frac{dy}{dx}\right) = 0$$

Integrating, we get $\mathbf{F - F_{y'}\frac{dy}{dx} = c.}$

This is a first order differential equation. This equation is solved to get the extremals.

Case II. If y is missing from the function 'F' *i.e.*, the functional is of the form $\int_{x_1}^{x_2} F(x, y')\,dx$, then $F_y = 0$.

$\therefore$ Euler's equation reduces to $0 - \frac{d}{dx}(F_{y'}) = 0$. Integrating, we get $F_{y'} = c$.

On solving this, we get an equation of the form $\mathbf{y' = \phi(x, c)}$.

This differential equation is solved to get the extremals.

Case III. If y' is missing from the function 'F' *i.e.*, the functional is of the form $\int_{x_1}^{x_2} F(x, y)\,dx$, then $F_{y'} = 0$.

$\therefore$ Euler's equation reduces to $F_y - \frac{d}{dx}(0) = 0$.

$\Rightarrow$ $F_y = 0$. This is not a differential equation. This finite equation, when solved for y, does not involve any arbitrary constant. Thus, in general, it is not possible to find y satisfying the boundary conditions $y(x_1) = y_1$ and $y(x_2) = y_2$ and as such this variational problem does not, in general, admit of a solution.

Case IV. If x and y are both missing from the function 'F' *i.e.*, the functional is of the form $\int_{x_1}^{x_2} F(y')\,dx$, then, $F_x = 0$ and $F_y = 0$.

$\therefore$ Euler's equation reduces to $0 - \frac{d}{dx}(F_{y'}) = 0$.

$$\Rightarrow \quad \frac{d}{dx}(F_{y'}) = 0 \Rightarrow F_{y'x} + F_{y'y}\frac{dy}{dx} + F_{y'y'}\frac{dy'}{dx} = 0$$

$$\Rightarrow \quad F_{xy'} + F_{yy'}\frac{dy}{dx} + F_{y'y'}\frac{d^2y}{dx^2} = 0$$

$$\Rightarrow \quad \frac{\partial}{\partial y'}(F_x) + \frac{\partial}{\partial y'}(F_y)\,.\,\frac{dy}{dx} + F_{y'y'}\frac{d^2y}{dx^2} = 0$$

$$\Rightarrow \qquad 0 + 0 \,.\, \frac{dy}{dx} + F_{y'y'}\frac{d^2y}{dx^2} = 0 \;\Rightarrow\; F_{y'y'}\frac{d^2y}{dx^2} = 0.$$

If $F_{y'y'} \neq 0$, then we have $\dfrac{d^2y}{dx^2} = 0.$

$\Rightarrow$ **$y = c_1x + c_2$.** $\therefore$ The extremals are all non-vertical straight lines.

WORKING RULES FOR SOLVING PROBLEMS

Step I. *Identify the functional $\int_{x_1}^{x_2} F(x, y, y')\,dx$ and the boundary conditions $y(x_1) = y_1$, $y(x_2) = y_2$ in the given problem.*

Step II. *Write the Euler's equation $\dfrac{\partial F}{\partial y} - \dfrac{d}{dx}\left(\dfrac{\partial F}{\partial y'}\right) = 0$ and solve this equation.*

Step III. *The solutions of this equation are extremals of the given problem.*

Step IV. *Find the extremal that satisfy the given boundary conditions. This particular extremal is the stationary function of the given problem.*

Example 1. *Find the extremals of the functional* $\int_0^{\pi} (y'^2 - y^2 + 4y\cos x)\,dx$.

Sol. Let $\qquad J[y(x)] = \int_0^{\pi} (y'^2 - y^2 + 4y\cos x)\,dx.$

Let $\qquad F(x, y, y') = y'^2 - y^2 + 4y\cos x$

The **Euler's equation** is $\dfrac{\partial F}{\partial y} - \dfrac{d}{dx}\left(\dfrac{\partial F}{\partial y'}\right) = 0.$ ...(1)

Here $\qquad \dfrac{\partial F}{\partial y} = -2y + 4\cos x, \quad \dfrac{\partial F}{\partial y'} = 2y'$

$\therefore$ (1) $\Rightarrow \quad -2y + 4\cos x - \dfrac{d}{dx}(2y') = 0$

$$\Rightarrow \qquad \frac{d^2y}{dx^2} + y = 2\cos x$$

The A.E. is $\quad D^2 + 1 = 0. \quad \therefore \quad D = \pm i$

$\therefore \qquad$ C.F. $= e^{0x}(c_1\cos 1.x + c_2\sin 1.x) = c_1\cos x + c_2\sin x$

$$\text{P.I.} = \frac{1}{D^2+1}(2\cos x) = 2x \,.\, \frac{1}{2D}\cos x = x\sin x$$

$\therefore \qquad$ **$y = c_1\cos x + c_2\sin x + x\sin x.$**

This is the equation of the extremals.

Example 2. *Find the stationary function of the functional $\int_0^4 (xy' - y'^2)dx$ which is determined by the boundary conditions $y(0) = 0$ and $y(4) = 3$.*

Sol. Given functional is $\int_0^4 (xy' - y'^2)\,dx$

Let $\quad F(x, y, y') = xy' - y'^2$

The **Euler's equation** is $\frac{\partial F}{\partial y} - \frac{d}{dx}\left(\frac{\partial F}{\partial y'}\right) = 0$ ...(1)

Here $\frac{\partial F}{\partial y} = 0$, $\frac{\partial F}{\partial y'} = x - 2y'$

$\therefore$ (1) $\Rightarrow$ $0 - \frac{d}{dx}(x - 2y') = 0 \Rightarrow -1 + 2\frac{d^2y}{dx^2} = 0$

$\Rightarrow$ $\frac{d^2y}{dx^2} = \frac{1}{2} \Rightarrow \frac{dy}{dx} = \frac{1}{2}x + c_1 \Rightarrow y = \frac{x^2}{4} + c_1x + c_2.$

The boundary conditions are $y(0) = 0$, $y(4) = 3$.

$y(0) = 0 \Rightarrow 0 + c_1(0) + c_2 = 0 \Rightarrow c_2 = 0$

$y(4) = 3 \Rightarrow 4 + 4c_1 + c_2 = 3 \Rightarrow 4c_1 = 3 - c_2 - 4 = -1 \Rightarrow c_1 = -1/4$

$\therefore$ $y = \frac{x^2}{4} + \left(\frac{-1}{4}\right)x + 0$ *i.e.*, $\mathbf{y = \frac{x^2}{4} - \frac{x}{4}}.$

This is the stationary function of the given functional.

Remark. The extremals of the given functional are given by two-parameter family of curves $y = \frac{x^2}{4} + c_1x + c_2.$

Example 3. *Find the curve on which the functional* $\int_0^1 (y'^2 + 12xy)dx$ *with boundary conditions y(0) = 0 and y(1) = 1 can be extremised.*

Sol. Given functional is $\int_0^1 (y'^2 + 12xy)dx$

Let $F(x, y, y') = y'^2 + 12xy$

The **Euler's equation** is $F_y - \frac{d}{dx}F_{y'} = 0$ ...(1)

Here $F_y = 12x$, $F_{y'} = 2y'$

$\therefore$ (1) $\Rightarrow$ $12x - \frac{d}{dx}(2y') = 0 \Rightarrow \frac{d^2y}{dx^2} = 6x$

$\Rightarrow$ $\frac{dy}{dx} = 3x^2 + c_1 \Rightarrow y = x^3 + c_1x + c_2$ ...(2)

This is the equation of the extremals. The boundary conditions are $y(0) = 0$, $y(1) = 1$.

$y(0) = 0 \Rightarrow 0 + c_1(0) + c_2 = 0 \Rightarrow c_2 = 0$

$y(1) = 1 \Rightarrow 1 + c_1(1) + c_2 = 1 \Rightarrow 1 + c_1 + 0 = 1 \Rightarrow c_1 = 0$

$\therefore \Rightarrow$ (2) $\Rightarrow$ $y = x^3 + 0.x + 0$ *i.e.*, $y = x^3$

This is the stationary function of the given variational problem.

$\therefore$ The given functional with given boundary conditions can be extremised on the function $\mathbf{y = x^3}$.

Example 4. *Show that the functional* $\int_0^1 (xy + y^2 - 2y^2y')\,dx$, *y(0) = 1, y(1) = 2 cannot have any stationary function.*

Sol. Given functional is $\int_0^1 (xy + y^2 - 2y^2 y')\, dx$

Let $\quad F(x, y, y') = xy + y^2 - 2y^2y'$

The **Euler's equation** is $\dfrac{\partial F}{\partial y} - \dfrac{d}{dx}\left(\dfrac{\partial F}{\partial y'}\right) = 0 \quad \text{...(1)}$

Here $\dfrac{\partial F}{\partial y} = x + 2y - 4yy'$, $\quad \dfrac{\partial F}{\partial y'} = -2y^2$

$\therefore$ (1) $\Rightarrow$ $(x + 2y - 4yy') - \dfrac{d}{dx}(-2y^2) = 0$

$\Rightarrow$ $\quad x + 2y - 4yy' + 4yy' = 0 \Rightarrow x + 2y = 0.$

$\therefore$ There is only one extremal of the given functional.

$y(0) = 1 \Rightarrow 0 + 2(1) = 0$, which is not true.

$\therefore$ There is no stationary function of the given problem.

Example 5. *Find the extremals of the functional* $J[y(x)] = \int_0^{2\pi} (y'^2 - y^2)\, dx$ *that satisfy the boundary conditions* $y(0) = 1$, $y(2\pi) = 1$.

Sol. We have $\quad J[y(x)] = \int_0^{2\pi} (y'^2 - y^2)\, dx$

Let $\quad F(x, y, y') = y'^2 - y^2$

The **Euler's equation** is $\dfrac{\partial F}{\partial y} - \dfrac{d}{dx}\left(\dfrac{\partial F}{\partial y'}\right) = 0 \quad \text{...(1)}$

Here $\dfrac{\partial F}{\partial y} = -2y$, $\quad \dfrac{\partial F}{\partial y'} = 2y'$

$\therefore$ (1) $\Rightarrow$ $-2y - \dfrac{d}{dx}(2y') = 0 \Rightarrow \dfrac{d^2y}{dx^2} + y = 0$

The A.E. is $D^2 + 1 = 0$. $\therefore$ $D = 0 \pm i$

$\therefore$ $\quad y = e^{0.x}(c_1 \cos 1.x + c_2 \sin 1.x)$ *i.e.,* $y = c_1 \cos x + c_2 \sin x \quad \text{...(2)}$

This is the equation of the extremals.

The boundary conditions are $y(0) = 1$, $y(2\pi) = 1$.

$y(0) = 1 \Rightarrow c_1 \cos 0 + c_2 \sin 0 = 1 \Rightarrow c_1 = 1$

$y(2\pi) = 1 \Rightarrow c_1 \cos 2\pi + c_2 \sin 2\pi = 1 \Rightarrow c_1 = 1$

$\therefore$ (2) $\Rightarrow$ $y = 1.\cos x + c_2 \sin x$ *i.e.,* $\mathbf{y = \cos x + c_2 \sin x.}$

These are the extremals satisfying given boundary conditions. Here c_2 is arbitrary constant.

Example 6. *Find the extremals and the stationary function of the functional* $\int_1^e (xy'^2 + yy')\, dx$; $y(1) = 0$, $y(e) = 1$.

Sol. Given functional is $\int_1^e (xy'^2 + yy')\, dx$

Let $\quad F(x, y, y') = xy'^2 + yy'$

The **Euler's equation** is $\frac{\partial F}{\partial y} - \frac{d}{dx}\left(\frac{\partial F}{\partial y'}\right) = 0$ ...(1)

Here $\frac{\partial F}{\partial y} = y'$, $\frac{\partial F}{\partial y'} = 2xy' + y$

$\therefore$ (1) $\Rightarrow$ $y' - \frac{d}{dx}(2xy' + y) = 0 \Rightarrow y' - 2(xy'' + 1 \,.\, y') - y' = 0 \Rightarrow xy'' + y' = 0$

$\Rightarrow \quad x\frac{d^2y}{dx^2} + \frac{dy}{dx} = 0 \Rightarrow x\frac{dz}{dx} + z = 0$, where $z = \frac{dy}{dx}$

$\Rightarrow \quad \frac{dz}{z} + \frac{dx}{x} = 0 \Rightarrow \log z + \log x = \log c_1$

$\Rightarrow \quad zx = c_1 \Rightarrow x\frac{dy}{dx} = c_1 \Rightarrow dy = \frac{c_1}{x}dx$

$\Rightarrow \quad \mathbf{y = c_1 \log x + c_2}$. ...(2)

These are the extremals. Here c_1, c_2 are arbitrary constants.

The boundary conditions are $y(1) = 0$, $y(e) = 1$.

$y(1) = 0 \Rightarrow c_1 \log 1 + c_2 = 0 \Rightarrow 0 + c_2 = 0 \Rightarrow c_2 = 0$

$y(e) = 1 \Rightarrow c_1 \log e + c_2 = 1 \Rightarrow c_1 + 0 = 1 \Rightarrow c_1 = 1$.

$\therefore$ (2) $\Rightarrow$ $y = 1 \,.\, \log x + 0$ *i.e.*, $\mathbf{y = \log x}$.

This is the stationary function.

Example 7. *Find the extremal of the functional $J[y(x)] = \int_0^1 yy'^2\, dx$ that satisfy the boundary conditions $y(0) = 1$, $y(1) = \sqrt[3]{4}$.*

Sol. We have $\quad J[y(x)] = \int_0^1 yy'^2\, dx$

Let $\quad F(x, y, y') = yy'^2$

The **Euler's equation** is $\frac{\partial F}{\partial y} - \frac{d}{dx}\left(\frac{\partial F}{\partial y'}\right) = 0$ ...(1)

Here $\frac{\partial F}{\partial y} = y'^2$, $\frac{\partial F}{\partial y'} = 2yy'$

$\therefore$ (1) $\Rightarrow$ $y'^2 - \frac{d}{dx}(2yy') = 0 \Rightarrow y'^2 - 2(yy'' + y'y') = 0 \Rightarrow -2yy'' - y'^2 = 0$

$\Rightarrow y'^2 + 2yy'' = 0$

Multiplying by y', we get $y'^3 + 2yy'y'' = 0$.

$\Rightarrow \quad y'(y'^2) + y(2y'y'') = 0 \Rightarrow \frac{d}{dx}(yy'^2) = 0$

$\Rightarrow \quad yy'^2 = c_1 \Rightarrow y' = \pm\frac{\sqrt{c_1}}{\sqrt{y}} \Rightarrow \frac{dy}{dx} = \frac{c_2}{\sqrt{y}}$ (Taking $c_2 = \pm\sqrt{c_1}$)

$\Rightarrow \quad \sqrt{y}\, dy = c_2 dx \Rightarrow \frac{2y^{3/2}}{3} = c_2 x + c_3$ ...(2)

This is the equation of the extremals. The boundary conditions are $y(0) = 1$, $y(1) = \sqrt[3]{4}$.

$y(0) = 1 \;\Rightarrow\; \dfrac{2(1)^{3/2}}{3} = c_2(0) + c_3 \qquad \Rightarrow\; c_3 = \dfrac{2}{3}$

$y(1) = 4^{1/3} = \dfrac{2(4^{1/3})^{3/2}}{3} = c_2(1) + c_3 \qquad \Rightarrow\; c_2 + c_3 = \dfrac{4}{3} \;\Rightarrow\; c_2 + \dfrac{2}{3} = \dfrac{4}{3} \;\Rightarrow\; c_2 = \dfrac{2}{3}$

$\therefore$ (2) $\Rightarrow \qquad \dfrac{2y^{3/2}}{3} = \dfrac{2}{3}x + \dfrac{2}{3}$

or $\qquad y^{3/2} = x + 1$ or $\mathbf{y = (x + 1)^{2/3}}$.

This is the required extremal.

Alternative method of finding extremals. Since x is missing in $F(x, y, y')$, the Euler's equation reduces to $F - F_{y'} \dfrac{dy}{dx} = c$.

$\Rightarrow \qquad yy'^2 - (2yy')y' = c$

$\Rightarrow \qquad yy'^2 = -c \;\Rightarrow\; y' = \pm \dfrac{\sqrt{-c}}{\sqrt{y}} = \dfrac{k_1}{\sqrt{y}} \qquad$ (Putting $k_1 = \pm\sqrt{-c}$)

$\Rightarrow \qquad \sqrt{y}\, dy = k dx \;\Rightarrow\; \dfrac{2y^{3/2}}{3} = k_1 x + k_2.$

$\therefore$ We get the same family of curves as extremals.

Example 8. *Find the extremals and the stationary function of the functional* $J[y(x)] = \int_0^1 (y'^2 + 4y^2)\, dx$ *subject to the boundary conditions* $y(0) = e^2$, $y(1) = 1$.

Sol. We have $\qquad J[y(x)] = \int_0^1 (y'^2 + 4y^2)\, dx.$

Let $\qquad F(x, y, y') = y'^2 + 4y^2$

The **Euler's equation** is $\dfrac{\partial F}{\partial y} - \dfrac{d}{dx}\left(\dfrac{\partial F}{\partial y'}\right) = 0 \qquad$ ...(1)

Here $\dfrac{\partial F}{\partial y} = 8y$, $\dfrac{\partial F}{\partial y'} = 2y'$

$\therefore$ (1) $\Rightarrow \qquad 8y - \dfrac{d}{dx}(2y') = 0 \qquad \Rightarrow\; \dfrac{d^2y}{dx^2} - 4y = 0$

$\therefore$ The A.E. is $\qquad D^2 - 4 = 0. \qquad \therefore\; D = \pm 2$

$\therefore \qquad \mathbf{y = c_1 e^{2x} + c_2 e^{-2x}} \qquad$ ...(2)

This is the equation of the extremals.

The boundary conditions are $y(0) = e^2$ and $y(1) = 1$.

$y(0) = e^2 \;\Rightarrow\; c_1 e^0 + c_2 e^0 = e^2 \;\Rightarrow\; c_1 + c_2 = e^2$

$y(1) = 1 \;\Rightarrow\; c_1 e^2 + c_2 e^{-2} = 1 \;\Rightarrow\; c_1 e^2 + c_2 e^{-2} = 1$

Solving, we get $\qquad c_1 = 0,\; c_2 = e^2$

$\therefore$ (2) $\Rightarrow \qquad y = 0.e^{2x} + e^2.e^{-2x}$ *i.e.*, $\mathbf{y = e^{2(1-x)}}$.

This gives the stationary function.

Example 9. *Show that the general solution of the Euler's equation for the functional*

$$\int_a^b \frac{1}{y}\sqrt{1+\left(\frac{dy}{dx}\right)^2}\,dx \text{ is } (x-h)^2+y^2=k^2.$$

Sol. Given functional is $\int_a^b \frac{\sqrt{1+y'^2}}{y}\,dx$.

Let $$F(x, y, y') = \frac{\sqrt{1+y'^2}}{y}$$

The **Euler's equation** is $F_y - \frac{d}{dx}F_{y'} = 0$ ...(1)

Since x is missing in $F(x, y, y')$, the Euler's equation reduces to

$$F - F_{y'}\frac{dy}{dx} = c \qquad \text{...(2)}$$

Here $$F_{y'} = \frac{y'}{y\sqrt{1+y'^2}}$$

$$\therefore \quad (2) \Rightarrow \frac{\sqrt{1+y'^2}}{y} - \frac{y'}{y\sqrt{1+y'^2}}\,y' = c$$

$$\Rightarrow \frac{1+y'^2-y'^2}{y\sqrt{1+y'^2}} = c \qquad \Rightarrow \frac{1}{y\sqrt{1+y'^2}} = c$$

$$\Rightarrow y^2(1+y'^2) = c_1 \qquad \Rightarrow y' = \sqrt{\frac{c_1}{y^2}-1} = \frac{\sqrt{c_1-y^2}}{y}$$

$$\left(\text{Putting } c_1 = \frac{1}{c^2} \quad \therefore \quad c_1 > 0\right)$$

$$\Rightarrow \frac{y}{\sqrt{c_1-y^2}}\,dy = dx \qquad \Rightarrow -\frac{1}{2}\cdot\frac{(c_1-y^2)^{1/2}}{1/2} = x + c_2$$

$$\Rightarrow -\sqrt{c_1-y^2} = x + c_2 \qquad \Rightarrow c_1 - y^2 = (x+c_2)^2$$

$$\Rightarrow (x+c_2)^2 + y^2 = c_1$$

$$\Rightarrow \mathbf{(x-h)^2 + y^2 = k^2.} \qquad (\text{Putting } h = -c_2,\ k^2 = c_1)$$

This is the general solution of the Euler's equation of the given functional.

Example 10. *Find the extremals of the functional* $J[y(x)] = \int_{x_1}^{x_2} \frac{\sqrt{1+y'^2}}{x}\,dx$.

Sol. We have $$J[y(x)] = \int_{x_1}^{x_2} \frac{\sqrt{1+y'^2}}{x}\,dx.$$

Let $$F(x, y, y') = \frac{\sqrt{1+y'^2}}{x}$$

The **Euler's equation** is $\frac{\partial F}{\partial y} - \frac{d}{dx}\left(\frac{\partial F}{\partial y'}\right) = 0$ ...(1)

Here $\dfrac{\partial F}{\partial y} = 0$, $\dfrac{\partial F}{\partial y'} = \dfrac{1}{x} \cdot \dfrac{2y'}{2\sqrt{1+y'^2}} = \dfrac{y'}{x\sqrt{1+y'^2}}$

$$\therefore \ (1) \Rightarrow \quad 0 - \frac{d}{dx}\left(\frac{y'}{x\sqrt{1+y'^2}}\right) = 0 \Rightarrow \frac{y'}{x\sqrt{1+y'^2}} = c_1$$

$$\Rightarrow \quad y'^2 = c_1^2 x^2 (1 + y'^2) \Rightarrow y'^2 = \frac{c_1^2 x^2}{1 - c_1^2 x^2} \Rightarrow y' = \frac{c_1 x}{\sqrt{1 - c_1^2 x^2}}$$

$$\Rightarrow \quad dy = -\frac{1}{2c_1} \cdot \frac{(-2c_1^2 x)}{\sqrt{1 - c_1^2 x^2}} dx$$

$$\Rightarrow \quad y = -\frac{1}{2c_1} \cdot \frac{(1 - c_1^2 x^2)^{1/2}}{1/2} + c_2 \Rightarrow y = -\frac{1}{c_1}\sqrt{1 - c_1^2 x^2} + c_2$$

$$\Rightarrow \quad (y - c_2)^2 = \frac{1}{c_1^2}(1 - c_1^2 x^2) = \frac{1}{c_1^2} - x^2$$

$$\Rightarrow \quad \mathbf{x^2 + (y - c_2)^2 = c_3^2}. \qquad \left(\text{Putting } c_3 = \frac{1}{c_1}\right)$$

This family of curves are the required extremals.

Example 11. *From among the curves connecting the points A(1, 3) and B(2, 5), find the curve on which an extremum of the functional*

$$J[y(x)] = \int_1^2 y'(1 + x^2 y')\, dx \text{ can be attained.}$$

Sol. We have $J[y(x)] = \int_1^2 y'(1 + x^2 y')\, dx$.

Let $F(x, y, y') = y'(1 + x^2 y')$

The **Euler's equation** is $\dfrac{\partial F}{\partial y} - \dfrac{d}{dx}\left(\dfrac{\partial F}{\partial y'}\right) = 0$ ...(1)

Here $\dfrac{\partial F}{\partial y} = 0$, $\dfrac{\partial F}{\partial y'} = 1 + 2x^2 y'$

$$\therefore \ (1) \Rightarrow \quad 0 - \frac{d}{dx}(1 + 2x^2 y') = 0 \Rightarrow 1 + 2x^2 y' = c_1$$

$$\Rightarrow \quad y' = \frac{c_1 - 1}{2x^2} \Rightarrow dy = \frac{c_1 - 1}{2} x^{-2} dx$$

$$\Rightarrow \quad y = \frac{c_1 - 1}{2} \cdot \frac{x^{-1}}{-1} + c_2 \Rightarrow y = \frac{c_3}{x} + c_2 \qquad ...(2)$$

$$\left(\text{Taking } c_3 = \frac{1 - c_1}{2}\right)$$

This is the equation of the extremals.

We are to find that extremal which passes through (1, 3) and (2, 5).

$\therefore$ The boundary conditions are $y(1) = 3$ and $y(2) = 5$.

$$y(1) = 3 \Rightarrow \frac{c_3}{1} + c_2 = 3 \Rightarrow c_2 + c_3 = 3$$

$$y(2) = 5 \Rightarrow \frac{c_3}{2} + c_2 = 5 \Rightarrow c_2 + \frac{c_3}{2} = 5$$

Solving, we get $c_3 = -4$ and $c_2 = 7$.

$\therefore$ (2) $\Rightarrow$ $y = -\dfrac{4}{x} + 7$

This is the stationary function of the given problem.

$\therefore$ The given functional can have an extremum on the curve $\mathbf{y = -\dfrac{4}{x} + 7}$.

8. PRACTICAL VARIATIONAL PROBLEMS

We know that if a function $y(x)$ extremises the functional $\int_{x_1}^{x_2} F(x, y, y')\, dx$, then $y(x)$ satisfies the Euler's equation $\dfrac{\partial F}{\partial y} - \dfrac{d}{dx}\left(\dfrac{\partial F}{\partial y'}\right) = 0$. This does not hold conversely. Thus, if $y(x)$ is a function satisfying the Euler's equation and the boundary conditions then this function may or may not extremise the given functional. In other words, the satisfaction of Euler's equation and boundary conditions is not a sufficient condition for a function to extremise the given functional. Therefore, an extremising function of a functional is a stationary function of the functional but a stationary function may not enjoy the status of an extremising function.

In case of extremising functions of one variable, we use second derivative to give sufficient conditions for deciding whether a particular stationary point is either a maximum point or a minimum point or a point of inflexion. Similar sufficient conditions are also available in the calculus of variations. In actual practice, the geometry or physics of the problem under discussion often makes it possible to determine whether a particular stationary function maximises or minimises the given functional (or neither).

Example 12. *By using Euler's equation find the curve passing through the points* (x_1, y_1) *and* (x_2, y_2) *such that its length between the given points is shortest.*

Sol. Let $y(x)$ be a function whose curve passes through $A(x_1, y_1)$ and $B(x_2, y_2)$. The length of the arc between A and B is given by $\int_{x_1}^{x_2} \sqrt{1 + y'^2}\, dx$.

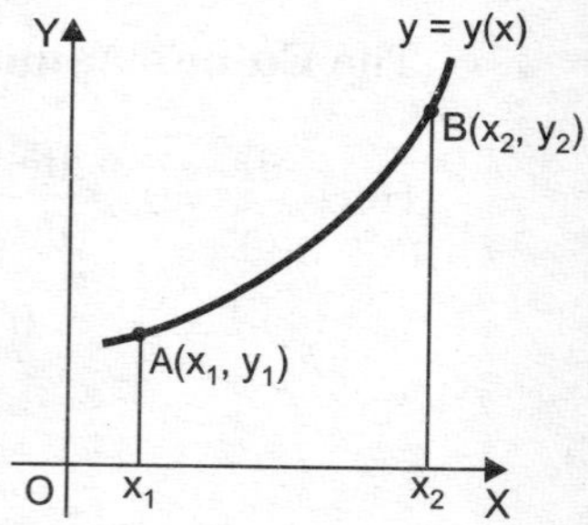

Let $\qquad F(x, y, y') = \sqrt{1 + y'^2}$

Let the length of the curve between given points be minimum for the curve $y(x)$.

$\therefore$ The functional $\int_{x_1}^{x_2} F(x, y, y')\, dx$ has a minimum value at the function $y(x)$.

$\therefore$ $y(x)$ satisfies the Euler's equation.

The **Euler's equation** is $\dfrac{\partial F}{\partial y} - \dfrac{d}{dx}\left(\dfrac{\partial F}{\partial y'}\right) = 0$. ...(1)

Here $\qquad \dfrac{\partial F}{\partial y} = 0, \quad \dfrac{\partial F}{\partial y'} = \dfrac{y'}{\sqrt{1 + y'^2}}$

$\therefore$ (1) $\Rightarrow$ $\qquad 0 - \dfrac{d}{dx}\left(\dfrac{y'}{\sqrt{1 + y'^2}}\right) = 0 \Rightarrow \dfrac{y'}{\sqrt{1 + y'^2}} = c_1$

$\Rightarrow \qquad y'^2 = c_1^2(1 + y'^2) \Rightarrow y' = \dfrac{c_1}{\sqrt{1 - c_1^2}} = c_2 \qquad \left(\text{Putting } c_2 = \dfrac{c_1}{\sqrt{1-c_1^2}}\right)$

$\Rightarrow \qquad dy = c_2 dx$

$\Rightarrow \qquad y = c_2 x + c_3$ ...(2)

This is the equation of the extremals. We are to find that extremal which passes through the points (x_1, y_1) and (x_2, y_2).

$\therefore$ The boundary conditions are $y(x_1) = y_1$, $y(x_2) = y_2$.

$$y(x_1) = y_1 \Rightarrow c_2 x_1 + c_3 = y_1$$
$$y(x_2) = y_2 \Rightarrow c_2 x_2 + c_3 = y_2$$

Solving, we get $\quad c_2 = \dfrac{y_2 - y_1}{x_2 - x_1}, \quad c_3 = \dfrac{x_2 y_1 - x_1 y_2}{x_2 - x_1}$

$\therefore$ (2) $\Rightarrow \qquad y = \left(\dfrac{y_2 - y_1}{x_2 - x_1}\right) x + \dfrac{x_2 y_1 - x_1 y_2}{x_2 - x_1}$ ...(2)

This is first degree equation.

$\therefore$ This represents a straight line passing through A and B.

$\therefore$ (2) minimises the functional *i.e.,* the length between given points.

$\therefore$ The shortest curve joining (x_1, y_1) and (x_2, y_2) is the straight line joining these points.

Example 13. *Find the curve passing through the points (x_1, y_1) and (x_2, y_2) such that its rotation about the x-axis give rise to a surface of revolution of minimum surface area.*

Sol. Let $y(x)$ be a function whose curve passes through $A(x_1, y_1)$ and $B(x_2, y_2)$. The area of the surface of revolution of the curve between A and B is given by

$$\int_{x_1}^{x_2} 2\pi y \sqrt{1 + y'^2}\, dx.$$

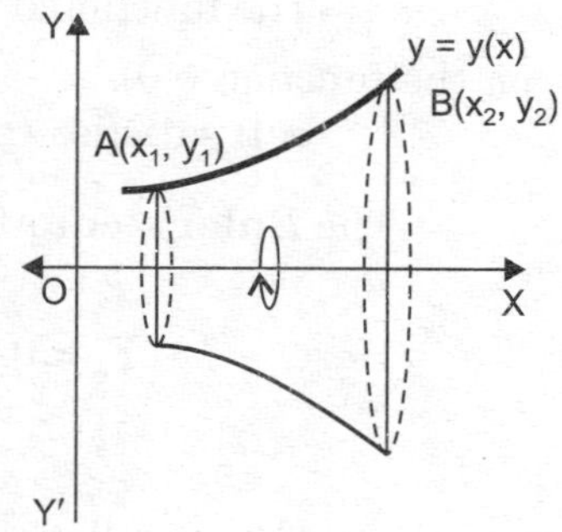

Let $\qquad F(x, y, y') = 2\pi y \sqrt{1 + y'^2}$

Let the area of surface of revolution of the curve between the given points has a minimum value for the curve $y(x)$.

$\therefore$ The functional $\displaystyle\int_{x_1}^{x_2} F(x, y, y')\, dx$ has a minimum value at the function $y(x)$.

$\therefore$ $y(x)$ satisfies the Euler's equation.

The **Euler's equation** is

$$\frac{\partial F}{\partial y} - \frac{d}{dx}\left(\frac{\partial F}{\partial y'}\right) = 0.$$

Since x is missing in $F(x, y, y')$, the Euler's equation reduces to $F - F_{y'} \dfrac{dy}{dx} = c.$

$\Rightarrow \qquad 2\pi y \sqrt{1 + y'^2} - \dfrac{2\pi y y'}{\sqrt{1 + y'^2}} \cdot y' = c \Rightarrow y\sqrt{1 + y'^2} - \dfrac{y y'^2}{\sqrt{1 + y'^2}} = \dfrac{c}{2\pi} = c_1$, say

$\Rightarrow \qquad y = c_1 \sqrt{1 + y'^2} \Rightarrow 1 + y'^2 = \dfrac{y^2}{c_1^2} \Rightarrow y' = \dfrac{\sqrt{y^2 - c_1^2}}{c_1}$

$\Rightarrow \qquad \dfrac{c_1 dy}{\sqrt{y^2 - c_1^2}} = dx \Rightarrow c_1 \cosh^{-1} \dfrac{y}{c_1} = x + c_2 \quad \therefore \; y = c_1 \cosh \dfrac{x + c_2}{c_1}.$

The values of c_1 and c_2 (provided they exist) are found by using the boundary conditions. This curve represents a catenary.

$\therefore \quad \mathbf{y = c_1 \cosh \frac{x + c_2}{c_1}}$ minimises the functional *i.e.*, the area of the surface of revolutions by the curve between the given points.

Example 14. *Show that the geodesics* on a sphere of radius a are its great circles.*

Sol. Let A and B be any two points on the given sphere with origin at the centre of the sphere. Let (a, θ_1, ϕ_1) and (a, θ_2, ϕ_2) be their spherical coordinates** respectively. Let $\phi(\theta)$ be a function whose curve passes through A and B and lying itself on the surface of the given sphere.

$\therefore$ The length of the arc between A and B is given by $\int_{\theta_1}^{\theta_2} ds$.

Since, we are dealing with spherical coordinates, we have

$$ds = \sqrt{(dr)^2 + (r\,d\theta)^2 + (r \sin\theta\, d\phi)^2} \qquad \textbf{(Note this step)}$$

$$\Rightarrow \quad ds = \sqrt{0 + a^2 (d\theta)^2 + a^2 \sin^2\theta \,.\, \phi'^2 (d\theta)^2} \qquad (\because \; r = a \Rightarrow dr = 0)$$

$$= a\sqrt{1 + \phi'^2 \sin^2\theta}\; d\theta$$

$$\therefore \quad \text{Length of arc} = \int_{\theta_1}^{\theta_2} a\sqrt{1 + \phi'^2 \sin^2\theta}\; d\theta$$

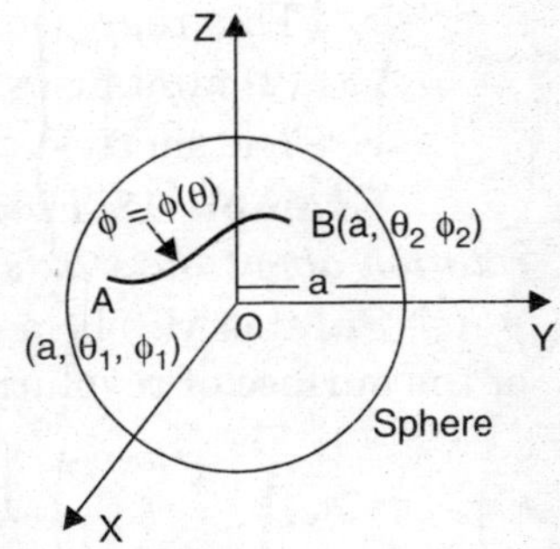

Let $\quad F(\theta, \phi, \phi') = a\sqrt{1 + \phi'^2 \sin^2\theta}$

Let the length of the curve between A and B has a minimum value for this curve. In other words, let $\phi(\theta)$ be a geodesic between A and B.

$\therefore$ The functional $\int_{\theta_1}^{\theta_2} F(\theta, \phi, \phi')\, d\theta$ has a minimum value on the function $\phi(\theta)$.

$\therefore$ $\phi(\theta)$ satisfies the Euler's equation.

The **Euler's equation** is $F_\phi - \frac{d}{d\theta} F_{\phi'} = 0 \qquad$ ...(1)

Here $\quad F_\phi = 0, \quad F_{\phi'} = \frac{a\phi' \sin^2\theta}{\sqrt{1 + \phi'^2 \sin^2\theta}}$

$$\therefore \quad (1) \Rightarrow 0 - \frac{d}{d\theta}\left(\frac{a\phi' \sin^2\theta}{\sqrt{1 + \phi'^2 \sin^2\theta}}\right) = 0 \Rightarrow \frac{d}{d\theta}\left(\frac{a\,\phi' \sin^2\theta}{\sqrt{1 + \phi'^2 \sin^2\theta}}\right) = 0$$

---

*A *geodesic* is a curve of shortest length lying on a given surface and joining two given points of the surface.

The equations of geodesics may be obtained as Euler's equations corresponding to the variational problem of finding the shortest distance on a surface between two given points.

** **Spherical coordinates.** Let $P(x, y, z)$ be any point in space and let Q be the foot of perpendicular from P on XY- plane.

If $\quad OP = r, \quad \angle ZOP = \theta, \quad \angle XOQ = \phi$, where $r \geq 0, \quad 0 \leq \theta \leq \pi, 0 \leq \phi \leq 2\pi$, then

(r, θ, ϕ) are called the **spherical coordinates** of the point $P(x, y, z)$. Also, $x = r \sin\theta \cos\phi$, $y = r \sin\theta \sin\phi$, $z = r\cos\theta$.

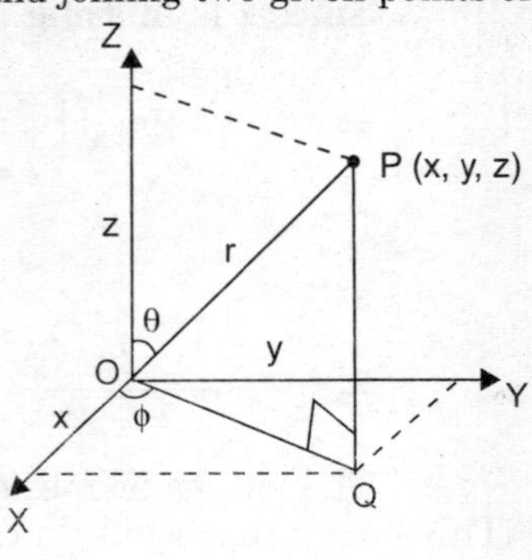

$$\Rightarrow \qquad \frac{\phi' \sin^2 \theta}{\sqrt{1 + \phi'^2 \sin^2 \theta}} = c_1$$

$$\Rightarrow \qquad \phi'^2 \sin^4 \theta = c_1^2 (1 + \phi'^2 \sin^2 \theta)$$

$$\Rightarrow \qquad \phi' = \frac{c_1}{\sin \theta \sqrt{\sin^2 \theta - c_1^2}} = \frac{c_1}{\sin^2 \theta \sqrt{1 - c_1^2 \operatorname{cosec}^2 \theta}}$$

$$\Rightarrow \qquad d\phi = \frac{-(-\operatorname{cosec}^2 \theta)\, c_1 \, d\theta}{\sqrt{(1 - c_1^2) - c_1^2 \cot^2 \theta}} = \frac{-d\,(c_1 \cot \theta)}{\sqrt{(1 - c_1^2) - c_1^2 \cot^2 \theta}}$$

Integrating, we get

$$\phi = \cos^{-1} \frac{c_1 \cot \theta}{\sqrt{1 - c_1^2}} + c_2 .$$

$$\Rightarrow \qquad \phi(\theta) = \cos^{-1} (c_3 \cot \theta) + c_2, \quad \text{say}$$

$$\Rightarrow \qquad c_3 \cot \theta = \cos (\phi(\theta) - c_2)$$

$$\Rightarrow \qquad c_3 \cot \theta = \cos \phi(\theta) \cos c_2 + \sin \phi(\theta) \sin c_2$$

$$\Rightarrow \qquad \cot \theta = \frac{\cos c_2}{c_3} \cos \phi(\theta) + \frac{\sin c_2}{c_3} \sin \phi(\theta)$$

$$\Rightarrow \qquad \cot \theta = c_4 \cos \phi(\theta) + c_5 \sin \phi(\theta), \quad \text{say}$$

Multiplying by $a \sin \theta$, we get

$$a \cos \theta = c_4(a \cos \phi(\theta) \sin \theta) + c_5(a \sin \phi(\theta) \sin \theta)$$

$$\Rightarrow \qquad z = c_4 x + c_5 y. \qquad \text{(Using cartesian coordinates)}$$

$z = c_4 x + c_5 y$ represents a plane passing through the centre (0, 0, 0) of the sphere.

The values of c_4 and c_5 are found by using the fact that this plane is to pass through A and B.

$\therefore$ The geodesic is the arc of the great circle passing through given points.

Example 15. *Show that the curve of shortest distance between any two points on a circular cylinder of radius a is a helix.*

Sol. Let the axis of the cylinder be taken along the z-axis. Let A and B be any two points on the given cylinder and let (a, ϕ_1, z_1) and (a, ϕ_2, z_2) be their cylindrical coordinates* respectively. Let $z(\phi)$ be a function whose curve passes through A and B and lying itself on the surface of the given cylinder.

---

*__Cylindrical coordinates.__ Let P(x, y, z) be any point in space and let Q be the foot of perpendicular from P on XY-plane.

If OQ $= r$, $\angle$XOQ $= \phi$, where $r \geq 0$, $0 \leq \phi \leq 2\pi$, $-\infty < z < \infty$, then (r, ϕ, z) are called the **cylindrical coordinates** of the point P(x, y, z).

Also, $x = r \cos \phi$, $y = r \sin \phi$, $z = z$.

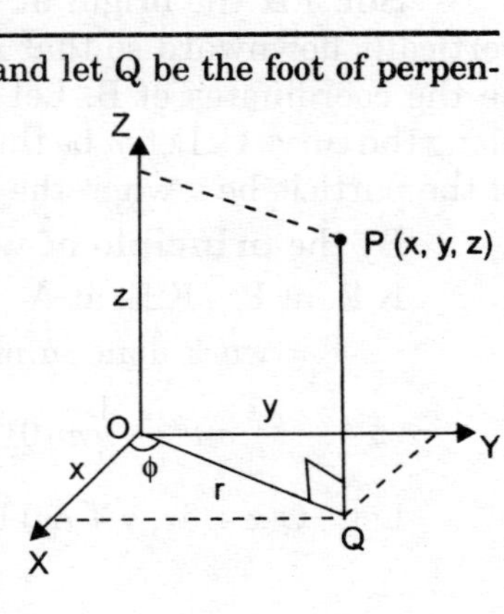

$\therefore$ The length of the arc between A and B is given by $\int_{\phi_1}^{\phi_2} ds$.

Since, we are dealing with cylindrical coordinates, we have

$$ds = \sqrt{(dr)^2 + (rd\phi)^2 + (dz)^2} \qquad \textbf{(Note this step)}$$

$$\Rightarrow \quad ds = \sqrt{0 + a^2(d\phi)^2 + (dz)^2} = \sqrt{a^2 + z'^2}\, d\phi \qquad (\because \; r = a \;\Rightarrow\; dr = 0)$$

$\therefore$ Length of arc $= \int_{\phi_1}^{\phi_2} \sqrt{a^2 + z'^2}\, d\phi$

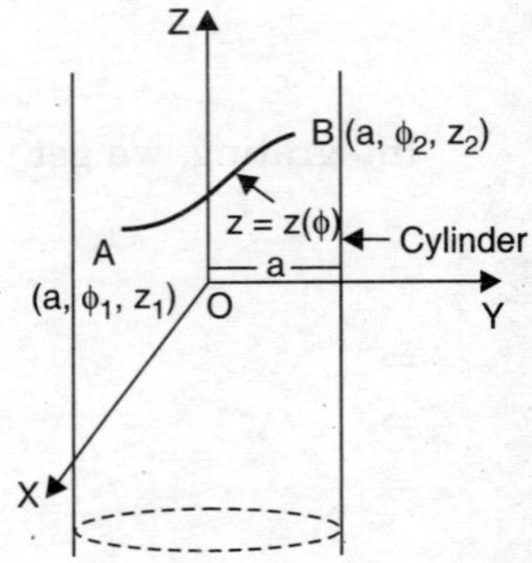

Let $\quad F(\phi, z, z') = \sqrt{a^2 + z'^2}$

Let the length of the curve between A and B has a minimum value for this curve.

$\therefore$ The functional $\int_{\phi_1}^{\phi_2} F(\phi, z, z')\, d\phi$ has a minimum value on the function $z(\phi)$.

$\therefore$ $z(\phi)$ satisfies the Euler's equation.

The **Euler's equation** is $F_z - \dfrac{d}{d\phi} F_{z'} = 0 \qquad ...(1)$

Here $F_z = 0, \quad F_{z'} = \dfrac{z'}{\sqrt{a^2 + z'^2}}$

$$\therefore \quad (1) \Rightarrow \quad 0 - \frac{d}{d\phi}\left(\frac{z'}{\sqrt{a^2 + z'^2}}\right) = 0 \;\Rightarrow\; \frac{z'}{\sqrt{a^2 + z'^2}} = c_1$$

$$\Rightarrow \quad z'^2 = c_1^2(a^2 + z'^2) \;\Rightarrow\; z' = \frac{c_1 a}{\sqrt{1 - c_1^2}} = c_2, \text{ say}$$

$$\Rightarrow \quad \mathbf{z = c_2\phi + c_3}.$$

The values of c_2 and c_3 are found by using the fact that this curve is to pass through A and B. This curve is called a *helix*.

Example 16. *Let A and B be two points in space but not in the same verticcl line. Find the curve joining A and B traversed by a particle starting from rest at A and moving under gravity from A to B in the shortest time and ignoring friction along the curve and the resistance of the medium.*

Sol. Fix the origin at A with x-axis horizontal and y-axis vertically downward so that point B is in the xy-plane. Let (h, k) be the coordinates of B. Let the particle traversed from A to B along the curve C. Let m be the mass of the particle. Let the velocity of the particle be v when the particle is at the point $P(x, y)$ on C.

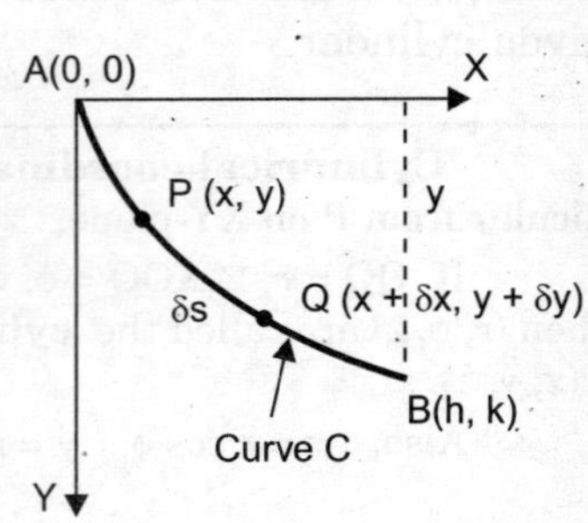

By the **principle of work and energy**,

K.E. at P – K.E. at A

= work done in moving the particle from A to P

$$\Rightarrow \quad \frac{1}{2}mv^2 - \frac{1}{2}m(0)^2 = mgy \;\Rightarrow\; v = \sqrt{2gy}.$$

Let $Q(x + \delta x, y + \delta y)$ be a neighbouring point on C such that $\widehat{PQ} = \delta s$.

$\therefore \quad \delta s = \sqrt{(\delta x)^2 + (\delta y)^2} = \sqrt{1+\left(\frac{\delta y}{\delta x}\right)^2}\,\delta x = \sqrt{1+y'^2}\,\delta x$, nearly

$\therefore$ Time of descent from P to Q $= \dfrac{\delta s}{v} = \dfrac{\sqrt{1+y'^2}}{\sqrt{2gy}}\,\delta x = \dfrac{1}{\sqrt{2g}}\cdot\dfrac{\sqrt{1+y'^2}}{\sqrt{y}}\,\delta x$

$\therefore$ Total time of transit from A to B

$$= \frac{1}{\sqrt{2g}}\int_0^h \frac{\sqrt{1+y'^2}}{\sqrt{y}}\,dx$$

Also, $y(0) = 0$ and $y(h) = k$.

$\therefore$ In order to find the curve of shortest time of descent, we have to minimise* the functional $\int_0^h \frac{\sqrt{1+y'^2}}{\sqrt{y}}\,dx$, subject to the fixed boundary conditions $y(0) = 0$ and $y(h) = k$.

Let $\qquad F(x, y, y') = \dfrac{\sqrt{1+y'^2}}{\sqrt{y}}$

The **Euler's equation** is $F_y - \frac{d}{dx}F_{y'} = 0.$

Since x is missing from $F(x, y, y')$, the Euler's equation reduces to

$$F - F_{y'}\frac{dy}{dx} = c.$$

$$\Rightarrow \quad \frac{\sqrt{1+y'^2}}{\sqrt{y}} - \left(\frac{1}{\sqrt{y}}\cdot\frac{2y'}{2\sqrt{1+y'^2}}\right)y' = c$$

$$\Rightarrow \quad \frac{1}{\sqrt{y}\,\sqrt{1+y'^2}}(1 + y'^2 - y'^2) = c$$

$$\Rightarrow \quad y(1 + y'^2) = \frac{1}{c^2} = c_1, \text{ say} \qquad \ldots(1)$$

Let $y' = \cot u$.

$$\therefore \ (1) \Rightarrow \quad y = \frac{c_1}{1+\cot^2 u} = c_1 \sin^2 u = \frac{c_1}{2}(1 - \cos 2u)$$

Also $y' = \cot u \Rightarrow$

$$dx = \frac{dy}{\cot u} = \frac{1}{\cot u}\cdot\frac{c_1}{2}(0 + 2\sin 2u)du = 2c_1 \sin^2 u\, du$$

$$\Rightarrow \quad \int dx = \int 2c_1 \sin^2 u\, du + c_2 \quad \Rightarrow \quad x = c_1\int(1 - \cos 2u)du + c_2$$

*From physical considerations, this functional has no maximum value.

$$\Rightarrow \qquad x = c_1\left(u - \frac{\sin 2u}{2}\right) + c_2 \quad \Rightarrow \quad x = \frac{c_1}{2}(2u - \sin 2u) + c_2$$

$\therefore$ The equations of the extremals are

$$x = \frac{c_1}{2}(t - \sin t) + c_2 \quad \text{and} \quad y = \frac{c_1}{2}(1 - \cos t). \qquad \text{(Putting } t = 2u\text{)}$$

The boundary conditions are $y(0) = 0$ and $y(h) = k$.

$$y = 0 \ \Rightarrow \ \frac{c_1}{2}(1 - \cos t) = 0 \ \Rightarrow \ \cos t = 1 \ \Rightarrow \ t = 0$$

$$\Rightarrow \qquad 0 = \frac{c_1}{2}(0 - \sin 0) + c_2 = c_2 \ \Rightarrow \ c_2 = 0.$$

$\therefore$ The required curve is

$$\mathbf{x = \frac{c_1}{2}(t - \sin t), \quad y = \frac{c_1}{2}(1 - \cos t).}$$

This represents a cycloid with $c_1/2$ as the radius of the rolling circle.

The value of c_1 is to be found by using $y(h) = k$ *i.e.*, by using the fact that the curve passes through B.

Note. This problem is known as **Brachistochrone problem.**

The name 'Brachistochrone is derived from the Greek words 'brachisto' meaning *shortest* and 'chrone' meaning *time.*

EXERCISE 4.1

1. On what curves can the functional $J[y(x)] = \int_1^2 (y'^2 - 2xy)\, dx$; $y(1) = 0$, $y(2) = -1$ attain an extremum ?
2. Find the stationary function of the functional:

   $J[y(x)] = \int_{-1}^{0} (12xy - y'^2)\, dx$ which is determined by the boundary conditions $y(-1) = 1$, $y(0) = 0$.
3. Find the extremal of the functional $J[y(x)] = \int_{-1}^{1} (y'^2 - 2xy)\, dx$ that satisfy the boundary conditions $y(-1) = -1$, $y(1) = 1$.
4. Find the extremal of the functional $J[y(x)] = \int_1^3 (3x - y)y\, dx$ that satisfy the boundary conditions $y(1) = 1$, $y(3) = 4\frac{1}{2}$.
5. Find the extremal of the functional $J[y(x)] = \int_1^2 \frac{x^3}{y'^2}\, dx$ that satisfy the boundary conditions $y(1) = 0$, $y(2) = 3$.
6. Find the extremals of the following functionals:

   (*i*) $\int_a^b \left(y + \frac{y^3}{3}\right) dx$ (*ii*) $\int_a^b (xy' + y'^2)\, dx.$

7. Find the extremal of the functional $\int_0^1 (2e^y - y^2)\, dx$ that satisfy the boundary conditions $y(0) = 1$, $y(1) = e$.

8. Find the extremals of the functional $\int_a^b (y^2 + y'^2 + 2ye^x)\, dx$.

9. Find the extremal of the functional $\int_0^{\pi/2} (y^2 - y'^2 - 8y \cosh x)\, dx$ that satisfies the boundary conditions $y(0) = 2$, $y\left(\frac{\pi}{2}\right) = 2 \cosh \frac{\pi}{2}$.

10. Find the extremal of the functional $\int_0^1 (x + y'^2)\, dx$ that satisfy the boundary conditions $y(0) = 1$, $y(1) = 2$.

11. Find the extremal of the functional $J[y(x)] = \int_0^{\pi} (4y \cos x + y'^2 - y^2)\, dx$ that satisfy the boundary conditions $y(0) = 0$, $y(\pi) = 0$.

12. Find the extremal of the functional $J[y(x)] = \int_0^{\pi/2} y(2x - y)\, dx$ that satisfy the boundary conditions $y(0) = 0$, $y(\pi/2) = \pi/2$.

13. Find the extremal of the functional $J[y(x)] = \int_0^1 (e^y + xy')\, dx$ that satisfy the boundary conditions $y(0) = 0$, $y(1) = 0$.

14. Find the extremal of the functional $J[y(x)] = \int_0^{\pi/4} (y'^2 - y^2)\, dx$ that satisfy the boundary conditions $y(0) = 1$, $y\left(\frac{\pi}{4}\right) = \frac{\sqrt{2}}{2}$.

15. Find the extremals and the stationary function of the functional $J[y(x)] = \int_0^{\pi} (y'^2 - y^2)\, dx$ that satisfy the boundary conditions $y(0) = 1$, $y(\pi) = -1$.

16. Find the stationary function of the functional $\int_0^{1} (y^2 + y'^2)\, dx$ subject to the boundary conditions $y(0) = 0$, $y(1) = 1$.

17. Show that the Euler's equation for the functional $\int_{x_1}^{x_2} (a(x)\, y'^2 + 2b(x)\, yy' + c(x) y^2)\, dx$ is a second order linear differential equation.

18. Find the extremal of the functional $\int_1^2 \frac{\sqrt{1 + y'^2}}{x}\, dx$ which satisfies the boundary conditions $y(1) = 0$, $y(2) = 1$.

19. Find the extremals of the functional $\int_a^b \frac{y'^2}{x^3}\, dx$.

20. By using Euler's equation, find the curve passing through (0, 3) and (4, 11) such that its length between the given points is shortest.

21. Find the geodesics on a right circular cone of semi-vertical angle α.

Answers

1. $y = \frac{x}{6}(1 - x^2)$ **2.** $y = -x^3$ **3.** $y = \frac{7}{6}x - \frac{1}{6}x^3$ **4.** No solution

5. $y = x^2 - 1$ **6.** (*i*) No extremals (*ii*) $y = -\frac{1}{4}x^2 + c_1x + c_2$

7. No solution **8.** $y = c_1e^x + c_2e^{-x} + \frac{1}{2}xe^x$, where c_1 and c_2 are arbitrary constants

9. $y = 2\cosh x$ **10.** $y = x + 1$ **11.** $y = (x + c)\sin x$, where c is an arbitrary constant.

12. $x - y = 0$ **13.** $y = 0$ **14.** $y = \cos x$

15. $y = c_1\cos x + c_2\sin x$, where c_1 and c_2 are arbitrary constants ; $y = \cos x + c_2\sin x$, where c_2 is an arbitrary constant

16. $y = \frac{\sinh x}{\sinh 1}$ **17.** $a(x)y'' + a'(x)y' + (b'(x) - c(x))y = 0$ **18.** $x^2 + (y - 2)^2 = 5$

19. $y = c_1x^4 + c_2$ **20.** $y = 2x + 3$ **21.** $c_1r = \sec(\phi\sin\alpha + c_2)$.

9. FUNCTIONALS DEPENDENT ON HIGHER ORDER DERIVATIVES

Theorem 1. *Let* $J[y(x)] = \int_{x_1}^{x_2} F(x, y(x), y'(x), y''(x))\,dx$ *be a functional defined on the set of functions* $y(x)$ *which are differentiable four times and satisfy the boundary conditions:* $y(x_1) = y_1$, $y'(x_1) = y_1'$, $y(x_2) = y_2$, $y'(x_2) = y_2'$, *where* y_1, y_1', y_2, y_2' *are prescribed at the fixed boundary points* x_1 *and* x_2. *Also, F is differentiable four times with respect to all its arguments.*

If the functional J [y(x)] has an extremum on a function y(x), then y(x) satisfy the differential equation

$$F_y - \frac{d}{dx}F_{y'} + \frac{d^2}{dx^2}F_{y''} = 0.$$

Proof. We have $$J[y(x)] = \int_{x_1}^{x_2} F(x, y(x), y'(x), y''(x))\,dx.$$

Let $y = y(x)$ be a function which extremise J[y(x)] and satisfying $y(x_1) = y_1$, $y'(x_1) = y_1'$, $y(x_2) = y_2$, $y'(x_2) = y_2'$.

Let $\eta(x)$ be any arbitrary but fixed function differentiable four times and

$$\eta(x_1) = \eta'(x_1) = \eta(x_2) = \eta'(x_2) = 0.$$

Let $\bar{y}(x) = y(x) + \alpha\eta(x)$, where α is a small parameter.

$\therefore$ $\bar{y}(x)$ is differentiable four times.

Now $$\bar{y}(x_1) = y_1,\ \bar{y}'(x_1) = y_1',\ \bar{y}(x_2) = y_2,\ \bar{y}'(x_2) = y_2'.$$

$\therefore$ $\bar{y}(x)$ also satisfy the boundary conditions.

$\therefore$ The functional J is defined at $\bar{y}(x)$.

$$\therefore \quad J[\bar{y}(x)] = \int_{x_1}^{x_2} F(x, \bar{y}(x), \bar{y}'(x), \bar{y}''(x))\,dx$$

$$= \int_{x_1}^{x_2} F(x, y(x) + \alpha\eta(x), y'(x), \alpha\eta'(x), y''(x) + \alpha\eta''(x))\,dx$$

On the curves of the family $y(x) + \alpha\eta(x)$, the right hand side is a function of α, say $\psi(\alpha)$.

$$\therefore \quad \psi(\alpha) = \int_{x_1}^{x_2} F(x, \bar{y}(x), \bar{y}'(x), \bar{y}''(x))\,dx$$

$$= \int_{x_1}^{x_2} F(x, y(x) + \alpha\eta(x), y'(x) + \alpha\eta'(x), y''(x) + \alpha\eta''(x))\,dx$$

Since the extremising curve $y = y(x)$ corresponds to $\alpha = 0$, it follows that $\psi(\alpha)$ is extremised at $\alpha = 0$.

$$\therefore \quad \frac{d\psi}{d\alpha} = 0 \quad \text{at} \quad \alpha = 0.$$

Differentiating $\psi(\alpha)$ under the integral sign, we have

$$\frac{d\psi}{d\alpha} = \int_{x_1}^{x_2} \frac{\partial}{\partial \alpha} F(x, \bar{y}(x), \bar{y}'(x), \bar{y}''(x))\, dx$$

$$= \int_{x_1}^{x_2} \left(\frac{\partial F}{\partial x}\frac{\partial x}{\partial \alpha} + \frac{\partial F}{\partial \bar{y}}\frac{\partial \bar{y}}{\partial \alpha} + \frac{\partial F}{\partial \bar{y}'}\frac{\partial \bar{y}'}{\partial \alpha} + \frac{\partial F}{\partial \bar{y}''}\frac{\partial \bar{y}''}{\partial \alpha} \right) dx$$

$$= \int_{x_1}^{x_2} \left(\frac{\partial F}{\partial x} . 0 + \frac{\partial F}{\partial \bar{y}} \eta(x) + \frac{\partial F}{\partial \bar{y}'} \eta'(x) + \frac{\partial F}{\partial \bar{y}''} \eta''(x) \right) dx$$

$$\therefore \quad \frac{\partial \psi}{\partial \alpha} = \int_{x_1}^{x_2} \left(\frac{\partial F}{\partial \bar{y}} \eta(x) + \frac{\partial F}{\partial \bar{y}'} \eta'(x) + \frac{\partial F}{\partial \bar{y}''} \eta''(x) \right) dx \qquad \ldots(1)$$

$$\text{Now} \qquad \frac{\partial \psi}{\partial \alpha} = 0 \quad \text{at} \quad \alpha = 0.$$

$\therefore$ Putting $\alpha = 0$ in (1), we get

$$\int_{x_1}^{x_2} \left(\frac{\partial F}{\partial y} \eta(x) + \frac{\partial F}{\partial y'} \eta'(x) + \frac{\partial F}{\partial y''} \eta''(x) \right) dx = 0 \qquad \ldots(2)$$

$$\Rightarrow \quad \int_{x_1}^{x_2} \frac{\partial F}{\partial y} \eta(x)\, dx + \left[\frac{\partial F}{\partial y'} \eta(x) \Big|_{x_1}^{x_2} - \int_{x_1}^{x_2} \frac{d}{dx}\left(\frac{\partial F}{\partial y'} \right) \eta(x)\, dx \right]$$

$$+ \left[\frac{\partial F}{\partial y''} \eta'(x) \Big|_{x_1}^{x_2} - \int_{x_1}^{x_2} \frac{d}{dx}\left(\frac{\partial F}{\partial y''} \right) \eta'(x)\, dx \right] = 0$$

$$\Rightarrow \quad \int_{x_1}^{x_2} \frac{\partial F}{\partial y} \eta(x)\, dx - \int_{x_1}^{x_2} \frac{d}{dx}\left(\frac{\partial F}{\partial y'} \right) \eta(x)\, dx - \int_{x_1}^{x_2} \frac{d}{dx}\left(\frac{\partial F}{\partial y''} \right) \eta'(x)\, dx = 0$$

$$\Rightarrow \quad \int_{x_1}^{x_2} \left(\frac{\partial F}{\partial y} \eta(x) - \frac{d}{dx}\left(\frac{\partial F}{\partial y'} \right) \eta(x) \right) dx - \left[\frac{d}{dx}\left(\frac{\partial F}{\partial y''} \right) \eta(x) \Big|_{x_1}^{x_2} - \int_{x_1}^{x_2} \frac{d^2}{dx^2}\left(\frac{\partial F}{\partial y''} \right) \eta(x)\, dx \right] = 0$$

$$\Rightarrow \quad \int_{x_1}^{x_2} \left(\frac{\partial F}{\partial y} - \frac{d}{dx}\left(\frac{\partial F}{\partial y'} \right) + \frac{d^2}{dx^2}\left(\frac{\partial F}{\partial y''} \right) \right) \eta(x) dx = 0.$$

This holds for any arbitrary continuous function $\eta(x)$.

$$\therefore \quad \text{We have} \qquad \frac{\partial F}{\partial y} - \frac{d}{dx}\left(\frac{\partial F}{\partial y'} \right) + \frac{d^2}{dx^2}\left(\frac{\partial F}{\partial y''} \right) = 0.$$

$\therefore$ The function $y(x)$ satisfies the equation

$$\mathbf{F_y - \frac{d}{dx} F_{y'} + \frac{d^2}{dx^2} F_{y''} = 0.}$$

This equation is known as the **Euler-Poisson equation.**

In the next theorem, we generalise the above result.

Theorem 2. *Let* $J[y(x)] = \int_{x_1}^{x_2} F(x, y(x), y'(x), \ldots, y^{(n)}(x))\, dx$ *be a functional defined on the set of functions y(x) which are differentiable 2n times and satisfy the boundary conditions:*

$$y(x_1) = y_1,\ y'(x_1) = y_1',\ \ldots,\ y^{(n-1)}(x_1) = y_1^{(n-1)},$$
$$y(x_2) = y_2,\ y'(x_2) = y_2',\ \ldots,\ y^{(n-1)}(x_2) = y_2^{(n-1)},$$

where $y_1, y_1', \ldots\ y_1^{(n-1)}, y_2, y_2', \ldots, y_2^{(n-1)}$ *are prescribed at the fixed boundary points* x_1 *and* x_2. *Also, F is differentiable n + 2 times with respect to all its arguments.*

If the functional J[y(x)] has an extremum on a function y(x) then y(x) satisfy the differential equation $F_y - \frac{d}{dx} F_{y'} + \frac{d^2}{dx^2} F_{y''} - \ldots\ldots + (-1)^n \frac{d^n}{dx^n} F_{y^{(n)}} = 0.$

Proof. We have $J[y(x)] = \int_{x_1}^{x_2} F(x, y(x), y'(x), \ldots, y^{(n)}(x))\, dx.$

Let $y = y(x)$ be a function which extremise J[y(x)] and satisfying

$$y(x_1) = y_1,\ y'(x_1) = y_1',\ \ldots, y^{(n-1)}(x_1) = y_1^{(n-1)},$$
$$y(x_2) = y_2,\ y'(x_2) = y_2',\ \ldots, y^{(n-1)}(x_2) = y_2^{(n-1)}.$$

Let $\eta(x)$ be any arbitrary but fixed function differentiable $2n$ times and

$$\eta(x_1) = \eta'(x_1) = \ldots\ldots = \eta^{(n-1)}(x_1) = 0$$
$$\eta(x_2) = \eta'(x_2) = \ldots\ldots = \eta^{(n-1)}(x_2) = 0.$$

Let $\bar{y}(x) = y(x) + \alpha\eta(x)$, where α is a small parameter.

∴ $\bar{y}(x)$ is differentiable $2n$ times.

Now
$$\bar{y}(x_1) = y_1,\ \bar{y}'(x_1) = y_1',\ \ldots,\ \bar{y}^{(n-1)}(x_1) = y_1^{(n-1)},$$
$$\bar{y}(x_2) = y_2,\ \bar{y}'(x_2) = y_2',\ \ldots,\ \bar{y}^{(n-1)}(x_2) = y_2^{(n-1)}.$$

∴ $\bar{y}(x)$ also satisfy the boundary conditions.

∴ The functional J is defined at $\bar{y}(x)$.

$$\therefore \quad J[\bar{y}(x)] = \int_a^b F(x, \bar{y}(x), \bar{y}'(x), \ldots, \bar{y}^{(n)}(x) dx$$

$$= \int_{x_1}^{x_2} F(x, y(x) + \alpha\eta(x), y'(x) + \alpha\eta'(x), \ldots, y^{(n)}(x) + \alpha\eta^{(n)}(x))\, dx$$

On the curves of the family $y(x) + \alpha\eta(x)$, the right hand side is a function of α, say $\psi(\alpha)$.

$$\therefore \quad \psi(\alpha) = \int_{x_1}^{x_2} F(x, \bar{y}(x), \bar{y}'(x), \ldots, \bar{y}^{(n)}(x))\, dx$$

$$= \int_{x_1}^{x_2} F(x, y(x) + \alpha\eta(x), y'(x) + \alpha\eta'(x), \ldots, y^{(n)}(x) + \alpha\eta^{(n)}(x))\, dx$$

Since the extremising curve $y = y(x)$ corresponds to $\alpha = 0$, it follows that $\psi(\alpha)$ is extremised for $\alpha = 0$.

$$\therefore \quad \frac{d\psi}{d\alpha} = 0 \text{ at } \alpha = 0.$$

Differentiating $\psi(\alpha)$ under the integral sign, we have

$$\frac{d\psi}{d\alpha} = \int_{x_1}^{x_2} \frac{\partial}{\partial \alpha} F(x, \bar{y}(x), \bar{y}'(x), ..., \bar{y}^{(n)}(x))\, dx$$

$$= \int_{x_1}^{x_2} \left(\frac{\partial F}{\partial x}\frac{\partial x}{\partial \alpha} + \frac{\partial F}{\partial \bar{y}}\frac{\partial \bar{y}}{\partial \alpha} + \frac{\partial F}{\partial \bar{y}'}\frac{\partial \bar{y}}{\partial \alpha} + ... + \frac{\partial F}{\partial \bar{y}^{(n)}}\frac{\partial \bar{y}^{(n)}}{\partial \alpha} \right) dx$$

$$= \int_{x_1}^{x_2} \left(\frac{\partial F}{\partial x} . 0 + \frac{\partial F}{\partial \bar{y}} \eta(x) + \frac{\partial F}{\partial \bar{y}'} \eta'(x) + ... + \frac{\partial F}{\partial \bar{y}^{(n)}} \eta^{(n)}(x) \right) dx$$

$$\therefore \quad \frac{d\psi}{d\alpha} = \int_{x_1}^{x_2} \left(\frac{\partial F}{\partial \bar{y}} \eta(x) + \frac{\partial F}{\partial \bar{y}'} \eta'(x) + ... + \frac{\partial F}{\partial \bar{y}^{(n)}} \eta^{(n)}(x) \right) dx \qquad ...(1)$$

Now $\frac{d\psi}{d\alpha} = 0$ at $\alpha = 0$.

$\therefore$ Putting $\alpha = 0$ in (1), we get

$$\int_{x_1}^{x_2} \left(\frac{\partial F}{\partial x} \eta(x) + \frac{\partial F}{\partial y'} \eta'(x) + ... + \frac{\partial F}{\partial y^{(n)}} \eta^{(n)}(x) \right) dx = 0 \qquad ...(2)$$

Integrating by parts the second term once, third term twice and so on and using boundary conditions, we get

$$\int_{x_1}^{x_2} \left(\frac{\partial F}{\partial x} - \frac{d}{dx}\left(\frac{\partial F}{\partial y'}\right) + \frac{d^2}{dx^2}\left(\frac{\partial F}{\partial y''}\right) - ...... + (-1)^n \frac{d^n}{dx^n}\left(\frac{\partial F}{\partial y^{(n)}}\right) \right) \eta(x)\, dx = 0.$$

This holds for any arbitrary continuous function $\eta(x)$.

$\therefore$ We have $\frac{\partial F}{\partial y} - \frac{d}{dx}\left(\frac{\partial F}{\partial y'}\right) + \frac{d^2}{dx^2}\left(\frac{\partial F}{\partial y''}\right) - ...... + (-1)^n \frac{d^n}{dx^n}\left(\frac{\partial F}{\partial y^{(n)}}\right) = 0$

$\therefore$ The function $y(x)$ satisfies the equation

$$\mathbf{F_y - \frac{d}{dx} F_{y'} + \frac{d^2}{dx^2} F_{y''} - ...... + (-1)^n \frac{d^n}{dx^n} F_{y^{(n)}} = 0.}$$

This equation is also known as the **Euler-Poisson equation.**

WORKING RULES FOR SOLVING PROBLEMS

Step I. *Identify the functional* $\int_{x_1}^{x_2} F(x, y, y', ..., y^{(n)})\, dx$ *and the boundary conditions in the given problem.*

Step II. *Write the Euler-Poisson equation*

$$F_y - \frac{d}{dx} F_{y'} + \frac{d^2}{dx^2} F_{y''} - ...... + (-1)^n \frac{d^n}{dx^n} F_{y^{(n)}} = 0.$$

Step III. *The solutions of this equation are extremals of the given problem.*

Step IV. *Find the extremal that satisfy the given boundary conditions. This particular extremal is the stationary function of the given problem.*

Example 1. *Find the stationary function of the functional* $\int_a^b (y'^2 + yy'')dx$; $y(a) = \lambda_1$, $y'(a) = \lambda_2$, $y(b) = \lambda_3$, $y'(b) = \lambda_4$.

Sol. Given functional is $\int_a^b (y'^2 + yy'')\,dx$.

Let $\quad F(x, y, y', y'') = y'^2 + yy''$

The **Euler-Poisson equation** is

$$F_y - \frac{d}{dx}F_{y'} + \frac{d^2}{dx^2}F_{y''} = 0 \qquad \text{...(1)}$$

Here $F_y = y'', F_{y'} = 2y', F_{y''} = y$

$$\therefore \quad (1) \Rightarrow \quad y'' - \frac{d}{dx}(2y') + \frac{d^2}{dx^2}(y) = 0$$

$$\Rightarrow \quad y'' - 2y'' + y'' = 0 \Rightarrow 0 = 0,$$

which is not a differential equation.

$\therefore$ There is no extremal and hence no stationary function.

Example 2. *Find the extremals of the functional* $\int_{x_0}^{x_1} (2xy + y'''^2)\,dx$.

Sol. Given functional is $\int_{x_0}^{x_1} (2xy + y'''^2)\,dx$.

Let $\quad F(x, y, y', y'', y''') = 2xy + y'''^2$

The **Euler-Poisson equation** is

$$F_y - \frac{d}{dx}F_{y'} + \frac{d^2}{dx^2}F_{y''} - \frac{d^3}{dx^3}F_{y'''} = 0 \qquad \text{...(1)}$$

Here $F_y = 2x, F_{y'} = 0, F_{y''} = 0, F_{y'''} = 2y'''$

$$\therefore \quad (1) \Rightarrow \quad 2x - \frac{d}{dx}(0) + \frac{d^2}{dx^2}(0) - \frac{d^3}{dx^3}(2y''') = 0$$

$$\Rightarrow \quad \frac{d^6y}{dx^6} = x \Rightarrow \frac{d^5y}{dx^5} = \frac{x^2}{2} + c_1$$

$$\Rightarrow \quad \frac{d^4y}{dx^4} = \frac{x^3}{6} + c_1x + c_2$$

$$\Rightarrow \quad \frac{d^3y}{dx^3} = \frac{x^4}{24} + c_1\frac{x^2}{2} + c_2x + c_3$$

$$\Rightarrow \quad \frac{d^2y}{dx^2} = \frac{x^5}{120} + c_1\frac{x^3}{6} + c_2\frac{x^2}{2} + c_3x + c_4$$

$$\Rightarrow \quad \frac{dy}{dx} = \frac{x^6}{720} + c_1\frac{x^4}{24} + c_2\frac{x^3}{6} + c_3\frac{x^2}{2} + c_4x + c_5$$

$$\Rightarrow \quad y = \frac{x^7}{5040} + c_1 \frac{x^5}{120} + c_2 \frac{x^4}{24} + c_3 \frac{x^3}{6} + c_4 \frac{x^2}{2} + c_5 x + c_6$$

$$\Rightarrow \quad y = \frac{x^7}{5040} + c_1' x^5 + c_2' x^4 + c_3' x^3 + c_4' x^2 + c_5 x + c_6,$$

where $c_1' = \frac{c_1}{120}, \quad c_2' = \frac{c_2}{24}, \quad c_3' = \frac{c_3}{6}, \quad c_4' = \frac{c_2}{2}$

$\therefore$ The equation of the extremals is

$$\mathbf{y = \frac{x^7}{5040} + c_1' x^5 + c_2' x^4 + c_3' x^3 + c_4' x^2 + c_5 x + c_6,}$$

where c_1', c_2', c_3', c_4', c_5, c_6 are arbitrary constants.

Example 3. *Find the curve that extremises the functional* $\int_0^{\pi/4} (y''^2 - y^2 + x^2)\, dx$ *under the boundary conditions* $y(0) = 0,\ y'(0) = 1,\ y(\pi/4) = y'(\pi/4) = 1/\sqrt{2}$.

Sol. Given functional is $\int_0^{\pi/4} (y''^2 - y^2 + x^2)\, dx$.

Let $\quad F(x, y, y', y'') = y''^2 - y^2 + x^2$

The **Euler-Poisson equation** is $\quad F_y - \frac{d}{dx} F_{y'} + \frac{d^2}{dx^2} F_{y''} = 0 \qquad$ ...(1)

Here $\quad F_y = -2y, \quad F_{y'} = 0, \quad F_{y''} = 2y''$

$\therefore$ (1) $\Rightarrow \quad -2y - \frac{d}{dx}(0) + \frac{d^2}{dx^2}(2y'') = 0 \quad \Rightarrow \quad \frac{d^4 y}{dx^4} - y = 0$

$\therefore$ The A.E. is $\quad D^4 - 1 = 0$

$\Rightarrow \quad D = \pm 1, \quad \pm i$

$\therefore \quad y = c_1 e^{1.x} + c_2 e^{-1.x} + e^{0.x}(c_3 \cos 1.x + c_4 \sin 1.x)$

or $\quad y = c_1 e^{x} + c_2 e^{-x} + c_3 \cos x + c_4 \sin x \qquad$ ...(2)

This is the equation of the extremals.

The boundary conditions are $y(0) = 0, \quad y'(0) = 1, \quad y(\pi/4) = y'(\pi/4) = 1/\sqrt{2}$.

(2) $\Rightarrow \quad y' = c_1 e^{x} - c_2 e^{-x} - c_3 \sin x + c_4 \cos x$

$y(0) = 0 \quad \Rightarrow \quad c_1 + c_2 + c_3 = 0 \qquad$ ...(3)

$y'(0) = 1 \quad \Rightarrow \quad c_1 - c_2 + c_4 = 1 \qquad$ ...(4)

$y\left(\frac{\pi}{4}\right) = \frac{1}{\sqrt{2}} \quad \Rightarrow \quad c_1 e^{\pi/4} + c_2 e^{-\pi/4} + \frac{c_3}{\sqrt{2}} + \frac{c_4}{\sqrt{2}} = \frac{1}{\sqrt{2}}$

$\Rightarrow \quad \sqrt{2}\, c_1 e^{\pi/4} + \sqrt{2}\, c_2 e^{-\pi/4} + c_3 + c_4 = 1 \qquad$ ...(5)

$y'\left(\frac{\pi}{4}\right) = \frac{1}{\sqrt{2}} \quad \Rightarrow \quad c_1 e^{\pi/4} - c_2 e^{-\pi/4} - \frac{c_3}{\sqrt{2}} + \frac{c_4}{\sqrt{2}} = \frac{1}{\sqrt{2}}$

$\Rightarrow \quad \sqrt{2}\, c_1 e^{\pi/4} - \sqrt{2}\, c_2 e^{-\pi/4} - c_3 + c_4 = 1 \qquad$ ...(6)

Solving (3), (4), (5), (6), we get

$$c_1 = c_2 = c_3 = 0, \quad c_4 = 1$$

$\therefore$ (2) $\Rightarrow \quad y = 0.e^{x} + 0.e^{-x} + 0.\cos x + 1.\sin x \quad$ *i.e.,* $\quad \mathbf{y = \sin x.}$

This is the equation of the required extremal.

Example 4. *Find the extremal of the functional* $\int_{-a}^{a} (\lambda y + \frac{1}{2}\mu y''^2)\, dx$ *which satisfy the boundary conditions* $y(-a) = 0, y'(-a) = 0, y(a) = 0, y'(a) = 0.$

Sol. Given functional is $\int_{-a}^{a} \left(\lambda y + \frac{1}{2} \mu y''^2 \right) dx.$

Let $$F(x, y, y', y'') = \lambda y + \frac{1}{2} \mu y''^2$$

The **Euler-Poisson equation** is $F_y - \frac{d}{dx} F_{y'} + \frac{d^2}{dx^2} F_{y''} = 0$ ...(1)

Here $F_y = \lambda,\quad F_{y'} = 0,\quad F_{y''} = \mu y''$

$$\therefore \quad (1) \Rightarrow \lambda - \frac{d}{dx}(0) + \frac{d^2}{dx^2}(\mu y'') = 0 \Rightarrow \mu \frac{d^4 y}{dx^4} + \lambda = 0 \Rightarrow \frac{d^4 y}{dx^4} = -\frac{\lambda}{\mu}$$

$$\Rightarrow \quad \frac{d^3 y}{dx^3} = -\frac{\lambda}{\mu} x + c_1$$

$$\Rightarrow \quad \frac{d^2 y}{dx^2} = -\frac{\lambda}{\mu} \frac{x^2}{2} + c_1 x + c_2$$

$$\Rightarrow \quad \frac{dy}{dx} = -\frac{\lambda}{\mu} \frac{x^3}{6} + c_1 \frac{x^2}{2} + c_2 x + c_3$$

$$\Rightarrow \quad y = -\frac{\lambda}{\mu} \frac{x^4}{24} + c_1 \frac{x^3}{6} + c_2 \frac{x^2}{2} + c_3 x + c_4 \quad \text{...(2)}$$

This is the equation of the extremals. The boundary conditions are

$$y(-a) = 0, \quad y'(-a) = 0, \quad y(a) = 0, \quad y'(a) = 0$$

$$y(-a) = 0 \Rightarrow -\frac{\lambda a^4}{24\mu} - \frac{c_1 a^3}{6} + \frac{c_2 a^2}{2} - c_3 a + c_4 = 0 \quad \text{...(3)}$$

$$y'(-a) = 0 \Rightarrow \frac{\lambda a^3}{6\mu} + \frac{c_1 a^2}{2} - c_2 a + c_3 = 0 \quad \text{...(4)}$$

$$y(a) = 0 \Rightarrow -\frac{\lambda a^4}{24\mu} + \frac{c_1 a^3}{6} + \frac{c_2 a^2}{2} + c_3 a + c_4 = 0 \quad \text{...(5)}$$

$$y'(a) = 0 \Rightarrow -\frac{\lambda a^3}{6\mu} + \frac{c_1 a^2}{2} + c_2 a + c_3 = 0 \quad \text{...(6)}$$

$$(3) - (5) \Rightarrow -\frac{c_1 a^3}{3} - 2c_3 a = 0 \Rightarrow c_1 a^2 + 6c_3 = 0 \quad \text{...(7)}$$

$$(4) + (6) \Rightarrow c_1 a^2 + 2c_3 = 0 \quad \text{...(8)}$$

Solving (7) and (8), we get $c_1 = 0, c_3 = 0$

$$(4) \Rightarrow \frac{\lambda a^3}{6\mu} + 0 - c_2 a + 0 = 0 \Rightarrow c_2 = \frac{\lambda a^2}{6\mu}$$

$$(3) \Rightarrow -\frac{\lambda a^4}{24\mu} - 0 + \frac{a^2}{2}\left(\frac{\lambda a^2}{6\mu}\right) - 0 + c_4 = 0 \Rightarrow c_4 = -\frac{\lambda a^4}{24\mu}.$$

$$\therefore \qquad c_1 = 0, \quad c_2 = \frac{\lambda a^2}{6\mu}, \quad c_3 = 0, \quad c_4 = -\frac{\lambda a^4}{24\mu}$$

$$\therefore \ (2) \Rightarrow \qquad y = -\frac{\lambda x^4}{24\mu} + 0 + \frac{\lambda a^2}{6\mu} \cdot \frac{x^2}{2} + 0 - \frac{\lambda a^4}{24\mu}$$

$$\therefore \qquad = -\frac{\lambda}{24\mu}(x^4 - 2a^2x^2 + a^4) = -\frac{\lambda}{24\mu}(x^2 - a^2)^2$$

$$\therefore \qquad \mathbf{y = -\frac{\lambda}{24\mu}(x^2 - a^2)^2.}$$

This is the equation of the required extremal.

EXERCISE 4.2

Find the extremals of the following functionals (Q. No. 1—4) :

1. $\int_{x_1}^{x_2} (16y^2 - y''^2 + x^2)\, dx$ **2.** $\int_{x_0}^{x_1} (y''^2 - y^2 + x^2)\, dx$

3. $\int_{x_0}^{x_1} (360x^2y - y''^2)\, dx$ **4.** $\int_{x_1}^{x_2} (y^2 + 2y'^2 + y''^2)\, dx.$

5. Find the extremal of the functional $\int_0^{\pi/2} (y''^2 - y^2 + x^2)\, dx$ that satisfy the boundary conditions: $y(0) = 1, y(\pi/2) = 0, y'(0) = 0, y'(\pi/2) = -1.$

6. Determine the extremal of the functional $J[y(x)] = \int_{-l}^{l} (\tfrac{1}{2}\mu y''^2 + ky)\, dx$ subject to $y(-l) = 0$, $y'(-l) = 0, y(l) = 0, y'(l) = 0.$

7. Find the extremals and the stationary function of the functional $\int_0^1 (y'^2 + y''^2)\, dx$ with the boundary conditions : $y(0) = 0, y(1) = \sinh 1, y'(0) = 1, y'(1) = \cosh 1.$

8. Find the extremals and the stationary function of the functional $\frac{1}{2}\int_0^1 (y'')^2\, dx$. The boundary conditions are $y(0) = 0, y(1) = \frac{1}{2}, y'(0) = 0, y'(1) = 1.$

9. Find the extremal of the functional $\int_a^b (y + y'')\, dx$ that satisfies the boundary conditions $y(a) = y_0$, $y(b) = y_1, y'(a) = y_0', y'(b) = y_1'.$

10. Find the extremal of the functional $J[y(x)] = \int_{-1}^{0} (240y + y'''^2)\, dx$ subject to the conditions: $y(-1) = 1, y(0) = 0, y'(-1) = -4.5, y'(0) = 0, y''(-1) = 16, y''(0) = 0.$

Answers

1. $y = c_1e^{2x} + c_2e^{-2x} + c_3 \cos 2x + c_4 \sin 2x$ **2.** $y = c_1e^{x} + c_2e^{-x} + c_3 \cos x + c_4 \sin x$

3. $y = \frac{x^6}{2} + c_1x^3 + c_2x^2 + c_3x + c_4$ **4.** $y = (c_1 + c_2x)e^{x} + (c_3 + c_4x)e^{-x}$

5. $y = \cos x$ **6.** $y = -\frac{k}{24\mu}(x^2 - l^2)^2$

7. $y = c_1 + c_2x + c_3e^x + c_4e^{-x}$, $y = \sinh x$

8. $y = c_1 + c_2x + c_3x^2 + c_4x^3$, $y = \frac{1}{2}x^2$

9. No extremal

10. $y = \dfrac{x^3}{6}(x^3 + 6x + 1)$.

10. FUNCTIONALS DEPENDENT ON MORE THAN ONE INDEPENDENT VARIABLE

Till now we have been discussing functionals depending on functions of one independent variable *i.e.*, on curves. In the present section we shall consider functionals depending on functions of more than one independent variable *i.e.*, on surfaces. We shall restrict ourselves only two independent variables.

Theorem. *Let* $J[z(x, y)] = \iint_D F(x, y, z, z_x, z_y)\, dx\, dy$ *be a functional defined on the set of functions* $z(x, y)$ *which admits of continuous derivatives upto the second order and assumes given values on the boundary* Γ *of the region D. Also, F is differentiable three times with respect to all its arguments.*

If the functional $J[z(x, y)]$ *has an extremum on a function* $z(x, y)$ *then* $z(x, y)$ *satisfy the differential equation*

$$F_z - \frac{\partial}{\partial x} F_{z_x} - \frac{\partial}{\partial y} F_{z_y} = 0.$$

Proof. We have $J[z(x, y)] = \iint_D F(x, y, z, z_x, z_y)\, dx\, dy.$

Let $z = z(x, y)$ be a function which extremise $J[z(x, y)]$ and assumes given values on the boundary Γ of the region D.

Let $\eta(x, y)$ be any arbitrary but fixed function admiting continuous derivatives upto the second order and taking value zero on the boundary of the region D.

Let $\bar{z}(x, y) = z(x, y) + \alpha\eta(x, y)$, where α is a small parameter.

$\therefore$ $\bar{z}(x, y)$ admits of continuous derivatives upto the second order and assumes given values on the boundary of the region D.

$\therefore$ The functional J is defined at $\bar{z}(x, y)$.

$$\therefore \quad J[\bar{z}(x, y)] = \iint_D F(x, y, \bar{z}, \bar{z}_x, \bar{z}_y)\, dx\, dy$$

$$= \iint_D F(x, y, z + \alpha\eta, z_x + \alpha\eta_x, z_y + \alpha\eta_y)\, dxdy$$

On the regions of the family $z(x, y) + \alpha\eta(x, y)$, the right hand side is a function of α, say $\psi(\alpha)$.

$$\therefore \quad \psi(\alpha) = \iint_D F(x, y, \bar{z}, \bar{z}_x, \bar{z}_y)\, dx\, dy$$

$$= \iint_D F(x, y, z + \alpha\eta, z_x + \alpha\eta_x, z_y + \alpha\eta_y)\, dx\, dy$$

Since the extremising surface $z = z(x, y)$ corresponds to $\alpha = 0$, it follows that $\psi(\alpha)$ is extremised for $\alpha = 0$.

$$\therefore \quad \frac{d\psi}{d\alpha} = 0 \quad \text{at} \quad \alpha = 0.$$

Differentiating $\psi(\alpha)$ under the integral sign, we have

$$\frac{d\psi}{d\alpha} = \iint_D \frac{\partial}{\partial \alpha} F(x, y, \bar{z}, \bar{z}_x, \bar{z}_y)\, dx\, dy$$

$$= \iint_D \left(\frac{\partial F}{\partial x}\frac{\partial x}{\partial \alpha} + \frac{\partial F}{\partial y}\frac{\partial y}{\partial \alpha} + \frac{\partial F}{\partial \bar{z}}\frac{\partial \bar{z}}{\partial \alpha} + \frac{\partial F}{\partial \bar{z}_x}\frac{\partial \bar{z}_x}{\partial \alpha} + \frac{\partial F}{\partial \bar{z}_y}\frac{\partial \bar{z}_y}{\partial \alpha} \right) dx\,dy$$

$$= \iint_D \left(\frac{\partial F}{\partial x}.0 + \frac{\partial F}{\partial y}.0 + \frac{\partial F}{\partial \bar{z}}\eta + \frac{\partial F}{\partial \bar{z}_x}\eta_x + \frac{\partial F}{\partial \bar{z}_y}\eta_y \right) dx\,dy$$

$$\therefore \quad \frac{d\psi}{d\alpha} = \iint_D \left(\frac{\partial F}{\partial \bar{z}}\eta + \frac{\partial F}{\partial \bar{z}_x}\eta_x + \frac{\partial F}{\partial \bar{z}_y}\eta_y \right) dx \cdot dy \qquad \ldots(1)$$

Now $\dfrac{d\psi}{d\alpha} = 0$ at $\alpha = 0$.

Putting $\alpha = 0$ in (1), we get

$$\iint_D \left(\frac{\partial F}{\partial z}\eta + \frac{\partial F}{\partial z_x}\eta_x + \frac{\partial F}{\partial z_y}\eta_y \right) dx\,dy = 0$$

$$\Rightarrow \quad \iint_D \frac{\partial F}{\partial z}\eta\,dx\,dy + \iint_D \left(\frac{\partial F}{\partial z_x}\eta_x + \frac{\partial F}{\partial z_y}\eta_y \right) dx\,dy = 0$$

$$\Rightarrow \quad \iint_D \frac{\partial F}{\partial z}\eta\,dx\,dy + \int_\Gamma \eta \left(\frac{\partial F}{\partial z_x}dy - \frac{\partial F}{\partial z_y}dx \right) - \iint_D \left(\frac{\partial}{\partial x}\left(\frac{\partial F}{\partial z_x}\right) + \frac{\partial}{\partial y}\left(\frac{\partial F}{\partial z_y}\right) \right) \eta\,dx\,dy = 0$$

(By using Green's theorem)

$$\Rightarrow \quad \iint_D \left(\frac{\partial F}{\partial z} - \frac{\partial}{\partial x}\left(\frac{\partial F}{\partial z_x}\right) - \frac{\partial}{\partial y}\left(\frac{\partial F}{\partial z_y}\right) \right) \eta\,dx\,dy + \int_\Gamma \eta \left(\frac{\partial F}{\partial z_x}dy - \frac{\partial F}{\partial z_y}dx \right) = 0$$

$$\Rightarrow \quad \iint_D \left(\frac{\partial F}{\partial z} - \frac{\partial}{\partial x}\left(\frac{\partial F}{\partial z_x}\right) - \frac{\partial}{\partial y}\left(\frac{\partial F}{\partial z_y}\right) \right) \eta\,dx\,dy = 0 \qquad (\because \ \eta = 0 \text{ on } \Gamma)$$

This holds for any arbitrary continuous function $\eta(x)$.

$\therefore$ We have $\dfrac{\partial F}{\partial z} - \dfrac{\partial}{\partial x}\left(\dfrac{\partial F}{\partial z_x}\right) - \dfrac{\partial}{\partial y}\left(\dfrac{\partial F}{\partial z_y}\right) = 0.$

$\therefore$ The function $z(x, y)$ satisfies the equation

$$\mathbf{F_z - \frac{\partial}{\partial x}F_{z_x} - \frac{\partial}{\partial y}F_{z_y} = 0.}$$

This equation is known as the **Euler-Ostrogradsky equation.**

In the above equation $\dfrac{\partial}{\partial x}F_{z_x}$ and $\dfrac{\partial}{\partial y}F_{z_y}$ are the total partial derivatives w.r.t. x and y respectively.

If $p = z_x$ and $q = z_y$, then

$$\frac{\partial}{\partial x}F_{z_x} = \frac{\partial}{\partial x}F_p = F_{px} + F_{pz}\frac{\partial z}{\partial x} + F_{pp}\frac{\partial p}{\partial x} + F_{pq}\frac{\partial q}{\partial x}$$

and

$$\frac{\partial}{\partial y}F_{z_y} = \frac{\partial}{\partial y}F_q = F_{qy} + F_{qz}\frac{\partial z}{\partial y} + F_{qp}\frac{\partial p}{\partial y} + F_{qq}\frac{\partial q}{\partial y}.$$

Putting the values of $\dfrac{\partial}{\partial x}F_{z_x}$ and $\dfrac{\partial}{\partial y}F_{z_y}$, the Euler-Ostrogradsky equation comes out to be a second order partial differential equation.

Example 1. *Find the Euler-Ostrogradsky equation for the functional* $J[z(x, y)]$ $= \iint_D \left[\left(\frac{\partial z}{\partial x}\right)^2 + \left(\frac{\partial z}{\partial y}\right)^2\right] dxdy$, *where the value of z are prescribed on the boundary of the domain D.*

Sol. We have $J[z(x, y)] = \iint \left[\left(\frac{\partial z}{\partial x}\right)^2 + \left(\frac{\partial z}{\partial y}\right)^2\right] dx\,dy$ *i.e.,* $\iint (z_x^{\,2} + z_y^{\,2})\,dx\,dy.$

Let $F(x, y, z, z_x, z_y) = z_x^2 + z_y^2$

The **Euler-Ostrogradsky equation** is

$$F_z - \frac{\partial}{\partial x} F_{z_x} - \frac{\partial}{\partial y} F_{z_y} = 0 \qquad \ldots(1)$$

Here $F_z = 0$, $F_{z_x} = 2z_x$, $F_{z_y} = 2z_y$

$$\therefore \quad (1) \Rightarrow \quad 0 - \frac{\partial}{\partial x}(2z_x) - \frac{\partial}{\partial y}(2z_y) = 0$$

$$\Rightarrow \quad \frac{\partial}{\partial x} z_x + \frac{\partial}{\partial y} z_y = 0 \Rightarrow \frac{\partial^2 \mathbf{z}}{\partial \mathbf{x}^2} + \frac{\partial^2 \mathbf{z}}{\partial \mathbf{y}^2} = \mathbf{0}$$

This is the required equation.

11. VARIATIONAL PROBLEMS IN PARAMETRIC FORM

Theorem. *Let* $J[x(t), y(t)] = \int_{t_1}^{t_2} F(x(t), y(t), \dot{\mathbf{x}}(t), \dot{\mathbf{y}}(t))dt$ *be a functional defined on the set of parametric functions* $x = x(t)$, $y = y(t)$ *which admits of continuous first order derivatives and satisfy the boundary conditions :* $x(t_1) = x_1$, $y(t_1) = y_1$, $x(t_2) = x_2$, $y(t_2) = y_2$, *where* x_1, y_1, x_2, y_2 *are prescribed at the fixed boundary points* t_1 *and* t_2. *Also F is a positive homogeneous function of first degree in* $\dot{x}(t)$ *and* $\dot{y}(t)$ *and F is differentiable three times with respect to all its arguments.*

If the functional $J[x(t), y(t)]$ *has an extremum on a parametric function* $x = x(t)$, $y = y(t)$ *then the functions* $x(t)$ *and* $y(t)$ *satisfy the Euler's equations*

$$F_x - \frac{d}{dt} F_{\dot{x}} = 0 \text{ and } F_y - \frac{d}{dt} F_{\dot{y}} = 0 \text{ respectively.}$$

Note. We accept this theorem without proof.

Remarks 1. One of the Euler's equations $\mathbf{F_x - \frac{d}{dt}F_{\dot{x}} = 0}$ **and** $\mathbf{F_y - \frac{d}{dt}F_{\dot{y}} = 0}$ **is a consequence of the other.**

2. The Weierstrarsian form of Euler's equations is

$$\frac{1}{r} = \frac{F_{x\dot{y}} - F_{y\dot{x}}}{F_1(\dot{x}^2 + \dot{y}^2)^{3/2}},$$

where r is the radius of curvature of the extremal and F_1 is the common value of the ratios

$$\frac{F_{\dot{x}\dot{x}}}{\dot{y}^2} = \frac{F_{\dot{y}\dot{y}}}{\dot{x}^2} = \frac{F_{\dot{x}\dot{y}}}{-\dot{x}\dot{y}}.$$

Example 2. *Find the extremals of the functional* $\int_{t_1}^{t_2} ((\dot{x}^2 + \dot{y}^2)^{1/2} + a^2 (x\dot{y} - y\dot{x}))\, dt$.

Sol. Given functional is $\int_{t_1}^{t_2} ((\dot{x}^2 + \dot{y}^2)^{1/2} + a^2 (x\dot{y} - y\dot{x}))\, dt$

Let $\quad F(x, y, \dot{x}, \dot{y}) = (\dot{x}^2 + \dot{y}^2)^{1/2} + a^2 (x\dot{y} - y\dot{x})$

$\therefore \quad F(x, y, \lambda\dot{x}, \lambda\dot{y}) = ((\lambda\dot{x})^2 + (\lambda\dot{y})^2)^{1/2} + a^2 (x(\lambda\dot{y}) - y(\lambda\dot{x}))$

$$= \lambda(\dot{x}^2 + \dot{y}^2)^{1/2} + \lambda a^2(x\dot{y} - y\dot{x}) \qquad (\because \lambda > 0)$$

$$= \lambda F(x, y, \dot{x}, \dot{y}).$$

$\therefore$ F is positive homogeneous function of degree one in $\dot{x}$ and $\dot{y}$.

Weierstrarsian form of Euler's equations is

$$\frac{1}{r} = \frac{F_{x\dot{y}} - F_{y\dot{x}}}{F_1 (\dot{x}^2 + \dot{y}^2)^{3/2}}. \qquad ...(1)$$

Also,
$$F_1 = \frac{F_{\dot{x}\dot{x}}}{\dot{y}^2} = \frac{F_{\dot{y}\dot{y}}}{\dot{x}^2} = \frac{F_{\dot{x}\dot{y}}}{-\dot{x}\dot{y}} \qquad ...(2)$$

Now,
$$F_x = a^2\dot{y}, \quad F_y = -a^2\dot{x}, \quad F_{\dot{x}} = \frac{\dot{x}}{\sqrt{\dot{x}^2 + \dot{y}^2}} - a^2 y$$

$\therefore$
$$F_{x\dot{y}} = a^2, \quad F_{y\dot{x}} = -a^2,$$

$$F_{\dot{x}\dot{x}} = \frac{\sqrt{\dot{x}^2 + \dot{y}^2}\,.\,1 - \dot{x}\dfrac{\dot{x}}{\sqrt{\dot{x}^2 + \dot{y}^2}}}{\dot{x}^2 + \dot{y}^2} - 0 = \frac{1}{(\dot{x}^2 + \dot{y}^2)^{3/2}}(\dot{x}^2 + \dot{y}^2 - \dot{x}^2) = \frac{\dot{y}^2}{(\dot{x}^2 + \dot{y}^2)^{3/2}}$$

$\therefore$ (2) $\Rightarrow$
$$F_1 = \frac{F_{\dot{x}\dot{x}}}{\dot{y}^2} = \frac{1}{\dot{y}^2}\cdot\frac{\dot{y}^2}{(\dot{x}^2 + \dot{y}^2)^{3/2}} = \frac{1}{(\dot{x}^2 + \dot{y}^2)^{3/2}}$$

$\therefore$ (1) $\Rightarrow$
$$\frac{1}{r} = \frac{a^2 - (-a^2)}{\dfrac{1}{(\dot{x}^2 + \dot{y}^2)^{3/2}}\cdot(\dot{x}^2 + \dot{y}^2)^{3/2}} = 2a^2$$

$\therefore$ The curvature $(1/r)$ of the extremals is $2a^2$, a constant.

$\therefore$ Extremals are arcs of circles of fixed radius, r, given by $1/2a^2$.

Remark. Extremals are full circles if $x(t_1) = x(t_2)$, $y(t_1) = y(t_2)$.

Example 3. *Find the extremals of the functional* $\int_{t_1}^{t_2} y\sqrt{\dot{x}^2 + \dot{y}^2}\, dt$.

Sol. Given functional is $\int_{t_1}^{t_2} y\sqrt{\dot{x}^2 + \dot{y}^2}\, dt$

Let $\quad F(x, y, \dot{x}, \dot{y}) = y\sqrt{\dot{x}^2 + \dot{y}^2}$.

F is a positive homogeneous function of degree one in $\dot{x}$ and $\dot{y}$

The **Euler's equations** are

$$F_x - \frac{d}{dt}F_{\dot{x}} = 0 \quad ...(1) \quad \text{and} \quad F_y - \frac{d}{dt}F_{\dot{y}} = 0 \quad ...(2)$$

We use (1) to find the extremals.

Here $F_x = 0, \quad F_{\dot{x}} = y \cdot \dfrac{2\dot{x}}{2\sqrt{\dot{x}^2 + \dot{y}^2}} = \dfrac{y\dot{x}}{\sqrt{\dot{x}^2 + \dot{y}^2}}$

$$\therefore \quad (1) \Rightarrow \quad 0 - \frac{d}{dt}\left(\frac{y\dot{x}}{\sqrt{\dot{x}^2 + \dot{y}^2}}\right) = 0 \Rightarrow \frac{\dot{x}y}{\sqrt{\dot{x}^2 + \dot{y}^2}} = c_1 \qquad ...(3)$$

Case I. $c_1 = 0$

$$(3) \Rightarrow \quad \dot{x}y = 0 \quad \Rightarrow \quad \dot{x} = 0, y = 0 \quad \Rightarrow \quad x = c_2, y = 0.$$

Case II. $c_1 \neq 0$

$$(3) \Rightarrow \frac{y}{\sqrt{1 + (\dot{y}/\dot{x})^2}} = c_1 \quad \Rightarrow \quad \frac{y}{\sqrt{1 + y'^2}} = c_1 \Rightarrow \frac{y^2}{c_1^2} = 1 + y'^2$$

$$\Rightarrow \quad y'^2 = \frac{y^2 - c_1^2}{c_1^2} \quad \Rightarrow \quad \frac{c_1 dy}{\sqrt{y^2 - c_1^2}} = dx \Rightarrow c_1 \cosh^{-1}\frac{y}{c_1} = x + c_3$$

$$\Rightarrow \quad \frac{y}{c_1} = \cosh\frac{x + c_3}{c_1} \quad \Rightarrow \quad y = c_1 \cosh\frac{x + c_3}{c_1}.$$

$\therefore$ The extremals are $\mathbf{x = c_2, y = 0}$ **and** $\mathbf{y = c_1 \cosh \dfrac{x + c_3}{c_1}}$.

Remarks. (2) $\Rightarrow \sqrt{\dot{x}^2 + \dot{y}^2} - \dfrac{d}{dt}\left(\dfrac{y\dot{y}}{\sqrt{\dot{x}^2 + \dot{y}^2}}\right) = 0 \Rightarrow \dfrac{d}{dt}\left(\dfrac{y\dot{y}}{\sqrt{\dot{x}^2 + \dot{y}^2}}\right) = \sqrt{\dot{x}^2 + \dot{y}^2}$

$$\Rightarrow \frac{1}{\sqrt{\dot{x}^2 + \dot{y}^2}}\frac{d}{dt}\left(\frac{y\dot{y}}{\sqrt{\dot{x}^2 + \dot{y}^2}}\right) = 1 \quad \Rightarrow 2\left(\frac{y\dot{y}}{\sqrt{\dot{x}^2 + \dot{y}^2}}\right)\frac{d}{dt}\left(\frac{y\dot{y}}{\sqrt{\dot{x}^2 + \dot{y}^2}}\right) = (2y)\,\dot{y}$$

$$\Rightarrow \quad \frac{d}{dt}\left(\frac{(y\dot{y})^2}{\dot{x}^2 + \dot{y}^2}\right) = \frac{d}{dt}(y^2) \Rightarrow \frac{y^2\dot{y}^2}{\dot{x}^2 + \dot{y}^2} = y^2 + \lambda \Rightarrow \frac{y^2\dot{y}^2}{\dot{x}^2 + \dot{y}^2} - y^2 = \lambda$$

$$\Rightarrow \quad \frac{\dot{x}^2 y^2}{\dot{x}^2 + \dot{y}^2} = \lambda \quad \Rightarrow \frac{\dot{x}y}{\sqrt{\dot{x}^2 + \dot{y}^2}} = \sqrt{\lambda} \Rightarrow \frac{\dot{x}y}{\sqrt{\dot{x}^2 + \dot{y}^2}} = c_1, \text{ where } c_1 = \sqrt{\lambda}.$$

This is same as (3).

$\therefore$ Equations (2) and (1) give the same solution.

12. FUNCTIONALS DEPENDENT ON MORE THAN ONE FUNCTION

Theorem. *Let* $J[y_1(x), y_2(x), ..., y_n(x)] = \int_{x_1}^{x_2} F(x, y_1(x), y_2(x), ... y_n(x), y_1'(x), y_2'(x), ...,$ $y_n'(x))dx$ *be a functional defined on the set of n functions* $y_1(x), y_2(x), ..., y_n(x)$ *which admits of continuous first order derivative and satisfy the boundary conditions :*

$$y_1(x_1) = Y_1, \quad y_2(x_1) = Y_2, ... , y_n(x_1) = Y_n,$$
$$y_1(x_2) = Z_1, \quad y_2(x_2) = Z_2, ... , y_n(x_2) = Z_n,$$

where $Y_1, Y_2, ..., Y_n, Z_1, Z_2 ......, Z_n$ *are prescribed at the fixed boundary points* x_1 *and* x_2. *Also, F is differentiable three times with respect to all its arguments.*

If the functional $J[y_1(x), y_2(x), ..., y_n(x)]$ has an extremum on functions $y_1(x), y_2(x), ...$ $y_n(x)$, then $y_1(x), y_2(x), ..., y_n(x)$ satisfy the Euler's equations :

$$F_{y_1} - \frac{d}{dx} F_{y_1'} = 0,\quad F_{y_2} - \frac{d}{dx} F_{y_2'} = 0,\ ......,\ F_{y_n} - \frac{d}{dx} F_{y_n'} = 0.$$

Note. Proof of this theorem is beyond the scope of this book.

Example 4. *Show that the functional* $\int_0^1 \left(2x + \left(\frac{dx}{dt}\right)^2 + \left(\frac{dy}{dt}\right)^2\right) dt$ such that $x(0) = 1$, $y(0) = 1$, $x(1) = 1.5$, $y(1) = 1$ *is stationary for* $x = \frac{2+t^2}{2}$, $y = 1$.

Sol. Given functional is $\int_0^1 \left(2x + \left(\frac{dx}{dt}\right)^2 + \left(\frac{dy}{dt}\right)^2\right) dt$ *i.e.,* $\int_0^1 (2x + x'^2 + y'^2)\, dt$

Let $\quad F(t, x, y, x', y') = 2x + x'^2 + y'^2$

The **Euler's equations** are

$$F_x - \frac{d}{dt} F_{x'} = 0 \qquad ...(1)$$

and

$$F_y - \frac{d}{dt} F_{y'} = 0 \qquad ...(2)$$

Here $F_x = 2,\ F_y = 0,\ F_{x'} = 2x',\ F_{y'} = 2y'$

$\therefore$ (1) $\Rightarrow$ $2 - \frac{d}{dt}(2x') = 0 \Rightarrow \frac{d^2x}{dt^2} = 1 \Rightarrow \frac{dx}{dt} = t + c_1$

$$\Rightarrow \quad x = \frac{t^2}{2} + c_1 t + c_2 \qquad ...(3)$$

(2) $\Rightarrow$ $0 - \frac{d}{dt}(2y') = 0 \Rightarrow \frac{dy'}{dt} = 0 \Rightarrow y' = c_3$

$$\Rightarrow \quad y = c_3 t + c_4 \qquad ...(4)$$

$\therefore$ (3) and (4) are the equations of the extremals.

The boundary conditions are $x(0) = 1,\ y(0) = 1,\ x(1) = 1.5,\ y(1) = 1$

$x(0) = 1 \Rightarrow 0 + c_1(0) + c_2 = 1 \Rightarrow c_2 = 1$

$y(0) = 1 \Rightarrow c_3(0) + c_4 = 1 \Rightarrow c_4 = 1$

$x(1) = 1.5 \Rightarrow \frac{1}{2} + c_1(1) + c_2 = 1.5 \Rightarrow \frac{1}{2} + c_1 + 1 = 1.5 \Rightarrow c_1 = 0$

$y(1) = 1 \Rightarrow c_3(1) + c_4 = 1 \Rightarrow c_3 + 1 = 1 \Rightarrow c_3 = 0$

$\therefore \quad c_1 = 0,\ c_2 = 1,\ c_3 = 0,\ c_4 = 1$

$\therefore$ (3) $\Rightarrow$ $x = \frac{t^2}{2} + 0.t + 1$ *i.e.,* $x = \frac{2+t^2}{2}$

(4) $\Rightarrow$ $y = 0.t + 1$ *i.e.,* $y = 1$

$\therefore$ The stationary functions are $\mathbf{x = \frac{2+t^2}{2}}$, $\mathbf{y = 1}$.

Remark. Since, the function $2x + x'^2 + y'^2$ is not a homogeneous functions of x' and y', we have treated the given functional as dependent on two functions.

Example 5. *Find the extremals of the functional* $J[y(x), z(x)] = \int_0^{\pi} (2yz - 2y^2 + y'^2 - z'^2)\, dx.$ *Also find that extremal which satisfy the boundary conditions :*

$$y(0) = 0, \quad y(\pi) = 1, \quad z(0) = 0, \quad z(\pi) = 1.$$

Sol. We have $J[y(x), z(x)] = \int_0^{\pi} (2yz - 2y^2 + y'^2 - z'^2)\, dx$

Let $F(x, y, z, y', z') = 2yz - 2y^2 + y'^2 - z'^2$

The **Euler's equations** are

$$F_y - \frac{d}{dx} F_{y'} = 0 \quad \text{...(1)}$$

and

$$F_z - \frac{d}{dx} F_{z'} = 0 \quad \text{...(2)}$$

Here $F_y = 2z - 4y$, $F_z = 2y$, $F_{y'} = 2y'$, $F_{z'} = -2z'$

$$\therefore \quad (1) \Rightarrow \quad 2z - 4y - \frac{d}{dx}(2y') = 0 \Rightarrow \frac{d^2y}{dx^2} + 2y - z = 0 \quad \text{...(3)}$$

$$(2) \Rightarrow \quad 2y - \frac{d}{dx}(-2z') = 0 \Rightarrow \frac{d^2z}{dx^2} + y = 0 \quad \text{...(4)}$$

$$(3) \Rightarrow \quad (D^2 + 2)y - z = 0 \quad \text{...(5)}$$

$$(4) \Rightarrow \quad y + D^2 z = 0 \quad \text{...(6)}$$

Operating (5) by D^2 and adding to (6), we get

$$D^2(D^2 + 2)y + y = 0$$

$$\Rightarrow \quad (D^4 + 2D^2 + 1)y = 0 \Rightarrow (D^2 + 1)^2 y = 0$$

The A.E. is $(D^2 + 1)^2 = 0 \quad \therefore \quad D = \pm i, \ \pm i$

$$\therefore \quad y = e^{0.x}((c_1 + c_2 x) \cos 1.x + (c_3 + c_4 x) \sin 1.x)$$

i.e.,

$$y = (c_1 + c_2 x) \cos x + (c_3 + c_4 x) \sin x \quad \text{...(7)}$$

$$(3) \Rightarrow \quad z = \frac{d^2y}{dx^2} + 2y \quad \text{...(8)}$$

$$(7) \Rightarrow \quad \frac{dy}{dx} = c_2 \cos x - (c_1 + c_2 x) \sin x + c_4 \sin x + (c_3 + c_4 x) \cos x$$

$$= (c_2 + c_3 + c_4 x) \cos x + (c_4 - c_1 - c_2 x) \sin x$$

$$\therefore \quad \frac{d^2y}{dx^2} = c_4 \cos x - (c_2 + c_3 + c_4 x) \sin x - c_2 \sin x + (c_4 - c_1 - c_2 x) \cos x$$

$$= (2c_4 - c_1 - c_2 x) \cos x - (2c_2 + c_3 + c_4 x) \sin x$$

$$\therefore \quad (8) \Rightarrow \quad z = (2c_4 - c_1 - c_2 x) \cos x - (2c_2 + c_3 + c_4 x) \sin x$$

$$+ 2(c_1 + c_2 x) \cos x + 2(c_3 + c_4 x) \sin x$$

$$= (2c_4 + c_1 + c_2 x) \cos x + (c_3 - 2c_2 + c_4 x) \sin x$$

$\therefore$ The family of extremals is

$$\mathbf{y = (c_1 + c_2 x) \cos x + (c_3 + c_4 x) \sin x,}$$

$$\mathbf{z = (2c_4 + c_1 + c_2 x) \cos x + (c_3 - 2c_2 + c_4 x) \sin x,}$$

where c_1, c_2, c_3, c_4 are arbitrary constants.

The boundary conditions are:

$$y(0) = 0, \quad y(\pi) = 1, \quad z(0) = 0, \quad z(\pi) = 1$$

$y(0) = 0 \quad \Rightarrow (c_1 + c_2.0)\cos 0 + (c_3 + c_4.0)\sin 0 = 0 \quad \Rightarrow \quad c_1 = 0$

$y(\pi) = 1 \quad \Rightarrow (c_1 + \pi c_2)\cos \pi + (c_3 + \pi c_4)\sin \pi = 1$

$\Rightarrow (0 + \pi c_2)(-1) = 1 \quad \Rightarrow \quad c_2 = -1/\pi.$

$z(0) = 0 \quad \Rightarrow (2c_4 + c_1 + c_2.0)\cos 0 + (c_3 - 2c_2 + c_4 . 0)\sin 0 = 0$

$\Rightarrow (2c_4 + 0 + 0) . 1 + \left(c_3 + \dfrac{2}{\pi} + 0\right) . 0 = 0 \quad \Rightarrow \quad c_4 = 0$

$z(\pi) = 1 \quad \Rightarrow (2(0) + 0 + \pi(-1/\pi))\cos \pi \ + (c_3 - 2(-1/\pi) + 0.\pi)\sin \pi = 1$

$\Rightarrow (-1)(-1) = 1$, which is true.

$\therefore \quad c_1 = 0, \quad c_2 = -1/\pi, \quad c_4 = 0$

$$\therefore \quad y = \left(0 - \frac{x}{\pi}\right)\cos x + (c_3 + 0.x)\sin x = c_3 \sin x - \frac{x}{\pi}\cos x$$

$$z = \left(2(0) + 0 - \frac{x}{\pi}\right)\cos x + \left(c_3 - 2\left(-\frac{1}{\pi}\right) + 0.x\right)\sin x$$

$$= c_3 \sin x + \frac{1}{\pi}(2 \sin x - x \cos x)$$

$\therefore$ The required extremals are $\mathbf{y = c_3 \sin x - \dfrac{x}{\pi} \cos x}$

$$\mathbf{z = c_3 \sin x + \frac{1}{\pi}(2 \sin x - x \cos x)},$$

where c_3 is any arbitrary constant.

Example 6. *Find the extremal of the functional* $\int_0^1 \sqrt{1 + y'^2 + z'^2}\, dx$ *that satisfy the boundary conditions* $y(0) = 0, y(1) = 2, z(0) = 0, z(1) = 4$.

Sol. Given functional is $\int_0^1 \sqrt{1 + y'^2 + z'^2}\, dx$.

Let $\quad F(x, y, z, y', z') = \sqrt{1 + y'^2 + z'^2}$

The **Euler's equations** are $F_y - \dfrac{d}{dx} F_{y'} = 0$ ...(1)

and $\quad F_z - \dfrac{d}{dx} F_{z'} = 0$ ...(2)

Here $F_y = 0, F_z = 0, F_{y'} = \dfrac{y'}{\sqrt{1 + y'^2 + z'^2}}, F_{z'} = \dfrac{z'}{\sqrt{1 + y'^2 + z'^2}}$

$$\therefore \quad (1) \quad \Rightarrow \quad 0 - \frac{d}{dx}\left(\frac{y'}{\sqrt{1 + y'^2 + z'^2}}\right) = 0$$

$$\Rightarrow \quad \frac{y'}{\sqrt{1 + y'^2 + z'^2}} = c_1 \quad \text{...(3)}$$

$$(2) \quad \Rightarrow \quad 0 - \frac{d}{dx}\left(\frac{z'}{\sqrt{1+y'^2+z'^2}}\right) = 0$$

$$\Rightarrow \quad \frac{z'}{\sqrt{1+y'^2+z'^2}} = c_2 \qquad \text{...(4)}$$

Dividing (3) by (4), we get $\frac{y'}{z'} = \frac{c_1}{c_2}, = c_3$ say

$\therefore \quad y' = c_3 z'$

Putting the value of y' in (4), we get $\dfrac{z'}{\sqrt{1+(1+c_3^2)\,z'^2}} = c_2$

$$\Rightarrow \quad z'^2 = c_2^2(1 + (1 + c_3^2)\,z'^2) \Rightarrow z' = \frac{c_2}{\sqrt{1-c_2^2-c_2^2\,c_3^2}} = c_4, \text{ (say)}$$

$$\Rightarrow \quad z = c_4 x + c_5 \qquad \text{...(5)}$$

$$\therefore \quad y' = c_3 z' \Rightarrow y' = c_3 c_4 = c_6, \text{ say}$$

$$\Rightarrow \quad y = c_6 x + c_7 \qquad \text{...(6)}$$

$\therefore$ The extremals are given by (5) and (6).

The boundary conditions are

$$y(0) = 0, \quad y(1) = 2, \quad z(0) = 0, \quad z(1) = 4$$

$$\therefore \quad c_6(0) + c_7 = 0, \quad c_6(1) + c_7 = 2, \; c_4(0) + c_5 = 0, \quad c_4(1) + c_5 = 4$$

Solving these equations, we get

$$c_6 = 2, \; c_7 = 0, \; c_4 = 4, \; c_5 = 0$$

(6) $\Rightarrow$ $y = 2.x + 0 \Rightarrow y = 2x$

(5) $\Rightarrow$ $z = 4.x + 0 \Rightarrow z = 4x$

$\therefore$ The required extremals are **y = 2x, z = 4x.**

EXERCISE 4.3

1. Find the Euler-Ostrogradsky equation for the functional $J[z(x, y)] = \iint_D \left[\left(\frac{\partial z}{\partial x}\right)^2 - \left(\frac{\partial z}{\partial y}\right)^2\right] dx\, dy$, where the values of z are prescribed on the boundary of the domain D.

2. Find the Euler-Ostrogradsky equation for the functional $J[z(x, y)] = \iint_D \left[\left(\frac{\partial z}{\partial x}\right)^4 + \left(\frac{\partial z}{\partial y}\right)^4 + 12z\, f(x, y)\right] dx\, dy$, where the values of z are prescribed on the boundary of the domain D.

3. Find the extremals of the functional $\int_{t_1}^{t_2} \sqrt{\dot{x}^2 + \dot{y}^2}\, dt$.

4. Find the stationary function of the functional $\int_{(-1,0)}^{(1,0)} (k\sqrt{\dot{x}^2 + \dot{y}^2} - \dot{x}y)\, dt, \; k > 0$.

5. Find the extremal of the functional $\int_0^1 (y'^2 + z'^2)\, dx$ that satisfy the boundary conditions $y(0) = 0$, $z(0) = 0$, $y(1) = 1$, $z(1) = 2$.

6. Find the extremal of the functional $\int_0^1 (y'^2 + z'^2 + 4z)\, dx$ that satisfy the boundary conditions $y(0) = 0$, $y(1) = 1$, $z(0) = 0$, $z(1) = 0$.

7. Find the extremal of the functional $\int_0^{\pi/2} (y'^2 + z'^2 + 2yz)\, dx$ that satisfy the boundary conditions $y(0) = 0$, $y(\pi/2) = -1$, $z(0) = 0$, $z(\pi/2) = 1$.

Answers

1. $\dfrac{\partial^2 z}{\partial x^2} - \dfrac{\partial^2 z}{\partial y^2} = 0$

2. $\left(\dfrac{\partial z}{\partial x}\right)^2 \dfrac{\partial^2 z}{\partial x^2} + \left(\dfrac{\partial z}{\partial y}\right)^2 \dfrac{\partial^2 z}{\partial y^2} = f(x, y)$

3. $y = c_1 x + c_2$, $x = c_3$

4. Arc of circle joining $(-1, 0)$ to $(1, 0)$ and of radius k.

5. $y = x$, $z = 2x$

6. $y = x$, $z = x^2 - x$

7. $y = -\sin x$, $z = \sin x$.

13. INVARIANCE OF EULER'S EQUATION UNDER COORDINATES TRANSFORMATION

Let $\quad J[y(x)] = \int_{x_1}^{x_2} F(x, y, y')\, dx$ be a functional to be extremised.

Let u and v be two variables defined by $x = x(u, v)$, $y = y(u, v)$ and $\begin{vmatrix} x_u & x_v \\ y_u & y_v \end{vmatrix} \neq 0$. Then the curve given by the equation $y = y(x)$ in the xy-plane corresponds to the curve given by the equation $v = v(u)$ in the uv-plane.

When we change the variables x and y, the functional

$$J[y(x)] = \int_{x_1}^{x_2} F(x, y, y')\, dx$$

changes into the functional

$$J_1[v(u)] = \int_{u_1}^{u_2} F\left(x(u, v), y(u, v), \frac{y_u + y_v v'}{x_u + x_v v'}\right)(x_u + x_v v')\, du$$

$$\left(\because \frac{dy}{dx} = \frac{dy/du}{dx/du} = \frac{y_u + y_v v'}{x_u + x_v v'}\right)$$

$$= \int_{u_1}^{u_2} G(u, v, v')\, du, \quad \text{(say)}$$

If $y = y(x)$ satisfies the Euler's equation $F_y - \dfrac{d}{dx} F_{y'} = 0$ corresponding to the original functional $J[y(x)]$ then it can be proved that the function $v = v(u)$ satisfies the Euler's equation $G_v - \dfrac{d}{du} G_{v'} = 0$ corresponding to the new functional $J_1[v(u)]$. In other words, $v(u)$ is an extremal of $J_1[v(u)]$ if $y(x)$ is an extremal of $J[y(x)]$. Thus, whether or not a curve is an extremal is a property which is independent of the choice of the coordinate system.

$\therefore$ The extremals of the functional J[y(x)] can be found by solving Euler's equation of the transformed functional $J_1[v(u)]$. This is called the **principle of invariance of Euler's equation under coordinates transformations.**

In solving practical problems, the change of variables is made directly in the integral representing the functional. We then write and solve the Euler's equation for the new integral involving new variables. In the extremals so obtained, the new variables are changed to the original variables to get the desired extremals.

Example. *Find the extremals of the functional* $\int_{\theta_1}^{\theta_2} \sqrt{r^2 + r'^2}\, d\theta$, *where* $r = r(\theta)$.

Sol. Let $$J[r(\theta)] = \int_{\theta_1}^{\theta_2} \sqrt{r^2 + r'^2}\, d\theta.$$

Let $$x = r\cos\theta, \quad y = r\sin\theta.$$

$\therefore$ $$r = \sqrt{x^2 + y^2}, \quad \theta = \tan^{-1}\frac{y}{x}$$

$$r' = \frac{dr}{d\theta} = \frac{dr/dx}{d\theta/dx} = \frac{r_x + r_y y'}{\theta_x + \theta_y y'} = \frac{\dfrac{x}{\sqrt{x^2+y^2}} + \dfrac{y}{\sqrt{x^2+y^2}}\, y'}{\dfrac{1}{1+\dfrac{y^2}{x^2}}\left(-\dfrac{y}{x^2}\right) + \dfrac{1}{1+\dfrac{y^2}{x^2}}\left(\dfrac{1}{x}\right) y'}$$

$$= \frac{\dfrac{x + yy'}{\sqrt{x^2+y^2}}}{\dfrac{xy' - y}{x^2 + y^2}} = \frac{\sqrt{x^2+y^2}\,(x + yy')}{xy' - y}$$

Also, $$d\theta = \frac{d\theta}{dx}\,.\,dx = (\theta_x + \theta_y y')dx = \frac{xy' - y}{x^2 + y^2}\, dx$$

$\therefore$ $$r^2 + r'^2 = x^2 + y^2 + \frac{(x^2+y^2)(x+yy')^2}{(xy'-y)^2}$$

$$= (x^2 + y^2)\left[\frac{(xy'-y)^2 + (x+yy')^2}{(xy'-y)^2}\right] = \frac{(x^2+y^2)(1+y'^2)\,(x^2+y^2)}{(xy'-y)^2}$$

$\therefore$ $$\sqrt{r^2 + r'^2} = \frac{(x^2+y^2)\sqrt{1+y'^2}}{xy' - y}$$

$\therefore$ $$\int \sqrt{r^2 + r'^2}\, d\theta = \int \frac{(x^2+y^2)\sqrt{1+y'^2}}{xy'-y}\,.\,\frac{xy'-y}{x^2+y^2}\, dx = \int \sqrt{1+y'^2}\, dx$$

Let $\theta = \theta_1 \Rightarrow x = x_1$ and $\theta = \theta_2 \Rightarrow x = x_2$.

Let the transformed functional be denoted by $J_1[y(x)]$

$\therefore$ $$J_1[y(x)] = \int_{x_1}^{x_2} \sqrt{1 + y'^2}\, dx \qquad ...(1)$$

Let $\qquad G(x, y, y') = \sqrt{1+y'^2}$

The **Euler's equation** for (1) is $\quad G_y - \dfrac{d}{dx} G_{y'} = 0 \qquad$ ...(2)

Here $\quad G_y = 0, \quad G_{y'} = \dfrac{y'}{\sqrt{1+y'^2}}$

$\therefore \quad (2) \Rightarrow \qquad 0 - \dfrac{d}{dx}\left(\dfrac{y'}{\sqrt{1+y'^2}}\right) = 0$

$\Rightarrow \qquad \dfrac{y'}{\sqrt{1+y'^2}} = k \Rightarrow y'^2 = k^2(1+y'^2) \Rightarrow y' = \dfrac{k}{\sqrt{1-k^2}} = c_1, \text{ (say)}$

$\Rightarrow \qquad y = c_1 x + c_2$

$\therefore$ The extremals of the given functional are $\mathbf{r \sin \theta = c_1 r \cos \theta + c_2}$.

Alternative method

We have $\qquad x = r\cos\theta, \quad y = r \sin \theta.$

$\therefore \qquad \dfrac{dx}{d\theta} = -r\sin\theta + r'\cos\theta, \qquad \dfrac{dy}{d\theta} = r\cos\theta + r'\sin\theta, \quad \text{where } r' = \dfrac{dr}{d\theta}$

$\therefore \qquad \left(\dfrac{dx}{d\theta}\right)^2 + \left(\dfrac{dy}{d\theta}\right)^2 = (-r\sin\theta + r'\cos\theta)^2 + (r\cos\theta + r'\sin\theta)^2 = r^2 + r'^2$

$$\therefore \quad \int \sqrt{r^2+r'^2}\, d\theta = \int \sqrt{\left(\frac{dx}{d\theta}\right)^2 + \left(\frac{dy}{d\theta}\right)^2}\, d\theta = \int \sqrt{(dx)^2 + (dy)^2}$$

$$= \int \sqrt{1 + \left(\frac{dy}{dx}\right)^2}\, dx = \int \sqrt{1+y'^2}\, dx.$$

Remark. The Euler's equation of $J[r(\theta)]$ is

$$\frac{r}{\sqrt{r^2+r'^2}} - \frac{d}{d\theta}\left(\frac{r'}{\sqrt{r^2+r'^2}}\right) = 0,$$

which is difficult to solve.

EXERCISE 4.4

1. Find the extremals of the functional $J[y(x)] = \displaystyle\int_0^{\log 2} (e^{-x}y'^2 - e^x y^2)\, dx$ by using the transformations:

$x = \log u, \quad y = v.$

2. Find the extremals of the functional $\displaystyle\int_{\theta_1}^{\theta_2} r\sin\theta\sqrt{r^2+r'^2}\, d\theta$ by using the transformations:

$x = r\cos\theta, \quad y = r\sin\theta.$

Answers

1. $y = c_1 \cos e^x + c_2 \sin e^x$
2. $c_1 \log (r \sin \theta + \sqrt{r^2 \sin^2 \theta - c_1^2}) = r \cos \theta + c_2$.

Hint

1. Use: $y' = \dfrac{dy}{dx} = \dfrac{dy}{dv}\dfrac{dv}{du}\dfrac{du}{dx} = 1.\dfrac{dv}{du}.e^x = u\dfrac{dv}{du}$.

14. CONDITIONAL EXTREMUM

Theorem. *If a curve $y = y(x)$ extremises the functional $\int_{x_1}^{x_2} F(x, y, y')\,dx$ subject to boundary conditions: $y(x_1) = y_1, y(x_2) = y_2$ and one more condition $\int_{x_1}^{x_2} G(x, y, y')\,dx = L$ and if $y = y(x)$ is not an extremal of the functional $\int_{x_1}^{x_2} G(x, y, y')\,dx$, then there exists a constant λ such that the curve $y = y(x)$ satisfy the Euler's equation $F_y - \dfrac{d}{dx}F_{y'} + \lambda\left(G_y - \dfrac{d}{dx}G_{y'}\right) = 0$.*

Note. The proof of this theorem is beyond the scope of this book.

Example 1. *Find the extremal of the functional $\int_0^1 y'^2\,dx$, $y(0) = 1, y(1) = 6$* subject to the condition $\int_0^1 y\,dx = 3$.

Sol. Given functional is $\int_0^1 y'^2\,dx$.

Given condition is $\int_0^1 y\,dx = 3$ ...(1)

Let $F(x, y, y') = y'^2$ and $G(x, y, y') = y$

The **Euler's equation** is $F_y - \dfrac{d}{dx}F_{y'} + \lambda\left(G_y - \dfrac{d}{dx}G_{y'}\right) = 0$ ...(2)

Here $F_y = 0, F_{y'} = 2y', G_y = 1, G_{y'} = 0$

$$\therefore \quad (2) \Rightarrow \quad 0 - \frac{d}{dx}(2y') + \lambda\left(1 - \frac{d}{dx}(0)\right) = 0$$

$$\Rightarrow \quad -2\frac{d^2y}{dx^2} + \lambda = 0 \quad \Rightarrow \quad \frac{d^2y}{dx^2} = \frac{\lambda}{2}$$

$$\Rightarrow \quad \frac{dy}{dx} = \frac{\lambda x}{2} + c_1 \quad \Rightarrow \quad y = \frac{\lambda x^2}{4} + c_1 x + c_2 \qquad ...(3)$$

The boundary conditions are $y(0) = 1, \; y(1) = 6$

$y(0) = 1 \Rightarrow 0 + c_1(0) + c_2 = 1 \Rightarrow c_2 = 1$

$$y(1) = 6 \Rightarrow \frac{\lambda}{4} + c_1 + c_2 = 6 \quad \Rightarrow \frac{\lambda}{4} + c_1 + 1 = 6 \Rightarrow c_1 = 5 - \frac{\lambda}{4}$$

$$\therefore \ (3) \Rightarrow \quad y = \frac{\lambda x^2}{4} + \left(5 - \frac{\lambda}{4}\right)x + 1 \qquad ...(4)$$

Putting the value of y in (1), we get

$$\int_0^1 \left(\frac{\lambda x^2}{4} + \left(5 - \frac{\lambda}{4}\right)x + 1\right)dx = 3$$

$$\Rightarrow \quad \left(\frac{\lambda x^3}{12} + \left(5 - \frac{\lambda}{4}\right)\frac{x^2}{2} + x\right)\Bigg|_0^1 = 3$$

$$\Rightarrow \frac{\lambda}{12}(1-0) + \left(\frac{5}{2} - \frac{\lambda}{8}\right)(1-0) + (1-0) = 3$$

$$\Rightarrow \quad \frac{\lambda}{12} + \frac{5}{2} - \frac{\lambda}{8} \div 1 = 3 \Rightarrow \lambda = 12$$

$$\therefore \ (4) \Rightarrow \quad y = \frac{12x^2}{4} + \left(5 - \frac{12}{4}\right)x + 1 \quad i.e., \quad \mathbf{y = 3x^2 + 2x + 1.}$$

This is the equation of the required extremal.

Example 2. *Find the extremal of the functional* $\int_0^{\pi} (y'^2 - y^2)dx$ *under the boundary conditions* $y(0) = 0$, $y(\pi) = 1$ *and subject to an additional condition* $\int_0^{\pi} y\,dx = 1.$

Sol. Given functional is $\int_0^{\pi} (y'^2 - y^2)\,dx$.

Given condition is $\int_0^{\pi} ydx = 1$ ...(1)

Let $F(x, y, y') = y'^2 - y^2$ and $G(x, y, y') = y$

The **Euler's equation** is

$$F_y - \frac{d}{dx}F_{y'} + \lambda\left(G_y - \frac{d}{dx}G_{y'}\right) = 0 \qquad ...(2)$$

Here $F_y = -2y$, $F_{y'} = 2y'$, $G_y = 1$, $G_{y'} = 0$

$$\therefore \ (2) \Rightarrow \quad -2y - \frac{d}{dx}(2y') + \lambda\left(1 - \frac{d}{dx}(0)\right) = 0$$

$$\Rightarrow \quad -2y - 2\frac{d^2y}{dx^2} + \lambda = 0 \Rightarrow \frac{d^2y}{dx^2} + y = \frac{\lambda}{2}$$

The A.E. is $D^2 + 1 = 0$. $\therefore$ $D = \pm i$

$$\therefore \quad \text{C.F.} = e^{0.x}(c_1 \cos 1.x + c_2 \sin 1.x) = c_1 \cos x + c_2 \sin x$$

$$\text{P.I.} = \frac{1}{D^2+1}\left(\frac{\lambda}{2}\right) = (1 + D^2)^{-1}\left(\frac{\lambda}{2}\right)$$

$$= (1 - D^2 + ...)\left(\frac{\lambda}{2}\right) = \frac{\lambda}{2} - 0 + ... = \frac{\lambda}{2}$$

$$\Rightarrow \qquad y = c_1 \cos x + c_2 \sin x + \frac{\lambda}{2} \qquad ...(3)$$

The boundary conditions are $y(0) = 0$, $y(\pi) = 1$.

$$y(0) = 0 \Rightarrow c_1(1) + c_2(0) + \frac{\lambda}{2} = 0 \Rightarrow c_1 + \frac{\lambda}{2} = 0 \qquad ...(4)$$

$$y(\pi) = 1 \Rightarrow c_1(-1) + c_2(0) + \frac{\lambda}{2} = 1 \Rightarrow -c_1 + \frac{\lambda}{2} = 1 \qquad ...(5)$$

Solving (4) and (5), we get $c_1 = -\frac{1}{2}$, $\lambda = 1$

$$\therefore \quad (3) \Rightarrow \qquad y = -\frac{1}{2} \cos x + c_2 \sin x + \frac{1}{2} \qquad ...(6)$$

Putting the value of y in (1), we get

$$\int_0^{\pi} \left(-\frac{1}{2} \cos x + c_2 \sin x + \frac{1}{2} \right) dx = 1$$

$$\Rightarrow \qquad \left(-\frac{1}{2} \sin x - c_2 \cos x + \frac{x}{2} \right) \Bigg|_0^{\pi} = 1$$

$$\Rightarrow \qquad \left(-\frac{1}{2}(0) - c_2(-1) + \frac{\pi}{2} \right) - \left(-\frac{1}{2}(0) - c_2(1) + 0 \right) = 1$$

$$\Rightarrow \qquad 0 + c_2 + \frac{\pi}{2} + c_2 = 1 \Rightarrow c_2 = \frac{2 - \pi}{4}$$

$$\therefore \quad (6) \Rightarrow \qquad \mathbf{y = -\frac{1}{2} \cos x + \frac{2-\pi}{4} \sin x + \frac{1}{2}}.$$

This equation is of the required extremal.

Example 3. *Find the extremal of the functional* $\int_0^1 (x^2 + y'^2)\, dx$, *$y(0) = 0$, $y(1) = 0$ subject to the condition* $\int_0^1 y^2\, dx = 2$.

Sol. Given functional is $\int_0^1 (x^2 + y'^2)\, dx$.

Given condition is $\int_0^1 y^2 dx = 2$ ...(1)

Let $F(x, y, y') = x^2 + y'^2$ and $G(x, y, y') = y^2$

The **Euler's equation** is

$$F_y - \frac{d}{dx} F_{y'} + \lambda \left(G_y - \frac{d}{dx} G_{y'} \right) = 0 \qquad ...(2)$$

Here $F_y = 0$, $F_{y'} = 2y'$, $G_y = 2y$, $G_{y'} = 0$

$$\therefore \quad (2) \Rightarrow \qquad 0 - \frac{d}{dx}(2y') + \lambda \left(2y - \frac{d}{dx}(0) \right) = 0$$

$$\Rightarrow \quad -2\frac{d^2y}{dx^2} + 2\lambda y = 0 \quad \Rightarrow \quad \frac{d^2y}{dx^2} - \lambda y = 0 \qquad \text{...(3)}$$

The A.E. is $\quad D^2 - \lambda = 0 \qquad$ ...(4)

Case I. $\lambda > 0$, (4) $\Rightarrow \quad D = \pm\sqrt{\lambda}$

$$\therefore \quad (3) \Rightarrow \quad y = c_1 e^{\sqrt{\lambda}x} + c_2 e^{-\sqrt{\lambda}x} \qquad \text{...(5)}$$

The boundary condition are $y(0) = 0, y(1) = 0$.

$$y(0) = 0 \quad \Rightarrow \quad c_1e^0 + c_2e^0 = 0 \quad \Rightarrow \quad c_1 + c_2 = 0 \qquad \text{...(6)}$$

$$y(1) = 0 \quad \Rightarrow \quad c_1e^{\sqrt{\lambda}} + c_2 e^{-\sqrt{\lambda}} = 0 \qquad \text{...(7)}$$

Solving (6) and (7), we get $c_1 = 0, c_2 = 0$.

$$\therefore \quad y = 0.e^{\sqrt{\lambda}x} + 0.e^{-\sqrt{\lambda}x} \quad i.e., \quad y = 0$$

Now $\quad \int_0^1 y^2 dx = \int_0^1 0.dx = 0 \neq 2$

$\therefore$ Given condition is not satisfied.

$\therefore$ $\lambda > 0$ is impossible.

Case II. $\lambda = 0$, $(3) \Rightarrow \quad \frac{d^2y}{dx^2} = 0 \quad \Rightarrow \quad \frac{dy}{dx} = c_1 \quad \Rightarrow \quad y = c_1x + c_2$

$$y(0) = 0 \quad \Rightarrow \quad c_1(0) + c_2 = 0 \quad \Rightarrow \quad c_2 = 0$$

$$y(1) = 0 \quad \Rightarrow \quad c_1(1) + c_2 = 0 \quad \Rightarrow \quad c_1 + 0 = 0 \quad \Rightarrow \quad c_1 = 0$$

$$\therefore \quad y = 0.x + 0 \quad i.e., \quad y = 0$$

Here also $\quad \int_0^{\pi} y^2 dx = \int_0^{\pi} 0\, dx = 0 \neq 2$

$\therefore$ $\lambda = 0$ is impossible.

Case III. $\lambda < 0$, (4) $\Rightarrow \quad D = \pm\sqrt{-\lambda}\, i$

$$\therefore \quad (3) \Rightarrow \quad y = e^{0x}(c_1 \cos\sqrt{-\lambda}\,x + c_2 \sin\sqrt{-\lambda}\,x)$$

i.e.,

$$y = c_1 \cos\sqrt{-\lambda}\,x + c_2 \sin\sqrt{-\lambda}\,x \qquad \text{...(8)}$$

$$y(0) = 0 \quad \Rightarrow \quad c_1 \cos 0 + c_2 \sin 0 = 0 \quad \Rightarrow \quad c_1 = 0$$

$$y(1) = 0 \quad \Rightarrow \quad c_1 \cos\sqrt{-\lambda} + c_2 \sin\sqrt{-\lambda} = 0 \quad \Rightarrow \sin\sqrt{-\lambda} = 0, \text{ for } c_2 \neq 0$$

$$\Rightarrow \quad \sqrt{-\lambda} = n\pi, \quad n \in \mathbf{N}$$

$$\therefore \quad (8) \Rightarrow \quad y = c_2 \sin n\pi x, \quad n \in \mathbf{N}$$

Also, $\quad \int_0^1 y^2\, dx = {c_2}^2 \int_0^1 \sin^2 n\pi x\, dx$

$$= \frac{{c_2}^2}{2}\int_0^1 (1 - \cos 2n\pi x)\, dx = \frac{{c_2}^2}{2}\left[x - \frac{\sin 2n\pi x}{2n\pi}\right]_0^1$$

$$= \frac{{c_2}^2}{2}[(1-0) - (0-0)] = \frac{{c_2}^2}{2}$$

$\therefore \quad \dfrac{c_2^{\,2}}{2} = 2$ (Using (1))

$\therefore \quad c_2^{\,2} = 4 \quad \text{or} \quad c_2 = \pm 2$

$\therefore \quad \mathbf{y = \pm 2 \sin n\pi x, \quad n \in N.}$

This is the equation of the required extremals.

Example 4. *Find the plane curve of fixed perimeter so that the area covered by the curve, ordinates and x-axis is maximum.*

Sol. Let $y = y(x)$ be a curve of given length l between given points $A(x_1, y_1)$ and $B(x_2, y_2)$.

$$\therefore \quad l = \int_{x_1}^{x_2} \sqrt{1 + y'^2}\, dx$$

Also, area enclosed by the curve, ordinates and x-axis is $\int_{x_1}^{x_2} y\, dx$. We are to maximise the functional $\int_{x_1}^{x_2} y\, dx$ under the condition $\int_{x_1}^{x_2} \sqrt{1+y'^2}\, dx = l$

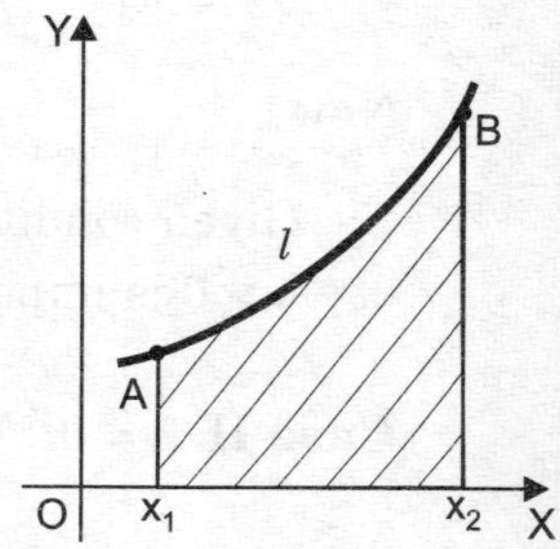

Let $\quad F(x, y, y') = y \quad$ and $\quad G(x, y, y') = \sqrt{1+y'^2}$.

The **Euler's equation** is

$$F_y - \frac{d}{dx} F_{y'} + \lambda \left(G_y - \frac{d}{dx} G_{y'} \right) = 0 \qquad \text{...(1)}$$

Here, $\quad F_y = 1, F_{y'} = 0, G_y = 0, G_{y'} = \dfrac{y'}{\sqrt{1+y'^2}}$

$$\therefore \quad (1) \Rightarrow \quad 1 - \frac{d}{dx}(0) + \lambda \left(0 - \frac{d}{dx} \left(\frac{y'}{\sqrt{1+y'^2}} \right) \right) = 0$$

$$\Rightarrow \quad \frac{d}{dx} \left(\frac{\lambda y'}{\sqrt{1+y'^2}} \right) = 1 \quad \Rightarrow \quad \frac{\lambda y'}{\sqrt{1+y'^2}} = x + c_1$$

$$\Rightarrow \quad \lambda^2 y'^{\,2} = (x + c_1)^2 (1 + y'^2)$$

$$\Rightarrow \quad y'^2 = \frac{(x+c_1)^2}{\lambda^2 - (x+c_1)^2}$$

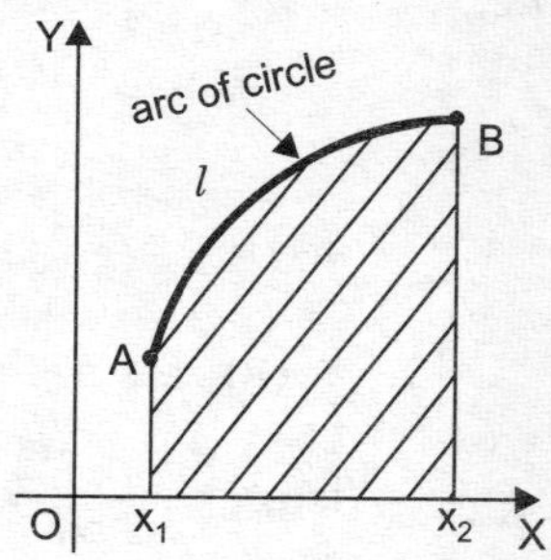

$$\Rightarrow \quad y' = \frac{x + c_1}{\sqrt{\lambda^2 - (x+c_1)^2}}$$

$$\Rightarrow \quad dy = -\frac{1}{2} \cdot \frac{-2(x+c_1)}{\sqrt{\lambda^2 - (x+c_1)^2}}\, dx$$

$$\Rightarrow \quad y = -\frac{1}{2} \cdot \frac{(\lambda^2 - (x+c_1)^2)^{1/2}}{1/2} + c_2$$

$$\Rightarrow \quad (y - c_2)^2 = \lambda^2 - (x + c_1)^2$$

$$\Rightarrow \quad (x + c_1)^2 + (y - c_2)^2 = \lambda^2 \qquad \text{...(2)}$$

Now $$\int_{x_1}^{x_2} \sqrt{1+y'^2}\, dx = \int_{x_1}^{x_2} \sqrt{1+\frac{(x+c_1)^2}{\lambda^2-(x+c_1)^2}}\, dx$$

$$= \int_{x_1}^{x_2} \frac{\lambda}{\sqrt{\lambda^2-(x+c_1)^2}}\, dx = \lambda \sin^{-1} \frac{x+c_1}{\lambda}\Bigg|_{x_1}^{x_2}$$

$\therefore$ $$\lambda\left[\sin^{-1}\frac{x_2+c_1}{\lambda} - \sin^{-1}\frac{x_1+c_1}{\lambda}\right] = l \qquad ...(3) \quad \left(\because \int_{x_1}^{x_2} \sqrt{1+y'^2}\, dx = l\right)$$

(3) is solved to find the value of λ.

$\therefore$ The required curves is an arc of the circle $\mathbf{(x+c_1)^2 + (y-c_2)^2} = \lambda^2$. The values of c_1 and c_2 are found by using the fact that A and B are on this arc. The position of the arc is shown in the adjoining figure.

Remarks 1. In particular, if $l = \pi\left(\frac{AB}{2}\right)$, then AB would be a diameter of the maximising circle.

2. This type of problems are called **isoperimetric problems.**

Example 5. *Prove that sphere is a solid figure of revolution, which for a given surface area, has a maximum volume.*

Sol. Let curve OPA rotate about x-axis as shown in the figure. Let coordinates of A be $(a, 0)$. Let S be the surface area of the solid figure of revolution.

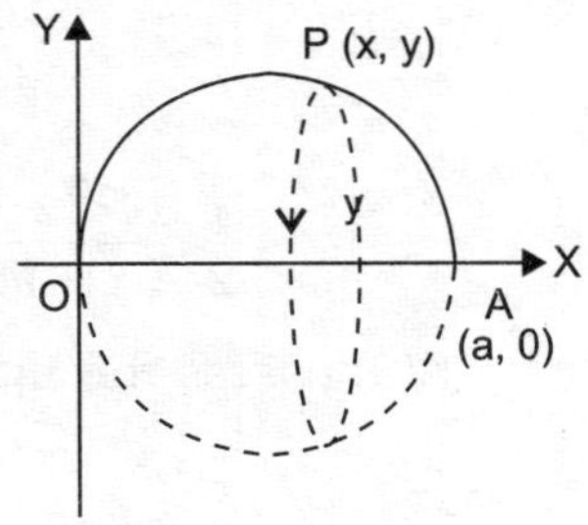

$\therefore$ $$\int_0^a 2\pi y\, ds = S$$

$\Rightarrow$ $$\int_0^a 2\pi y \sqrt{1+y'^2}\, dx = S$$

The volume of the solid figure of revolution is $\int_0^a \pi y^2\, dx$.

$\therefore$ We have to maximise the functional $\int_0^a \pi y^2\, dx$ with boundary conditions $y(0) = 0$, $y(a) = 0$ and under the condition $\int_0^a 2\pi y \sqrt{1+y'^2}\, dx = S$.

Let $F(x, y, y') = \pi y^2$ and $G(x, y, y') = 2\pi y\sqrt{1+y'^2}$.

The **Euler's equation** is $F_y - \frac{d}{dx} F_{y'} + \lambda\left(G_y - \frac{d}{dx} G_{y'}\right) = 0$

Here x is missing from both functions F and G.

$\therefore$ The Euler's equation reduces to

$$F + \lambda G - (F_{y'} + \lambda G_{y'}) \frac{dy}{dx} = c \qquad ...(1)$$

Here $F_{y'} = 0$, $G_{y'} = \dfrac{2\pi y y'}{\sqrt{1+y'^2}}$

$\therefore$ (1) $\Rightarrow$ $\pi y^2 + 2\pi\lambda y\sqrt{1+y'^2} - \left(0 + \dfrac{2\pi\lambda yy'}{\sqrt{1+y'^2}}\right)y' = c$

$\Rightarrow$ $\pi y^2 + \left(\dfrac{2\pi\lambda y(1+y'^2) - 2\pi\lambda yy'^2}{\sqrt{1+y'^2}}\right) = c$

$\Rightarrow$ $\pi y^2 + \dfrac{2\pi\lambda y}{\sqrt{1+y'^2}} = c$ ...(2)

Since the curve passes through (0, 0), we have $0 + 0 = c$ *i.e.,* $c = 0$

$\Rightarrow$ $\pi y^2 + \dfrac{2\pi\lambda y}{\sqrt{1+y'^2}} = 0$ $\Rightarrow$ $y + \dfrac{2\lambda}{\sqrt{1+y'^2}} = 0$

$\Rightarrow$ $\sqrt{1+y'^2} = -\dfrac{2\lambda}{y}$ $\Rightarrow$ $y'^2 = \dfrac{4\lambda^2}{y^2} - 1 = \dfrac{4\lambda^2 - y^2}{y^2}$

$\Rightarrow$ $y' = \dfrac{\sqrt{4\lambda^2 - y^2}}{y}$ $\Rightarrow$ $\dfrac{y}{\sqrt{4\lambda^2 - y^2}}\,dy = dx$

$\Rightarrow$ $-\dfrac{1}{2}\cdot\dfrac{(4\lambda^2 - y^2)^{1/2}}{1/2} = x + c_1$ $\Rightarrow$ $-\sqrt{4\lambda^2 - y^2} = x + c_1$ ...(3)

(3) passes through (0, 0) $\therefore$ $-2\lambda = c_1$

$\therefore$ $-\sqrt{4\lambda^2 - y^2} = x - 2\lambda$

$\Rightarrow$ $4\lambda^2 - y^2 = (x - 2\lambda)^2$ $\Rightarrow$ $(x - 2\lambda)^2 + y^2 = 4\lambda^2$ ...(4)

(4) passes through $(a, 0)$

$\therefore$ $(a - 2\lambda)^2 + 0 = 4\lambda^2$

$\Rightarrow$ $a^2 + 4\lambda^2 - 4a\lambda = 4\lambda^2$

$\Rightarrow$ $a^2 - 4a\lambda = 0$ $\Rightarrow$ $\lambda = a/4$

$\therefore$ (4) $\Rightarrow$ $\left(x - \dfrac{a}{2}\right)^2 + y^2 = \dfrac{a^2}{4}$ ...(5)

(5) represents a circle with centre at $(a/2, 0)$ and radius $a/2$.

$\therefore$ The curve OPA is a semi-circle.

$\therefore$ The solved figure revolution is a sphere.

Hence the result holds.

EXERCISE 4.5

1. Find the extremals of the functional $\int_{x_1}^{x_2} y'^2\,dx$ subject to the condition $\int_{x_1}^{x_2} y\,dx = c$, a constant.

2. Show that the extremal of the functional $\int_1^4 y'^2\,dx$ with boundary conditions $y(1) = 3$, $y(4) = 24$ and subject to an additional condition $\int_1^4 y dx = 36$ is a parabola.

3. Find the extremal of the functional $\int_0^\pi y'^2\,dx$, $y(0) = 0$, $y(\pi) = 0$ subject to the condition $\int_0^\pi y^2 dx = 1$.

4. Find the extremal of the functional $\int_0^1 y'^2\,dx$, $y(0) = 0$, $y(1) = \frac{1}{4}$ subject to the condition
$$\int_0^1 (y - y'^2)\,dx = \frac{1}{12}.$$

5. Among all curves of length $= l$ ($> 2a$) in the upper half plane passing through the points $(-a, 0)$ and $(a, 0)$, find the one which together with the segment $-a \le x \le a$ encloses the maximum area.

Answers

1. $y = \dfrac{3}{x_2^3 - x_1^3}(2c - 2c_2(x_2 - x_1) - c_1(x_2^2 - x_1^2))x^2 + c_1x + c_2.$

3. $y = \pm\sqrt{\dfrac{2}{\pi}}\sin nx,\ n \in \mathbf{N}$

4. $y = -\dfrac{x^2}{4} + \dfrac{x}{2}.$

5. $x^2 + (y + \sqrt{\lambda_0^2 - a^2})^2 = \lambda_0^2$, where λ_0 is a solution of the equation $\sin\dfrac{l}{2\lambda} = \dfrac{a}{\lambda}.$

5

Variational Problems with Moving Boundaries

1. INTRODUCTION

We know that a functional associates a unique number to each function belonging to a certain class of functions. In the preceding chapter, we studied the methods of finding stationary function of the functional of the form $\int_{x_1}^{x_2} F(x, y(x), y'(x))\, dx$, where the boundary points (x_1, y_1) and (x_2, y_2) were fixed points. In the present chapter, we shall allow one or both boundary points to move. In this case the class of admissible functions is extended because, in addition to the comparison curves with fixed boundary points, we have to admit functions with variable boundary points.

2. FUNCTIONALS DEPENDENT ON ONE FUNCTION

Theorem. *Let* $J[y(x)] = \int_{x_1}^{x_2} F(x, y(x), y'(x))\, dx$ *be a functional defined on the set of functions* $y = y(x)$ *which admits of continuous first order derivative and the end points of which* (x_1, y_1) *and* (x_2, y_2) *lie on the given curves* $y = \phi(x)$ *and* $y = \psi(x)$ *respectively so that* $y_1 = \phi(x_1)$ *and* $y_2 = \psi(x_2)$, *where* $y = \phi(x)$ *and* $y = \psi(x)$ *are functions from the class* $C_1[x_1, x_2]$. *Also, F is differentiable three times with respect to all its arguments.*

If the functional $J[y(x)]$ *has an extremum on a function* $y = y(x)$ *joining two arbitrary points of the given curves* $y = \phi(x)$ *and* $y = \psi(x)$, *then* $y(x)$ *satisfy the Euler's equation*

$$F_y - \frac{d}{dx} F_{y'} = 0$$

and the transversality conditions

$$(F + (\phi' - y') F_{y'})|_{x = x_1} = 0 \quad \text{and} \quad (F + (\psi' - y') F_{y'})|_{x = x_2} = 0.$$

Note. Proof of this theorem is beyond the scope of this book.

Remark. If one of the boundaries say, (x_1, y_1) is fixed, then we need not introduce the function $\phi(x)$. In this case, we shall have only one transversality condition involving the function $\psi(x)$.

WORKING RULES FOR SOLVING PROBLEMS

Step I. *Identify the functions* F, ϕ, ψ *in the given problem.*

Step II. *Solve the Euler's equation* $F_y - \frac{d}{dx} F_{y'} = 0$ *and obtain a family of extremals* $y = f(x, c_1, c_2)$ *dependent on parameters* c_1 *and* c_2.

Step III. *Using $y_1 = \phi(x_1)$ and $y_2 = \psi(x_2)$ form the equations*
$$f(x_1, c_1, c_2) = \phi(x_1) \text{ and } f(x_2, c_1, c_2) = \psi(x_2).$$
Step IV. *Simplify the transversality equations.*
Step V. *Solve the equations obtained in the above two steps to find the values of c_1, c_2, x_1, x_2.*
Step VI. *Find the extremum value of the given functional.*

Example 1. *Prove that for a functional of the form $J[y(x)] = \int_{x_1}^{x_2} h(x, y)\sqrt{1+y'^2}\,dx$, where $h(x, y) \neq 0$ at the moving boundary points $(x_1, \phi(x_1))$ and $(x_2, \psi(x_2))$, the transversality conditions are of the form $y'(x) = -\dfrac{1}{\phi'(x)}$ at $x = x_1$ and $y'(x) = -\dfrac{1}{\psi'(x)}$ at $x = x_2$.*

Sol. We have $J[y(x)] = \int_{x_1}^{x_2} h(x, y)\sqrt{1+y'^2}\,dx$.

Let $F(x, y, y') = h(x, y)\sqrt{1+y'^2}$

The transversality conditions are :

$$(F + (\phi' - y')F_{y'})|_{x=x_1} = 0 \quad ...(1)$$

and

$$(F + (\psi' - y')F_{y'})|_{x=x_2} = 0 \quad ...(2)$$

$$(1) \Rightarrow \left[h(x, y)\sqrt{1+y'^2} + (\phi' - y')\frac{h(x, y)y'}{\sqrt{1+y'^2}}\right]_{x=x_1} = 0$$

$$\Rightarrow \quad h(x, y)\,[1 + y'^2 + (\phi' - y')y']|_{x=x_1} = 0$$

$$\Rightarrow \quad h(x, y)(1 + \phi' y')|_{x=x_1} = 0 \Rightarrow (1 + \phi' y')|_{x=x_1} = 0 \qquad [\because h(x, y) \neq 0 \text{ at } x = x_1]$$

$$\therefore \quad y'(x) = -\frac{1}{\phi'(x)} \text{ at } x = x_1.$$

Similarly, (2) implies $y'(x) = -\dfrac{1}{\psi'(x)}$ at $x = x_2$.

Example 2. *Find the shortest distance between the point A(1, 0) and the ellipse*
$$4x^2 + 9y^2 = 36.$$

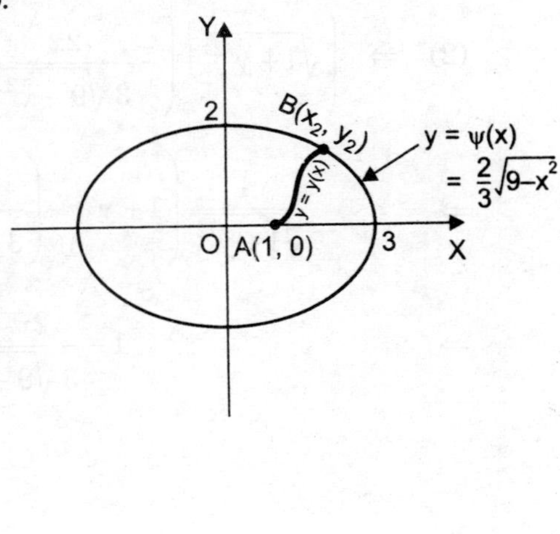

Sol. The given ellipse is

$$\frac{x^2}{9} + \frac{y^2}{4} = 1.$$

$$\Rightarrow \quad y = \pm\frac{2}{3}\sqrt{9 - x^2}$$

Let $y = \psi(x) = \dfrac{2}{3}\sqrt{9 - x^2}$ for the upper half ellipse. Let $B(x_2, y_2)$ be an arbitrary point on the curve

$y = \psi(x)$. Let $y = y(x)$ be a function whose curve passes through A and B. The length of the arc between A and B is given by the functional $\int_1^{x_2} \sqrt{1+y'^2}\, dx$.

Let $F(x, y, y') = \sqrt{1+y'^2}$

Let the length of the curve between A and B be minimum on the curve of $y = y(x)$.

$\therefore$ The functional $\int_1^{x_2} F(x, y, y')\, dx$ has a minimum value on $y = y(x)$.

$\therefore$ $y(x)$ satisfies the equations :

$$F_y - \frac{d}{dx} F_{y'} = 0 \qquad \text{...(1)}$$

and

$$(F + (\psi' - y')F_{y'})|_{x = x_2} = 0 \qquad \text{...(2)}$$

Here $F_y = 0$, $F_{y'} = \dfrac{y'}{\sqrt{1+y'^2}}$, $\psi' = \dfrac{2}{3} \cdot \dfrac{1}{2} (9 - x^2)^{-1/2}(-2x) = -\dfrac{2x}{3\sqrt{9-x^2}}$

$$(1) \Rightarrow \quad 0 - \frac{d}{dx}\left(\frac{y'}{\sqrt{1+y'^2}}\right) = 0 \quad \Rightarrow \quad \frac{y'}{\sqrt{1+y'^2}} = c_1$$

$$\Rightarrow \quad y'^2 = c_1^2 (1 + y'^2) \quad \Rightarrow \quad y' = \frac{c_1}{\sqrt{1-c_1^2}} = c_2, \text{ say}$$

$$\Rightarrow \quad y = c_2 x + c_3 \qquad \text{...(3)}$$

This is the equation of the extremals. The boundary conditions are $y(1) = 0$,

$y(x_2) = y_2 = \psi(x_2) = \dfrac{2}{3}\sqrt{9 - x_2^2}$.

$$y(1) = 0 \quad \Rightarrow \quad c_2(1) + c_3 = 0 \quad \Rightarrow \quad c_2 + c_3 = 0 \qquad \text{...(4)}$$

$$y(x_2) = \frac{2}{3}\sqrt{9-x_2^2} \quad \Rightarrow \quad c_2 x_2 + c_3 = \frac{2}{3}\sqrt{9-x_2^2} \qquad \text{...(5)}$$

$$\Rightarrow \quad c_2 x_2 - c_2 = \frac{2}{3}\sqrt{9-x_2^2} \qquad \text{(Using (4))}$$

$$\Rightarrow \quad c_2(x_2 - 1) = \frac{2}{3}\sqrt{9-x_2^2} \qquad \text{...(6)}$$

$$(2) \Rightarrow \quad \left(\sqrt{1+y'^2} + \left(-\frac{2x}{3\sqrt{9-x^2}} - y'\right)\frac{y'}{\sqrt{1+y'^2}}\right)\Bigg|_{x=x_2} = 0$$

$$\Rightarrow \quad \frac{1}{\sqrt{1+y'^2}}\left(1 + y'^2 - \left(\frac{2xy'}{3\sqrt{9-x^2}} + y'^2\right)\right)\Bigg|_{x=x_2} = 0$$

$$\Rightarrow \quad 1 - \frac{2x_2 c_2}{3\sqrt{9-x_2^2}} = 0 \Rightarrow 2c_2 x_2 = 3\sqrt{9-x_2^2} \qquad \text{...(7)}$$

$\therefore$ (6) $\Rightarrow$ $c_2(x_2 - 1) = \dfrac{2}{3}\left(\dfrac{2}{3}c_2x_2\right)$ $\Rightarrow$ $x_2 - 1 = \dfrac{4x_2}{9}$ $\Rightarrow$ $x_2 = \dfrac{9}{5}$

$\therefore$ (7) $\Rightarrow$ $2c_2\left(\dfrac{9}{5}\right) = 3\sqrt{9 - \dfrac{81}{25}}$ $\Rightarrow$ $\dfrac{18c_2}{5} = \dfrac{36}{5}$ $\Rightarrow$ $c_2 = 2$

$\therefore$ (4) $\Rightarrow$ $2 + c_3 = 0$ $\Rightarrow$ $c_3 = -2$

Using (3) the stationary function is $y = 2x - 2$.

$\therefore$ $y = 2x - 2$ minimises the functional *i.e.*, the distance between A and B.

$$\therefore \quad \text{Shortest distance} = \int_1^{9/5} \sqrt{1+(2)^2}\, dx = \sqrt{5}\left(\frac{9}{5} - 1\right) = \frac{4\sqrt{5}}{5} \textbf{ units.}$$

Example 3. *Find the shortest distance between the parabola $y = x^2$ and the straight line $x - y = 5$.*

Sol. Let $\phi(x) = x^2$ and $\psi(x) = x - 5$. Let $A(x_1, y_1)$ and $B(x_2, y_2)$ be arbitrary points on the curves $y = \phi(x)$ and $y = \psi(x)$ respectively. Let $y = y(x)$ be a function whose curve passes through A and B. The length of the arc between A and B is given by $\int_{x_1}^{x_2} \sqrt{1+y'^2}\, dx$.

Let $\quad F(x, y, y') = \sqrt{1+y'^2}$

Let the length of the curve between A and B be minimum on the curve of $y = y(x)$.

$\therefore$ The functional $\int_{x_1}^{x_2} F(x, y, y')dx$ has a minimum value on $y(x)$.

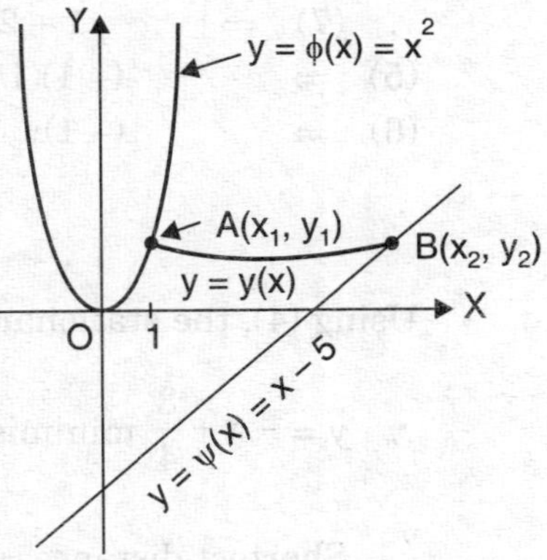

$\therefore$ $y(x)$ satisfies the equations :

$$F_y - \frac{d}{dx} F_{y'} = 0 \qquad ...(1)$$

$$(F + (\phi' - y')F_{y'})|_{x = x_1} = 0 \qquad ...(2)$$

and $$(F + (\psi' - y')F_{y'})|_{x = x_2} = 0 \qquad ...(3)$$

Here $F_y = 0$, $F_{y'} = \dfrac{y'}{\sqrt{1+y'^2}}$, $\phi' = 2x$, $\psi' = 1$

(1) $\Rightarrow$ $0 - \dfrac{d}{dx}\left(\dfrac{y'}{\sqrt{1+y'^2}}\right) = 0$ $\Rightarrow$ $\dfrac{y'}{\sqrt{1+y'^2}} = c_1$

$\Rightarrow$ $y'^2 = c_1^2(1 + y'^2)$ $\Rightarrow$ $y' = \dfrac{c_1}{\sqrt{1-c_1^2}} = c_2$, say

$\Rightarrow$ $y = c_2x + c_3$ ...(4)

This is the equation of the extremals.

The boundary conditions are:

$$y(x_1) = y_1 = \phi(x_1) = x_1^2, \quad y(x_2) = y_2 = \psi(x_2) = x_2 - 5.$$

$$y(x_1) = x_1^2 \quad \Rightarrow \quad c_2 x_1 + c_3 = x_1^2 \qquad \text{...(5)}$$

$$y(x_2) = x_2 - 5 \quad \Rightarrow \quad c_2 x_2 + c_3 = x_2 - 5 \qquad \text{...(6)}$$

$$(2) \Rightarrow \left.\left(\sqrt{1+y'^2} + (2x - y')\frac{y'}{\sqrt{1+y'^2}}\right)\right|_{x=x_1} = 0$$

$$\Rightarrow \left.\frac{1}{\sqrt{1+y'^2}}(1+2xy')\right|_{x=x_1} = 0$$

$$\Rightarrow 1 + 2x_1 c_2 = 0 \qquad \text{...(7)}$$

$$(3) \Rightarrow \left.\left(\sqrt{1+y'^2} + (1-y')\frac{y'}{\sqrt{1+y'^2}}\right)\right|_{x=x_2} = 0$$

$$\Rightarrow \left.\frac{1}{\sqrt{1+y'^2}}(1+y')\right|_{x=x_2} = 0$$

$$\Rightarrow 1 + c_2 = 0 \quad \Rightarrow \quad c_2 = -1 \qquad \text{...(8)}$$

$$\therefore \ (7) \Rightarrow 1 + 2x_1(-1) = 0 \quad \Rightarrow \quad x_1 = 1/2$$

$$(5) \Rightarrow (-1)(1/2) + c_3 = 1/4 \quad \Rightarrow \quad c_3 = 3/4$$

$$(6) \Rightarrow (-1)x_2 + (3/4) = x_2 - 5 \quad \Rightarrow \quad x_2 = 23/8$$

$$\therefore \quad c_2 = -1, \quad c_3 = \frac{3}{4}, \quad x_1 = \frac{1}{2}, \quad x_2 = \frac{23}{8}$$

Using (4), the stationary function is $y = (-1)x + \dfrac{3}{4}$.

$\therefore \quad y = -x + \dfrac{3}{4}$ minimises the functional *i.e.*, the distance between A and B.

$$\therefore \quad \text{Shortest distance} = \int_{1/2}^{23/8} \sqrt{1+(-1)^2}\, dx = \sqrt{2}\left(\frac{23}{8} - \frac{1}{2}\right) = \frac{19\sqrt{2}}{8} \textbf{ units.}$$

EXERCISE 5.1

1. Find the shortest distance between the point A(– 1, 3) and the straight line $y = 1 - 3x$.
2. Find the shortest distance between the point A(0, 0) and the straight line $x + y = \sqrt{2}$.
3. Find the shortest distance from the point A(– 1, 5) to the parabola $y^2 = x$.
4. Find the shortest distance between the circle $x^2 + y^2 = 1$ and the straight line $x + y = 4$.

Answers

1. $\dfrac{\sqrt{10}}{10}$ units **2.** 1 unit **3.** $2\sqrt{5}$ units **4.** $(2\sqrt{2} - 1)$ units.

3. FUNCTIONALS DEPENDENT ON TWO FUNCTIONS

Theorem. *Let $J[y(x), z(x)] = \int_{x_1}^{x_2} F[x, y(x), z(x), y'(x), z'(x)]dx$ be a functional defined on the set of functions $y = y(x)$, $z = z(x)$ which admits of continuous first order derivative. Let the point $A(x_1, y_1, z_1)$ corresponding to the lower limit in the above integral be fixed and the other boundary point $B(x_2, y_2, z_2)$ be in motion along a certain curve specified by the equations $y = \phi(x)$, $z = \psi(x)$ so that $y_2 = \phi(x_2)$, $z_2 = \psi(x_2)$, where $y = \phi(x)$ and $z = \psi(x)$ are functions from the class $C_1[x_1, x_2]$. Also, F is differentiable three times with respect to all its arguments.*

If the functional $J[y(x), z(x)]$ has an extremum on a curve $y = y(x)$, $z = z(x)$, then $y(x)$ and $z(x)$ satisfy the Euler's equations $F_y - \frac{d}{dx}F_{y'} = 0$, $F_z - \frac{d}{dx}F_{z'} = 0$ and the transversality condition $[F + (\phi' - y')F_{y'} + (\psi' - z')F_{z'}]\big|_{x = x_2} = 0$.

Note. The proof of this theorem is beyond the scope of this book.

Remark. If the boundary point $A(x_1, y_1, z_1)$ is fixed and the other boundary point $B(x_2, y_2, z_2)$ can move on the surface $z = \psi(x, y)$, then the transversality conditions are:

$$[\mathbf{F - y'F_{y'} + (\psi_x - z')F_{z'}}]\big|_{\mathbf{x = x_2}} = \mathbf{0}$$

and
$$[\mathbf{F_{y'} + F_{z'}\,\psi_y}]\big|_{\mathbf{x = x_2}} = \mathbf{0.}$$

Example 1. *Find the shortest distance from the point A(0, 2, 1) to the straight line $\frac{x}{1} = \frac{y}{2} = \frac{z}{3}$.*

Sol. The given line is $\frac{x}{1} = \frac{y}{2} = \frac{z}{3}$ or equivalently $y = 2x$, $z = 3x$. Let $y = \phi(x) = 2x$ and $z = \psi(x) = 3x$. Let $B(x_2, y_2, z_2)$ be an arbitrary point on the curve $y = \phi(x)$, $z = \psi(x)$. Let $y = y(x)$, $z = z(x)$ be functions whose curve passes through A and B. The length of the arc between A and B is given by the functional $\int_0^{x_2} \sqrt{1 + y'^2 + z'^2}\, dx$.

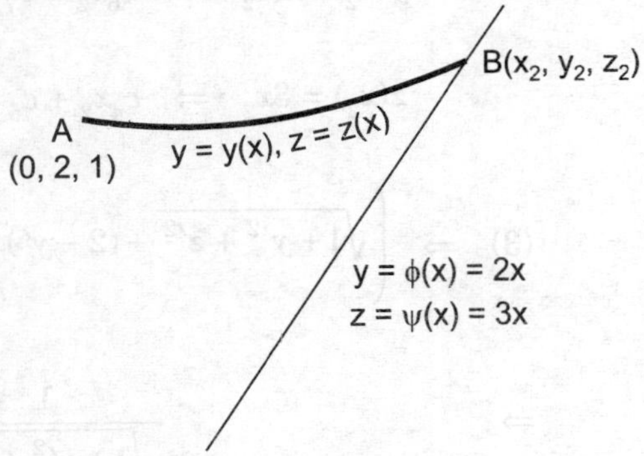

Let $\quad F(x, y, z, y', z') = \sqrt{1 + y'^2 + z'^2}$

Let the length of the curve between A and B be minimum on the curve given by $y = y(x)$, $z = z(x)$.

$\therefore$ The functional $\int_0^{x_2} F(x, y, z, y', z')\, dx$ has a minimum value on the functions $y = y(x)$, $z = z(x)$.

$\therefore$ $y(x)$, $z(x)$ satisfy the equations
$$F_y - \frac{d}{dx}F_{y'} = 0 \quad \text{...(1)}$$
$$F_z - \frac{d}{dx}F_{z'} = 0 \quad \text{...(2)}$$
and
$$[F + (\phi' - y')F_{y'} + (\psi' - z')F_{z'}]\big|_{x = x_2} = 0. \quad \text{...(3)}$$

Here $F_y = 0$, $F_z = 0$, $F_{y'} = \frac{y'}{\sqrt{1 + y'^2 + z'^2}}$, $F_{z'} = \frac{z'}{\sqrt{1 + y'^2 + z'^2}}$, $\phi' = 2$, $\psi' = 3$.

$\therefore$ (1) $\Rightarrow$ $$0 - \frac{d}{dx}\left(\frac{y'}{\sqrt{1+y'^2+z'^2}}\right) = 0 \Rightarrow \frac{y'}{\sqrt{1+y'^2+z'^2}} = c_1 \quad ...(4)$$

(2) $\Rightarrow$ $$0 - \frac{d}{dx}\left(\frac{z'}{\sqrt{1+y'^2+z'^2}}\right) = 0 \Rightarrow \frac{z'}{\sqrt{1+y'^2+z'^2}} = c_2 \quad ...(5)$$

Dividing (4) by (5), we get $\frac{y'}{z'} = \frac{c_1}{c_2} = c_3$, say

$\therefore$ $$y' = c_3 z'$$

Putting the value of y' in (5), we get $\frac{z'}{\sqrt{1+(1+c_3^2)\,z'^2}} = c_2$.

$\Rightarrow$ $$z'^2 = c_2^2\,(1 + (1+c_3^2)z'^2) \Rightarrow z' = \frac{c_2}{\sqrt{1-c_2^2-c_2^2\,c_3^2}} = c_4, \text{ say}$$

$\Rightarrow$ $$z = c_4 x + c_5 \quad ...(6)$$

$$y' = c_3 z' \Rightarrow y' = c_3 c_4 = c_6, \text{ say}$$

$\Rightarrow$ $$y = c_6 x + c_7 \quad ...(7)$$

$\therefore$ The extremals are given by (6) and (7)

The boundary conditions are:

$$y(0) = 2,\quad z(0) = 1,\ y(x_2) = y_2 = \phi(x_2) = 2x_2,\ z(x_2) = z_2 = \psi(x_2) = 3x_2$$

$$y(0) = 2 \Rightarrow c_6(0) + c_7 = 2 \Rightarrow c_7 = 2$$

$$z(0) = 1 \Rightarrow c_4(0) + c_5 = 1 \Rightarrow c_5 = 1$$

$$y(x_2) = 2x_2 \Rightarrow c_6 x_2 + c_7 = 2x_2 \Rightarrow c_6 x_2 + 2 = 2x_2 \Rightarrow c_6 = \frac{2x_2 - 2}{x_2} \quad ...(8)$$

$$z(x_2) = 3x_2 \Rightarrow c_4 x_2 + c_5 = 3x_2 \Rightarrow c_4 x_2 + 1 = 3x_2 \Rightarrow c_4 = \frac{3x_2 - 1}{x_2} \quad ...(9)$$

(3) $\Rightarrow$ $$\left(\sqrt{1+y'^2+z'^2} + (2-y')\frac{y'}{\sqrt{1+y'^2+z'^2}} + (3-z')\frac{z'}{\sqrt{1+y'^2+z'^2}}\right)\Bigg|_{x=x_2} = 0$$

$\Rightarrow$ $$\frac{1}{\sqrt{1+y'^2+z'^2}}(1+y'^2+z'^2+2y'-y'^2+3z'-z'^2)\Bigg|_{x=x_2} = 0$$

$\Rightarrow$ $$1 + 2c_6 + 3c_4 = 0$$

$\Rightarrow$ $$1 + 2\left(\frac{2x_2-2}{x_2}\right) + 3\left(\frac{3x_2-1}{x_2}\right) = 0$$

$\Rightarrow$ $$x_2 + 4x_2 - 4 + 9x_2 - 3 = 0 \Rightarrow x_2 = 1/2$$

$\therefore$ (8) $\Rightarrow$ $$c_6 = 2 - \frac{2}{1/2} = -2$$

(9) $\Rightarrow$ $$c_4 = 3 - \frac{1}{1/2} = 1$$

$\therefore$ $$c_4 = 1, c_5 = 1, c_6 = -2, c_7 = 2, x_2 = \frac{1}{2}$$

$\therefore$ (7) $\Rightarrow$ $y = -2x + 2$, (6) $\Rightarrow$ $z = x + 1$

$\therefore$ The stationary functions are $y = -2x + 2, z = x + 1$.

$\therefore$ The curve $y = -2x + 2, z = x + 1$ minimises the functional *i.e.*, the distance between A and B.

$\therefore$ Shortest distance $= \displaystyle\int_0^{1/2} \sqrt{1 + (-2)^2 + (1)^2}\, dx = \sqrt{6}\left(\frac{1}{2} - 0\right) = \frac{\sqrt{6}}{2}$ **units.**

Remark. The minimising curve is a straight line given by the equations:

$$\frac{x-0}{1} = \frac{y-2}{-2} = \frac{z-1}{1}.$$

Example 2. *Find the shortest distance from the point A(1, 1, 1) to the sphere* $x^2 + y^2 + z^2 = 1$.

Sol. The given sphere is $x^2 + y^2 + z^2 = 1$. $\therefore$ $z = \pm\sqrt{1 - x^2 - y^2}$

Let $z = \psi(x, y) = \sqrt{1 - x^2 - y^2}$ for the upper half sphere. Let B (x_2, y_2, z_2) be an arbitrary point on the surface $z = \psi(x, y)$. Let $y = y(x)$, $z = z(x)$ be functions whose curve passes through A and B. The length of the arc between A and B is given by the functional

$$\int_1^{x_2} \sqrt{1 + y'^2 + z'^2}\, dx.$$

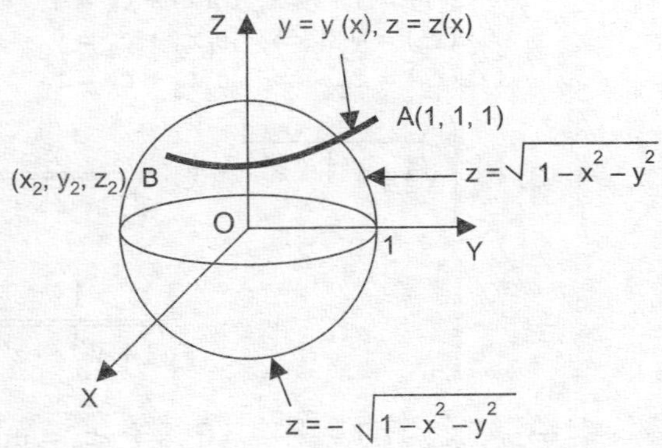

Let $F(x, y, z, y', z') = \sqrt{1 + y'^2 + z'^2}$

Let the length of the curve between A and B be minimum on the curve given by

$$y = y(x), z = z(x).$$

$\therefore$ The functional $\displaystyle\int_1^{x_2} F(x, y, z, y', z')\, dx$ has a minimum value on the functions $y = y(x)$, $z = z(x)$.

$\therefore$ $y(x)$, $z(x)$ satisfy the equations $$F_y - \frac{d}{dx} F_{y'} = 0 \quad \text{...(1)}$$

$$F_z - \frac{d}{dx} F_{z'} = 0 \quad \text{...(2)}$$

$$[F - y'F_{y'} + (\psi_x - z')F_{z'}]\,\big|_{x = x_2} = 0 \quad \text{...(3)}$$

and $$[F_{y'} + F_{z'}\,\psi_y]\,\big|_{x = x_2} = 0 \quad \text{...(4)}$$

Here $F_y = 0$, $F_z = 0$, $F_{y'} = \dfrac{y'}{\sqrt{1 + y'^2 + z'^2}}$, $F_{z'} = \dfrac{z'}{\sqrt{1 + y'^2 + z'^2}}$, $\psi_x = -\dfrac{x}{\sqrt{1 - x^2 - y^2}}$,

$$\psi_y = -\frac{y}{\sqrt{1 - x^2 - y^2}}$$

$$(1) \Rightarrow 0 - \frac{d}{dx}\left(\frac{y'}{\sqrt{1+y'^2+z'^2}}\right) = 0 \Rightarrow \frac{y'}{\sqrt{1+y'^2+z'^2}} = c_1 \quad ...(5)$$

$$(2) \Rightarrow 0 - \frac{d}{dx}\left(\frac{z'}{\sqrt{1+y'^2+z'^2}}\right) = 0 \Rightarrow \frac{z'}{\sqrt{1+y'^2+z'^2}} = c_2 \quad ...(6)$$

Dividing (4) by (5), we get $\frac{y'}{z'} = \frac{c_1}{c_2} = c_3$, say $y' = c_3 z'$

Putting the value of y' in (6), we get $\frac{z'}{\sqrt{1+(1+c_3^2)\,z'^2}} = c_2$.

$$\Rightarrow \quad z'^2 = c_2^2\,(1+(1+c_3^2)\,z'^2) \Rightarrow z' = \frac{c_2}{\sqrt{1-c_2^2-c_2^2\,c_3^2}} = c_4, \text{ say}$$

$$\Rightarrow \quad z = c_4 x + c_5 \quad ...(7)$$

$$y' = c_3 z' \Rightarrow y' = c_3 c_4 = c_5, \text{ say}$$

$$\Rightarrow \quad y = c_6 x + c_7 \quad ...(8)$$

∴ The extremals are given by (7) and (8).

The boundary conditions are:

$$y(1) = 1,\ \ z(1) = 1,\ \ y(x_2) = y_2,\ z(x_2) = z_2 \ \text{ and } \ z_2 = \sqrt{1-x_2^2-y_2^2}.$$

$$y(1) = 1 \Rightarrow c_6(1) + c_7 = 1 \Rightarrow c_6 + c_7 = 1 \quad ...(9)$$

$$z(1) = 1 \Rightarrow c_4(1) + c_5 = 1 \Rightarrow c_4 + c_5 = 1 \quad ...(10)$$

$$y(x_2) = y_2 \Rightarrow c_6 x_2 + c_7 = y_2 \quad ...(11)$$

$$z(x_2) = z_2 \Rightarrow c_4 x_2 + c_5 = z_2 \quad ...(12)$$

$$(3) \Rightarrow \left(\sqrt{1+y'^2+z'^2} - y'\cdot\frac{y'}{\sqrt{1+y'^2+z'^2}} + \left(-\frac{x}{\sqrt{1-x^2-y^2}} - z'\right)\frac{z'}{\sqrt{1+y'^2+z'^2}}\right)\Bigg|_{x=x_2} = 0$$

$$\Rightarrow \quad \frac{1}{\sqrt{1+y'^2+z'^2}}\left(1+y'^2+z'^2-y'^2-\frac{xz'}{\sqrt{1-x^2-y^2}}-z'^2\right)\Bigg|_{x=x_2} = 0$$

$$\Rightarrow \quad 1-\frac{x_2 c_4}{z_2} = 0 \Rightarrow z_2 = c_4 x_2 \quad ...(13)$$

$$(4) \Rightarrow \left(\frac{y'}{\sqrt{1+y'^2+z'^2}} + \frac{z'}{\sqrt{1+y'^2+z'^2}}\left(-\frac{y}{\sqrt{1-x^2-y^2}}\right)\right)\Bigg|_{x=x_2} = 0$$

$$\Rightarrow \quad \frac{1}{\sqrt{1+y'^2+z'^2}}\left(y' - \frac{yz'}{\sqrt{1-x^2-y^2}}\right)\Bigg|_{x=x_2} = 0$$

$$\Rightarrow \quad c_6 - \frac{y_2 c_4}{z_2} = 0 \Rightarrow c_6 z_2 = c_4 y_2 \quad ...(14)$$

$$(12) \Rightarrow z_2 + c_5 = z_2 \Rightarrow c_5 = 0 \quad \text{(Using (13))}$$

$$(10) \Rightarrow c_4 + 0 = 1 \Rightarrow c_4 = 1$$

(14) $\Rightarrow$ $c_6 z_2 = 1 \,.\, y_2$ *i.e.,* $y_2 = c_6 z_2$

$\Rightarrow$ $y_2 = c_6(c_4 x_2) = c_6 x_2$ $(\because\ c_4 = 1)$

$\therefore$ (11) $\Rightarrow$ $y_2 + c_7 = y_2$ $\Rightarrow$ $c_7 = 0$

(9) $\Rightarrow$ $c_6 + 0 = 1$ $\Rightarrow$ $c_6 = 1$

$\therefore$ $c_4 = 1,\quad c_5 = 0,\quad c_6 = 1,\quad c_7 = 0$

Also (13) $\Rightarrow$ $z_2 = x_2$ (14) $\Rightarrow$ $z_2 = y_2$

$\therefore$ $x_2 = y_2 = z_2$

Also $z_2 = \sqrt{1 - x_2^2 - y_2^2}$

$\Rightarrow$ $x_2^2 = 1 - x_2^2 - x_2^2 \Rightarrow x_2 = \dfrac{1}{\sqrt{3}}.$

$\therefore$ $B = (1/\sqrt{3}, 1/\sqrt{3}, 1/\sqrt{3})$

$\therefore$ (8) $\Rightarrow$ $y = 1.x + 0$ *i.e.,* $y = x$

(7) $\Rightarrow$ $z = 1.x + 0$ *i.e.,* $z = x$

$\therefore$ **The stationary functions are $y = x$, $z = x$.**

$\therefore$ **The curve $y = x$, $z = x$ minimises the functional *i.e.,* the distance between A and B.**

$\therefore$ Shortest distance

$$= \int_1^{1/\sqrt{3}} \sqrt{1 + (1)^2 + (1)^2}\, dx = \sqrt{3}\left(\frac{1}{\sqrt{3}} - 1\right) = (\sqrt{3} - 1) \text{ units.}$$

(Taking numerical value)

Remark. The minimising curve is a straight line given by the equations:

$$\frac{x-0}{1} = \frac{y-0}{1} = \frac{z-0}{1}.$$

EXERCISE 5.2

1. Find the shortest distance from the point (0, 2, 0) to the straight line $x = y = z$.
2. Find the shortest distance of the point (0, 2, 4) to the straight line $\dfrac{x-1}{1} = \dfrac{y}{3} = \dfrac{z}{4}$.

Answers

1. $2\sqrt{6}/3$ units **2.** $\sqrt{105/26}$ units.

6

Sufficient Conditions for an Extremum

1. INTRODUCTION

At an extremum, a functional may have either maximum value or minimum value. In the preceding two chapters, we have studied the necessary conditions for an extremum. At a stationary function, a functional may or may not have an extremum. In the present chapter, we shall confine ourselves to the study of sufficient conditions for extremums of variational problems with fixed boundaries only.

2. PROPER FIELD IN A GIVEN DOMAIN

A family of curves $y = y(x, c)$ is called a **proper field in a given domain** D of the xy-plane if through every point of D, there passes one and only one curve of the family $y = y(x, c)$.

In other words, the curves of the family $y = y(x, c)$ cover the entire domain D without self-intersections.

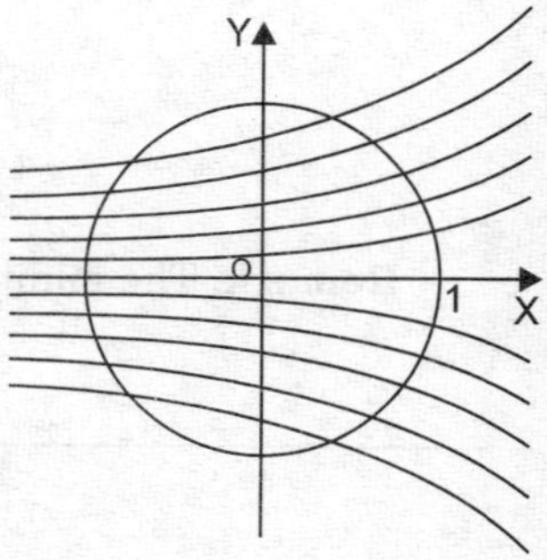

For example, the family of curves $y = ce^x$, where c is an arbitrary constant, form a proper field inside the circle $x^2 + y^2 = 1$, because through any point inside or on the circle there passes exactly one curve belonging to the family $y = ce^x$.

To take another example, the family of parabolas $y = (x + c)^2$, where c is an arbitrary constant, does not form a proper field inside the circle $x^2 + y^2 = 1$, because through any point inside or on the circle there may pass more than one parabola belonging to the family $y = (x + c)^2$. Moreover, there are infinitely many points inside the circle from where no parabola of the family $y = (x + c)^2$ pass.

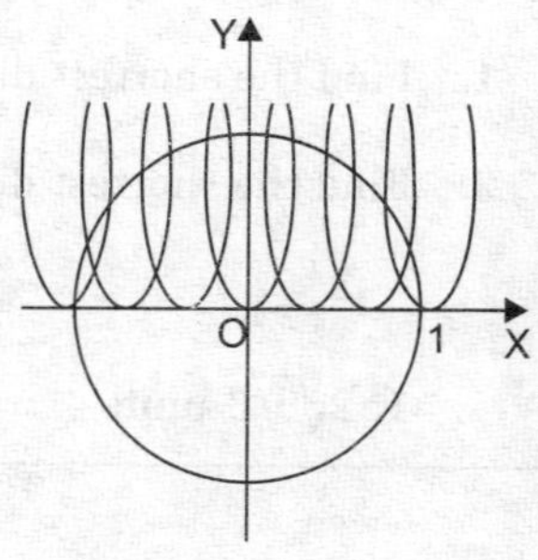

3. CENTRAL FIELD IN A GIVEN DOMAIN

A family of curves $y = y(x, c)$ is called a **central field in a given domain** D of the xy-plane if through every point of D there passes one and only one curve of the family $y = y(x, c)$ and all curves issue from a single point (x_0, y_0) lying outside D. The point (x_0, y_0) is called the **centre of the central field** or **centre of the pencil of curves.**

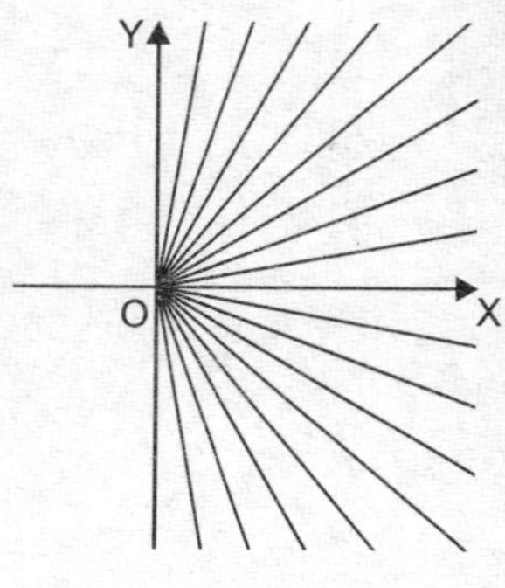

For example, the family of curves $y = cx$, where c is an arbitrary constant, form a central field with centre (0, 0) in the domain $x > 0$.

4. EXTREMAL FIELD

A field (proper or central) is called an **extremal field** (or a **field of extremals**) if it is formed by a family of extremals of some variational problem.

Example 1. *Find the proper and central field of extremals for the functional* $\int_0^1 (y^2 - y'^2)\, dx$.

Sol. Given functional is $\int_0^1 (y^2 - y'^2)\, dx$.

Let $F(x, y, y') = y^2 - y'^2$

The **Euler's equation** is

$$F_y - \frac{d}{dx} F_{y'} = 0 \qquad ...(1)$$

Here $F_y = 2y, \quad F_{y'} = -2y'$

$\therefore \quad (1) \Rightarrow \quad 2y - \dfrac{d}{dx}(-2y') = 0 \quad \Rightarrow \quad \dfrac{d^2y}{dx^2} + y = 0$

The A.E. is $D^2 + 1 = 0 \quad \therefore \quad D = \pm i$

$\therefore \qquad y = e^{0.x}(c_1 \cos 1.x + c_2 \sin 1.x)$

i.e., $\qquad y = c_1 \cos x + c_2 \sin x. \qquad ...(2)$

This is the equation of the extremals.

For $c_2 = 0$, (2) $\Rightarrow$ $\mathbf{y = c_1 \cos x}$. This forms a proper field of extremals in the domain $0 \le x \le 1$.

For $c_1 = 0$, (2) $\Rightarrow$ $\mathbf{y = c_2 \sin x}$. This forms a central field of extremals in the domain $0 < x \le 1$ with centre at (0, 0).

5. INCLUSION OF AN EXTREMAL IN A FIELD OF EXTREMALS

Let $y = y(x)$ be an extremal of the functional $\int_{x_1}^{x_2} F(x, y, y')\, dx$ with fixed boundary points (x_1, y_1) and (x_2, y_2). The extremal $y = y(x)$ is said to be **included in a proper field of extremals** if the given functional yields a family of extremals $y = y(x, c)$ that forms a proper field and contain the extremal $y = y(x)$ for some particular value of c. Also, this extremal $y = y(x)$ does not lie on the boundary of the domain D in which the family $y = y(x, c)$ form a proper field.

Similarly, the extremal $y = y(x)$ is said to be **included in a central field of extremals** if the given functional yields a family of extremals $y = y(x, c)$ that forms a central field and containing the extremal $y = y(x)$ for some particular value of c. Also, this extremal $y = y(x)$ does not lie on the boundary of the domain D in which the family $y = y(x, c)$ form a central field.

Example 2. *Show that the extremal of the variational problem* $\int_0^2 (y'^3 + \sin^2 x)\, dx$, $y(0) = 0$, $y(2) = 6$ *is included in a central field of extremals of the given functional.*

Sol. Given functional is $\int_0^2 (y'^3 + \sin^2 x)\, dx$.

Let $F(x, y, y') = y'^3 + \sin^2 x$

The **Euler's equation** is $F_y - \dfrac{d}{dx} F_{y'} = 0 \qquad ...(1)$

Here $F_y = 0$, $F_{y'} = 3y'^2$

$\therefore$ (1) $\Rightarrow$ $0 - \dfrac{d}{dx}(3y'^2) = 0 \Rightarrow \dfrac{dy'^2}{dx} = 0$

$\Rightarrow$ $y'^2 = c_1 \Rightarrow y' = \sqrt{c_1} = c_2$, say

$\Rightarrow$ $y = c_2 x + c_3$ ...(2)

This is the equation of the extremals. The boundary conditions are $y(0) = 0$, $y(2) = 6$.

$$y(0) = 0 \Rightarrow c_2(0) + c_3 = 0 \Rightarrow c_3 = 0$$

$$y(2) = 6 \Rightarrow c_2(2) + c_3 = 6 \Rightarrow 2c_2 = 6 \Rightarrow c_2 = 3$$

$\therefore$ (2) $\Rightarrow$ $y = 3x + 0$ *i.e.,* $y = 3x$

This is the extremal of the given variational problem.

$$c_3 = 0 \text{ in (2)} \Rightarrow y = c_2 x$$

This is a central field of extremals in the domain $0 < x \leq 1$ with centre at (0, 0).

For $c_2 = 3$, $y = c_2 x \Rightarrow y = 3x$

$\therefore$ The extremal $y = 3x$ is included in the central field of extremals $y = c_2 x$.

EXERCISE 6.1

1. Show that the family of parabolas $y = (x - c)^2$ is not a proper field inside the circle $x^2 + y^2 = 4$.
2. Find the proper and central fields of extremals for the functional $\int_0^4 y'^2\, dx$.
3. Find the proper and central fields of extremals for the functional $\int_0^{\pi/4} (y'^2 - y^2 + 2x^2 + 4)\, dx$.
4. Show that the extremal of the variational problem $\int_0^2 (y'^3 + \cos^2 x)\, dx$, $y(0) = 2$, $y(2) = 2$ is included in a proper field of extremals of the given functional.
5. Show that the extremal of the variational problem $\int_0^1 (y'^2 - 2xy)\, dx$, $y(0) = 0$, $y(1) = 0$ is included in a central field of extremals of the given functional.
6. Show that the extremal of the variational problem $\int_0^2 (y'^2 + x^2)\, dx$, $y(0) = 1$, $y(2) = 3$ is included in a proper field of extremals of the given functional.

Answers

2. $y = c_2$, $0 \leq x \leq 4$; $y = c_1 x$, (0, 0), $0 < x \leq 4$
3. $y = c_1 \cos x$, $0 \leq x \leq \pi/4$; $y = c_2 \sin x$, (0, 0), $0 < x \leq \pi/4$.

Hint

5. The extremal $y = -\dfrac{x^3}{6} + \dfrac{x}{6}$ is included in the central field of extremals $y = -\dfrac{x^3}{6} + c_1 x$.

6. ENVELOPE OF A FAMILY

Let $\phi(x, y, c) = 0$ be a one parameter family of plane curves. The **envelope** of the family $\phi(x, y, c) = 0$ is a curve, each point of which touches a certain curve of the given family and each portion of which is touched by infinitely many of curves of the family.

7. c-DISCRIMINANT OF THE FAMILY $\phi(x, y, c) = 0$

Let $\phi(x, y, c) = 0$ be a one parameter family of plane curves. The locus of points defined by the equations $\phi(x, y, c) = 0$, $\dfrac{\partial \phi}{\partial c} = 0$ is called the **c-discriminant** of the given family of plane curves $\phi(x, y, c) = 0$.

In general, the *c*-discriminant includes the envelopes of the family $\phi(x, y, c) = 0$, the locus of nodal points and the locus of cusps.

For example, if $\phi(x, y, c) = 0$ represents a pencil of curves with centre at the point A, then the centre at A of the pencil belongs to the *c*-discriminant of the family $\phi(x, y, c) = 0$.

Example 1. *Find the c-discriminant of the family of curves $y(c - x) - c^2 = 0$.*

Sol. Given family is $y(c - x) - c^2 = 0$. ...(1)

Let $\phi(x, y, c) = y(x - c) - c^2$

The equations of *c*-discriminant are $\phi(x, y, c) = 0$, $\dfrac{\partial \phi}{\partial c} = 0$.

$$\phi(x, y, c) = 0 \Rightarrow y(c - x) - c^2 = 0 \qquad ...(2)$$

$$\frac{\partial \phi}{\partial c} = 0 \Rightarrow y.1 - 0 - 2c = 0 \qquad ...(3)$$

(2) $\Rightarrow c = \dfrac{y}{2}$. Putting the value of *c* in (1), we get

$$y\left(\frac{y}{2} - x\right) - \frac{y^2}{4} = 0 \text{ or } \frac{y^2}{4} - xy = 0 \text{ or } \mathbf{y(y - 4x) = 0.}$$

This is the equation of the *c*-discriminant.

8. CONJUGATE POINT OF A POINT

Let $\phi(x, y, c) = 0$ be a pencil of curves with centre at A and $\psi(x, y) = 0$ represents the *c*-discriminant of the family $\phi(x, y, c) = 0$. Let $y = y(x)$ be a curve belonging to the pencil of curves $\phi(x, y, c) = 0$. If A* be a point, other than A, on the curve of $y = y(x)$ and also on the *c*-discriminant $\psi(x, y) = 0$, then the point A* is called a **conjugate point** of the point A.

Example 2. *Given a family of curves $y = c(x + 4)x$, find a point conjugate to the point (0, 0).*

Sol. Given family is $c(x + 4)x - y = 0$.

Given family of curves is a pencil of curves with centre (0, 0).

Let $\phi(x, y, c) = c(x + 4)x - y$

The equations of *c*-discriminant are $\phi(x, y, c) = 0$, $\dfrac{\partial \phi}{\partial c} = 0$.

$$\phi(x, y, c) = 0 \Rightarrow c(x + 4)x - y = 0 \qquad ...(1)$$

$$\frac{\partial \phi}{\partial c} = 0 \Rightarrow (x + 4)x.1 - 0 = 0 \qquad ...(2)$$

(2) $\Rightarrow$ $(x + 4)x = 0$ $\therefore$ (1) $\Rightarrow c.0 - y = 0 \Rightarrow y = 0$.

Also, $(x + 4)x = 0$ $\Rightarrow x = -4, 0$

$\therefore$ c-discriminant is the set $\{(-4, 0), (0, 0)\}$.

Let $y = c_0(x + 4)x$ be a curve belonging to the given pencil of curves. The curve of this function passes through $(-4, 0)$ which is also on the c-discriminant of the given family.

$\therefore$ **$(-4, 0)$** is a conjugate point of the centre $(0, 0)$.

EXERCISE 6.2

Find the c-discriminant of the following families of curves (Q. No. 1–3):

1. $y = (x - c)^2$ **2.** $y = cx + c^2$ **3.** $(x - c)^2 + y^2 = 1$

4. Given a family of curves $y = c(x - 1)x$, find a conjugate point of the point $(0, 0)$.

5. Given a family of curves $y = c \sinh x$, find a conjugate point of the point $(0, 0)$.

Answers

1. $y = 0$ **2.** $y = -\dfrac{x^2}{4}$ **3.** $y^2 - 1 = 0$ **4.** $(1, 0)$

5. No conjugate point.

9. JACOBI CONDITION

A given extremal may or may not be included in a central field of extremals. The following is the sufficient condition for an arc of an extremal to be included in a central field of extremals:

For an arc AB of an extremal to be included in a central field of extremals with centre at A, it is sufficient that no conjugate point of A, lie on the arc AB.

This sufficient condition for including an extremal in a central field of extremals is called the **Jacobi condition.**

Let $J[y(x)] = \int_{x_1}^{x_2} F(x, y, y')\, dx$ be a functional with fixed boundaries $A(x_1, y_1)$ and $B(x_2, y_2)$.

The equation $\left(\mathbf{F_{yy}} - \dfrac{\mathbf{d}}{\mathbf{dx}}\mathbf{F_{yy'}}\right)\mathbf{u} - \dfrac{\mathbf{d}}{\mathbf{dx}}(\mathbf{F_{y'y'}u'}) = \mathbf{0}$ is called the **Jacobi equation.**

Let $u = u(x)$ be the solution of the Jacobi equation satisfying the condition $u(x_1) = 0$. We have the following results:

(i) If $u(x) = 0$ for some $x \in (x_1, x_2)$ then a conjugate point of A lies on the arc AB.

$\therefore$ By the **Jacobi condition**, the arc AB of the extremal of the variational problem is not included in a central field of extremals with centre at A.

(ii) If $u(x) \neq 0$ for all $x \in (x_1, x_2]$ then no conjugate point of A lie on the arc AB.

$\therefore$ By the **Jacobi condition**, the arc AB of the extremal of the variational problem is included in a central field of extremals with centre at A.

Remark. In the Jacobi equation if any of F_{yy}, $F_{yy'}$, $F_{y'y'}$, contain y, then the value of y in terms of x is substituted from the extremal of the variational problem.

Example 1. *Show that the Jacobi condition is fulfilled for the extremal of the functional $\int_0^{\lambda} (x^2 + y^2 + y'^2)\, dx$ with fixed boundaries A(0, 0) and B (λ, 0).*

Sol. Given functional is $\int_0^{\lambda} (x^2 + y^2 + y'^2)\, dx$.

Let $F(x, y, y') = x^2 + y^2 + y'^2$

The **Euler's equation** is $F_y - \dfrac{d}{dx} F_{y'} = 0$ ...(1)

Here $F_y = 2y, \quad F_{y'} = 2y'$

$$\therefore \ (1) \Rightarrow \quad 2y - \frac{d}{dx}(2y') = 0 \ \Rightarrow \ \frac{d^2y}{dx^2} - y = 0$$

The A.E. is $\quad D^2 - 1 = 0 \quad \therefore \quad D = \pm 1$

$$\therefore \quad y = c_1e^x + c_2e^{-x} \qquad ...(2)$$

This is the equation of the extremals. The boundary conditions are $y(0) = 0$, $y(\lambda) = 0$.

$$y(0) = 0 \ \Rightarrow \ c_1e^0 + c_2e^0 = 0 \ \Rightarrow \ c_1 + c_2 = 0 \ \Rightarrow \ c_2 = -c_1$$

$$y(\lambda) = 0 \ \Rightarrow \ c_1e^{\lambda} + c_2e^{-\lambda} = 0 \ \Rightarrow \ c_1(e^{\lambda} - e^{-\lambda}) = 0 \ \Rightarrow \ c_1 = 0$$

$$\therefore \quad c_2 = -(0) = 0$$

$$\therefore \ (2) \Rightarrow \quad y = 0.e^x + 0.e^{-x} \ i.e., \ \ y = 0. \qquad ...(3)$$

This is the extremal of the given variational problem.

The **Jacobi equation** is

$$\left(F_{yy} - \frac{d}{dx}F_{yy'}\right)u - \frac{d}{dx}(F_{y'y'}\,u') = 0 \qquad ...(4)$$

Here $F_{yy} = 2, \ F_{yy'} = 0, \ \ F_{y'y'} = 2$

$$\therefore \ (4) \Rightarrow \quad \left(2 - \frac{d}{dx}(0)\right)u - \frac{d}{dx}(2u') = 0 \ \Rightarrow \ \frac{d^2u}{dx^2} - u = 0$$

The A.E. is $\quad D^2 - 1 = 0 \quad \therefore \quad D = \pm 1$

$$\therefore \quad u = k_1e^x + k_2e^{-x}$$

Also, $\quad u(0) = 0 \ \Rightarrow \ k_1e^0 + k_2e^0 = 0 \ \Rightarrow \ k_1 + k_2 = 0 \ \Rightarrow \ k_2 = -k_1$

Let $\quad u(x) = k_1e^x - k_1e^{-x} = k_1(e^x - e^{-x}), \ k_1 \neq 0$

Here $u(x) \neq 0$ for $x \in (0, \lambda]$

∴ No conjugate point of A(0, 0) lie on the arc AB *i.e.,* on the curve of $y = 0$.

∴ By the Jacobi condition, the curve of the extremal $y = 0$ is included in a central field of extremals with centre at A(0, 0).

Verification. The family of extremals is $y = c_1e^x + c_2e^{-x}$.

$$c_2 = -c_1 \ \Rightarrow \ y = c_1e^x - c_1e^{-x} = 2c_1\left(\frac{e^x - e^{-x}}{2}\right) = 2c_1 \sinh x\,.$$

This is a central field of extremals with centre at A(0, 0) in the domain $0 < x \leq \lambda$.

For $c_1 = 0, \ \ y = 2(0) \sinh x \Rightarrow \ y = 0.$

∴ The extremal $y = 0$ is included in the central field of extremals $y = 2c_1 \sinh x$ with centre at A(0, 0).

Example 2. *Show that the Jacobi condition is fulfilled for the extremal of the functional* $\int_0^a (y'^2 - 4y^2 + e^{-x^2})\,dx,\ a \neq \frac{n\pi}{2}$ *with fixed boundaries A(0, 0) and B(a, 0).*

Sol. Given functional is $\int_0^a (y'^2 - 4y^2 + e^{-x^2})dx$.

Let $F(x, y, y') = y'^2 - 4y^2 + e^{-x^2}$

The **Euler's equation** is $F_y - \frac{d}{dx} F_{y'} = 0$ ...(1)

Here $F_y = -8y,\ F_{y'} = 2y'$

$\therefore$ (1) $\Rightarrow$ $-8y - \frac{d}{dx}(2y') = 0 \Rightarrow \frac{d^2y}{dx^2} + 4y = 0$

The A.E. is $D^2 + 4 = 0 \quad \therefore \quad D = \pm 2i$

$\therefore$ $y = e^{0.x}(c_1 \cos 2x + c_2 \sin 2x)$

i.e., $y = c_1 \cos 2x + c_2 \sin 2x$ ...(2)

This is the equation of the extremals. The boundary conditions are $y(0) = 0$, $y(a) = 0$.

$y(0) = 0 \Rightarrow c_1 \cos 0 + c_2 \sin 0 = 0 \Rightarrow c_1 = 0$

$y(a) = 0 \Rightarrow c_1 \cos 2a + c_2 \sin 2a = 0 \Rightarrow 0.\cos 2a + c_2 \sin 2a = 0 \Rightarrow c_2 = 0$

$\left(\because a \neq \frac{n\pi}{2}\right)$

$\therefore$ (2) $\Rightarrow$ $y = 0.\cos 2x + 0.\sin 2x$ *i.e.,* $y = 0$. ...(3)

This is the extremal of the given variational problem.

The **Jacobi equation** is

$$\left(F_{yy} - \frac{d}{dx} F_{yy'}\right)u - \frac{d}{dx}(F_{y'y'}\, u') = 0 \qquad \text{...(4)}$$

Here $F_{yy} = -8,\ F_{yy'} = 0,\ F_{y'y'} = 2$

$\therefore$ (4) $\Rightarrow$ $\left(-8 - \frac{d}{dx}(0)\right)u - \frac{d}{dx}(2u') = 0 \Rightarrow \frac{d^2u}{dx^2} + 4u = 0$

$\therefore$ $u = k_1 \cos 2x + k_2 \sin 2x$

Also, $u(0) = 0 \Rightarrow k_1 \cos 0 + k_2 \sin 0 = 0 \Rightarrow k_1 = 0$

Let $u(x) = k_2 \sin 2x,\ k_2 \neq 0$.

For a, there are two possibilities.

Case I. a < π/2

Here $u(x) = k_2 \sin 2x \neq 0\ \forall\ x \in (0, a]$ $\left(\because 0 < a < \frac{\pi}{2} \Rightarrow 0 < 2a < \pi \Rightarrow \sin 2x \neq 0\right)$

$\therefore$ No conjugate point of A(0, 0) lie on the arc AB *i.e.,* on the curve of $y = 0$.

$\therefore$ By the Jacobi condition, the curve of the extremal $y = 0$ is included in a central field of extremals with centre at A(0, 0).

Case II. $a > \pi/2$

Here $u(x) = k_2 \sin 2x = 0$ for $x = \frac{\pi}{2}$, because $\frac{\pi}{2} \in (0, a)$

$\therefore$ A conjugate point of A(0, 0) lies on the arc AB *i.e.,* on the curve of $y = 0$.

$\therefore$ By the Jacobi condition, the curve of the extremal $y = 0$ is not included in a central field of extremals with centre at A(0, 0).

Verification. Case I. $a < \pi/2$.

The family of extremals is $y = c_1 \cos 2x + c_2 \sin 2x$.

$c_1 = 0 \Rightarrow y = c_2 \sin 2x$. This is a central field of extremals in the domain $0 < x \le a$ with centre at A(0, 0). $\left(\because 0 < a < \frac{\pi}{2} \Rightarrow 0 < 2a < \pi \Rightarrow \sin 2x \neq 0\right)$

For $c_2 = 0$, $y = c_2 \sin 2x \Rightarrow y = 0$.

$\therefore$ The extremal $y = 0$ is included in the central field of extremals $y = c_2 \sin 2x$ with centre at A(0, 0).

Case II. $a > \pi/2$. The family of extremals is $y = c_1 \cos 2x + c_2 \sin 2x$.

$c_1 = 0 \Rightarrow y = c_2 \sin 2x$. This is not a central field of extremals because each curve in this family passes through the points (0, 0) and $(\pi/2, 0)$

$c_2 = 0 \Rightarrow y = c_1 \cos 2x$. This is also not a central field of extremals because there is no centre of this family of curve.

$\therefore$ The extremal $y = 0$ is not included in a central field of extremals.

EXERCISE 6.3

1. Show that the Jacobi condition is fulfilled for the functional $\int_0^a (x^2 + y'^2)\, dx$, $a > 0$ with fixed boundaries A(0, 0) and B(a, 2).

2. Show that the Jacobi condition is fulfilled for the functional $\int_0^5 (9y^2 + y'^2 - 3x)\, dx$ with fixed boundaries A(0, 0) and B(5, 0).

3. Show that the Jacobi condition is fulfilled for the functional $\int_{-1}^1 (12xy + y'^2 + x^2)\, dx$ with fixed boundaries A(– 1, – 2) and B(1, 0).

4. Show that the Jacobi condition is fulfilled for the functional $\int_0^1 (1 + y'^2)\, dx$ with fixed boundaries A(0, 0) and B(1, 0).

Hint

1. The extremal is $y = \frac{2}{a}x$.

   $u(x) = k_1 x$, $(k_1 \neq 0)$ and $u(x) \neq 0\ \forall\ x \in (0, a]$.

   Also $y = c_1 x$ is a central field of extremals with centre at (0, 0) and containing $y = \frac{2}{a}x$.

10. LEGENDRE CONDITION

A given extremal may or may not give a maximum or minimum value of the functional under consideration. The following is the sufficient condition for an extremal of a functional to give a maximum or minimum value of the functional.

Let $J[y(x)] = \int_{x_1}^{x_2} F(x, y, y')\,dx$ be a functional with fixed boundaries $A(x_1, y_1)$ and $B(x_2, y_2)$. Let $F(x, y, y')$ has a continuous partial derivative $F_{y'y'}(x, y, y')$. Let C be the curve of the extremal of the given functional which passes through A and B. Also, let the extremal of the curve C be included in a field of extremals.

(*i*) **For a strong maximum (resp. minimum) to be attained on the curve C, it is sufficient to have $F_{y'y'} \le 0$ (resp. $F_{y'y'} \ge 0$) at the points close to the curve C and also for arbitrary values of y'.**

(*ii*) **For a weak maximum (resp. minimum) to be attained on the curve C, it is sufficient to have $F_{y'y'} < 0$ (resp. $F_{y'y'} > 0$) on the curve C.**

This sufficient condition for an extremal to give a maximum or minimum value of the corresponding functional is called the **Legendre condition.**

Remark. It may be recalled that every strong extremum is at the same time a weak extremum but not conversely.

Example 1. *Using Legendre condition, test the functional*

$$I[y(x)] = \int_0^2 (4 - e^{y'})\,dx,\ y(0) = 0,\ y(2) = 1 \text{ for extremal.}$$

Sol. We have $\qquad I[y(x)] = \int_0^2 (4 - e^{y'})\,dx.$

Let $\qquad F(x, y, y') = 4 - e^{y'}$

The **Euler's equation** is

$$F_y - \frac{d}{dx} F_{y'} = 0 \qquad \text{...(1)}$$

Here $\qquad F_y = 0,\ F_{y'} = -e^{y'}$

$$\therefore\ (1) \Rightarrow\quad 0 - \frac{d}{dx}(-e^{y'}) = 0 \quad\Rightarrow\quad \frac{de^{y'}}{dx} = 0 \Rightarrow e^{y'} = e^{c_1}$$

$$\Rightarrow\quad y' = c_1 \quad\Rightarrow\quad y = c_1 x + c_2 \qquad \text{...(2)}$$

This is the equation of the extremals.

The boundary conditions are $\quad y(0) = 0,\ y(2) = 1.$

$$y(0) = 0 \Rightarrow c_1(0) + c_2 = 0 \Rightarrow \quad c_2 = 0$$

$$y(2) = 1 \Rightarrow c_1(2) + c_2 = 1 \Rightarrow 2c_1 + 0 = 1 \Rightarrow c_1 = \frac{1}{2}$$

$$\therefore\ (2) \Rightarrow\quad y = \frac{1}{2}x + 0 \text{ i.e., } y = \frac{x}{2} \qquad \text{...(3)}$$

This is the extremal of the given variational problem.

$c_2 = 0$ in (2) $\Rightarrow$ $y = c_1 x$. This is a central field of extremals with centre (0, 0) and domain $0 < x \le 2$.

For $c_1 = \frac{1}{2}, y = c_1 x \Rightarrow y = \frac{x}{2}$

$\therefore$ The extremal $y = \frac{x}{2}$ is included in the central field of extremals $y = c_1x$.

Now,
$$F_{y'y'} = \frac{\partial}{\partial y'}\left(\frac{\partial}{\partial y'} F\right) = \frac{\partial}{\partial y'}\left(\frac{\partial}{\partial y'}(4 - e^{y'})\right) = \frac{\partial}{\partial y'}(-e^{y'}) = -e^{y'}$$

The function '$-e^{y'}$' is continuous.

Also $F_{y'y'} = -e^{y'} \leq 0$ for all values of y'.

$\therefore$ By Legendre condition, the given functional has a strong maximum on the function $y = x/2$.

Example 2. *Using Legendre condition, test the following functional for extremal:*

$$J[y(x)] = \int_2^3 \frac{x^3}{y'^2}\, dx\,, y(2) = 4\,, y(3) = 9.$$

Sol. We have
$$J[y(x)] = \int_2^3 \frac{x^3}{y'^2}\, dx.$$

Let
$$F(x, y, y') = \frac{x^3}{y'^2}$$

The **Euler's equation** is
$$F_y - \frac{d}{dx} F_{y'} = 0 \qquad \text{...(1)}$$

Here
$$F_y = 0,\ F_{y'} = -\frac{2x^3}{y'^3}$$

$\therefore$ (1)
$$0 - \frac{d}{dx}\left(-\frac{2x^3}{y'^3}\right) = 0 \Rightarrow \frac{d}{dx}\left(\frac{x^3}{y'^3}\right) = 0$$

$$\Rightarrow \quad \frac{x^3}{y'^3} = c_1 \Rightarrow y' = \left(\frac{1}{c_1}\right)^{1/3} x$$

$$\Rightarrow \quad y = \left(\frac{1}{c_1}\right)^{1/3} \frac{x^2}{2} + c_2 \Rightarrow y = c_3x^2 + c_2, \text{ say} \qquad \text{...(2)}$$

This is the equation of the extremals. The boundary conditions are $y(2) = 4$, $y(3) = 9$.

$$y(2) = 4 \Rightarrow c_3(2)^2 + c_2 = 4 \Rightarrow 4c_3 + c_2 = 4 \qquad \text{...(3)}$$
$$y(3) = 9 \Rightarrow c_3(3)^2 + c_2 = 9 \Rightarrow 9c_3 + c_2 = 9 \qquad \text{...(4)}$$

Solving, we get $c_2 = 0$, $c_3 = 1$

$\therefore$ (2) $\Rightarrow$ $y = 1.x^2 + 0$ *i.e.*, $y = x^2$. ...(5)

This is the extremal of the given variational problem.

$c_2 = 0$ in (2) $\Rightarrow$ $y = c_3x^2$.

This is a central field of extremals with centre at (0, 0) and domain $2 \leq x \leq 3$.

For $c_3 = 1, \quad y = c_3x^2 \quad \Rightarrow \quad y = x^2$

$\therefore$ The extremal $y = x^2$ is included in the central field of extremals $y = c_3x^2$.

Now $$F_{y'y'} = \frac{\partial}{\partial y'} F_{y'} = \frac{\partial}{\partial y'}\left(-\frac{2x^3}{y'^3}\right) = \frac{6x^3}{y'^4}$$

The function $\dfrac{6x^3}{y'^4}$ is continuous.

For $y' = 0$, the quantity $\dfrac{6x^3}{y'^4}$ is not defined.

$\therefore$ The function $y = x^2$ is neither a strong maximum nor a strong minimum.

$$y = x^2 \quad \Rightarrow \quad y' = 2x.$$

$\therefore$ On the extremal $y = x^2$,

$$F_{y'y'} = \frac{6x^3}{(2x)^4} = \frac{3}{8x}$$

Also, on [2, 3], $\dfrac{3}{8x} > 0$

$\therefore$ $F_{y'y'} > 0$ on the extremal $y = x^2$.

$\therefore$ By **Legendre condition,** the given functional has a weak minimum on the function $y = x^2$.

EXERCISE 6.4

Using Legendre condition, test the following functionals for extremal:

1. $J[y(x)] = \int_0^1 (x^2 + y'^2)\, dx$, $y(0) = -1$, $y(1) = 1$
2. $J[y(x)] = \int_0^2 (4 + e^{y'})\, dx$, $y(0) = 0$, $y(2) = 1$
3. $J[y(x)] = \int_0^2 (e^{y'} + 3)\, dx$, $y(0) = 0$, $y(2) = 1$
4. $J[y(x)] = \int_0^1 yy'^2\, dx$, $y(0) = 4$, $y(1) = 4$
5. $J[y(x)] = \int_1^2 (xy'^4 - 2yy'^3)\, dx$, $y(1) = 0$, $y(2) = 1$.

Answers

1. Strong minimum on the function $y = 2x - 1$
2. Strong minimum on the function $y = \dfrac{x}{2}$
3. Strong minimum on the function $y = \dfrac{x}{2}$
4. Strong minimum on the function $y = 4$
5. Weak minimum on the function $y = x - 1$.

Hints

1. The extremal $y = 2x - 1$ is included in the central field of extremals $y = cx - 1$ with centre at (0, – 1).
2. The extremal $y = x/2$ is included in the central field of extremal $y = cx$ with centre at (0, 0).

11. VARIATION OF A FUNCTIONAL

Let $J[y(x)]$ be a functional defined on a class A of functions $y(x)$. The increment of the argument $y(x)$ of the functional $J[y(x)]$ is given by the difference $y_1(x) - y(x)$ for some function $y_1(x)$ in A. This increment is denoted by $\delta y(x)$. The **variation of the functional** $J[y(x)]$ at the point $y(x)$ is defined as the value of the derivative of the functional $J[y(x) + \alpha\delta y(x)]$ with respect to the parameter α when $\alpha = 0$. The variation $J[y(x)]$ is denoted by δJ.

$$\therefore \quad \delta J = \frac{\partial}{\partial\alpha} J[y(x) + \alpha\delta y(x)]\Bigg|_{\alpha=0}$$

If the variation of a functional exists, then the functional is said to **differentiable.**

Example 1. *Find the variation of the functional* $J[y(x)] = \int_{x_1}^{x_2} y(x)\,dx$.

Sol. We have $$J[y(x)] = \int_{x_1}^{x_2} y(x)\,dx.$$

Let $\delta y(x)$ be an increment of $y(x)$.

$$\therefore \quad J[y(x) + \alpha\delta y(x)] = \int_{x_1}^{x_2} (y(x) + \alpha\delta y(x))\,dx$$

$$\therefore \quad \frac{\partial}{\partial\alpha} J[y(x) + \alpha\delta y(x)] = \frac{\partial}{\partial\alpha}\int_{x_1}^{x_2} (y(x) + \alpha\delta y(x))\,dx$$

$$= \int_{x_1}^{x_2} \frac{\partial}{\partial\alpha}(y(x) + \alpha\delta y(x))\,dx$$

$$= \int_{x_1}^{x_2} (0 + 1.\,\delta y(x))\,dx = \int_{x_1}^{x_2} \delta y(x)\,dx$$

$$\therefore \quad \frac{\partial}{\partial\alpha} J[y(x) + \alpha\delta y(x)]\Bigg|_{\alpha=0} = \int_{x_1}^{x_2} \delta y(x)\,dx$$

$\therefore$ Variation of $J[y(x)] = \int_{x_1}^{x_2} \delta y(x)\,dx$.

Example 2. *Find the variation of the functional*

$$J[y(x)] = y^2(0) + \int_0^1 (x\,.\,y(x) + (y'(x))^2)\,dx.$$

Sol. We have $$J[y(x)] = y^2(0) + \int_0^1 (x\,.\,y(x) + (y'(x))^2)\,dx$$

Let $\delta y(x)$ be an increment of $y(x)$.

$$\therefore \quad J[y(x) + \alpha\delta y(x)] = (y(0) + \alpha\delta y(0))^2 + \int_0^1 (x(y(x) + \alpha\delta y(x)) + (y'(x) + \alpha(\delta y)'(x)^2)\,dx$$

$$\therefore \quad \frac{\partial}{\partial\alpha} J[y(x)+\alpha\delta y(x)] = 2(y(0)+\alpha\delta y(0))\,\delta y(0) + \int_0^1 (0 + x\delta y(x) + 2(y'(x)+\alpha(\delta y)'(x))\,(\delta y)'(x))\,dx$$

$$\therefore \quad \left.\frac{\partial}{\partial\alpha} J[y(x)+\alpha\delta y(x)]\right|_{\alpha=0} = 2(y(0)+0)\,\delta y(0) + \int_0^1 (x\delta y(x) + 2(y'(x)+0)\,(\delta y)'(x))\,dx$$

$$= 2y(0)\,\delta y(0) + \int_0^1 (x\delta y(x) + 2y'(x)\,(\delta y)'(x))\,dx$$

$\therefore$ Variation of $\quad J[y(x)] = \mathbf{2y(0)\delta y(0) + \int_0^1 (x\delta y(x) + 2y'(x)\,(\delta y)'(x))\,dx}$.

12. SECOND VARIATION OF A FUNCTIONAL

Let $J[y(x)]$ be a functional defined on a class A of functions $y(x)$. The **second variation of the functional** $J[y(x)]$ is defined as the value of the second derivative of the functional $J[y(x) + \alpha\delta y(x)]$ with respect to the parameter α when $\alpha = 0$. The second variation of $J[y(x)]$ is denoted by $\delta^2 J$.

$$\therefore \quad \delta^2 \mathbf{J} = \left.\frac{\partial^2}{\partial\alpha^2}\mathbf{J\,[y(x)+\alpha\delta y(x)]}\right|_{\alpha=0}.$$

Example 3. *Find the second variation of the functional* $J[y(x)] = \int_{x_1}^{x_2} y(x)\,dx$.

Sol. We have $\quad J[y(x)] = \int_{x_1}^{x_2} y\,dx$

Let δy be an increment of y

$$\therefore \quad J[y+\alpha\delta y] = \int_{x_1}^{x_2} (y+\alpha\delta y)\,dx$$

$$\therefore \quad \frac{\partial}{\partial\alpha} J\,[y+\alpha\delta y] = \frac{\partial}{\partial\alpha}\int_{x_1}^{x_2} (y+\alpha\delta y)\,dx = \int_{x_1}^{x_2} \frac{\partial}{\partial\alpha}(y+\alpha\delta y)\,dx = \int_{x_1}^{x_2} (0+\delta y)\,dx$$

$$\therefore \quad \frac{\partial^2}{\partial\alpha^2} J\,[y+\alpha\delta y] = \frac{\partial}{\partial\alpha}\int_{x_1}^{x_2} \delta y\,dx = \int_{x_1}^{x_2} \frac{\partial}{\partial\alpha}(\delta y)\,dx = \int_{x_1}^{x_2} 0\,dx = 0$$

$$\therefore \quad \left.\frac{\partial^2}{\partial\alpha^2} J[y+\alpha\delta y]\right|_{\alpha=0} = 0$$

$$\therefore \quad \delta^2 J = \mathbf{0}.$$

Remark. For the sake of simplicity, the function $y(x)$ has been written as 'y'.

Example 4. *Find the second variation of the functional* $e^{F(y)}$, *where* $F(y)$ *is a twice differentiable function.*

Sol. Let $\quad J[y] = e^{F(y)}$

Let δy be an increment of y.

$$\therefore \quad J[y+\alpha\delta y] = e^{F(y+\alpha\delta y)}$$

$$\therefore \quad \frac{\partial}{\partial \alpha} J[y + \alpha\delta y] = e^{F(y+\alpha\delta y)}.F'(y + \alpha\delta y).\delta y$$

$$\therefore \quad \frac{\partial^2}{\partial \alpha^2} J[y + \alpha\delta y] = \delta y[e^{F(y+\alpha\delta y)} F'(y + \alpha\delta y)\delta y \,.\, F'(y + \alpha\delta y) + e^{F(y+\alpha\delta y)}.F''(y + \alpha\delta y)\delta y]$$

$$= (\delta y)^2 e^{F(y+\alpha\delta y)}[(F'(y + \alpha\delta y))^2 + F''(y + \alpha\delta y)]$$

$$\therefore \quad \left.\frac{\partial^2}{\partial \alpha^2} J[y + \alpha\delta y]\right|_{\alpha=0} = (\delta y)^2 e^{F(y+0)}[(F'(y + 0))^2 + F''(y + 0)]$$

$$= (\delta y)^2 e^{F(y)}[(F'(y))^2 + F''(y)] \qquad ...(1)$$

Let $\quad I[y] = F(y)$

$$\therefore \quad I[y + \alpha\delta y] = F(y + \alpha\delta y)$$

$$\therefore \quad \frac{\partial}{\partial \alpha} I[y + \alpha\delta y] = F'(y + \alpha\delta y).\delta y$$

and

$$\frac{\partial^2}{\partial \alpha^2} I[y + \alpha\delta y] = \delta y.F''(y + \alpha\delta y)\,.\,\delta y = F''(y + \alpha\delta y)\,.\,(\delta y)^2$$

$$\therefore \quad \left.\frac{\partial}{\partial \alpha} I[y + \alpha\delta y]\right|_{\alpha=0} = F'(y).\delta y \quad \text{and} \quad \left.\frac{\partial^2}{\partial \alpha^2} I[y + \alpha\delta y]\right|_{\alpha=0} = F''(y).(\delta y)^2$$

$$\therefore \quad \delta F = F'(y).\delta y \quad \text{and} \quad \delta^2 F = F''(y).(\delta y)^2$$

$$\therefore \quad (1) \Rightarrow \quad \left.\frac{\partial^2}{\partial \alpha^2} J[y + \alpha\delta y]\right|_{\alpha=0} = e^{F(y)}[F'(y)\delta y)^2 + F''(y).(\delta y)^2] = e^{F(y)}[(\delta F)^2 + \delta^2 F]$$

$$\therefore \quad \boldsymbol{\delta^2 J = e^{F(y)}[(\delta F)^2 + \delta^2 F].}$$

EXERCISE 6.5

1. Show that the functional $J[y(x)] = y(4)$, $y(x) \in C\,[4, 10]$ is differentiable.
2. Find the variation of the functional $\int_{x_1}^{x_2} (6y(x) + 7)\, dx$.
3. Find the variation of the functional $\int_3^5 y^2(x)\, dx$.
4. Find the variation of the functional $\int_{x_1}^{x_2} (x + y)\, dx$.
5. Find the variation of the functional $\int_a^b (y^2 - y'^2)\, dx$.
6. Find the variation of the functional $\int_0^{\pi} y' \sin y\ dx$.
7. Find the second variation of the functional $\int_a^b y^2(x) dx$.
8. Find the second variation of the functional $\int_0^1 (xy^2 + y'^3)\, dx$.

Answers

2. $6\int_{x_1}^{x_2} \delta y(x)\, dx$

3. $2\int_{3}^{5} y(x)\, \delta y(x)\, dx$

4. $\int_{x_1}^{x_2} \delta y(x)\, dx$

5. $2\int_{a}^{b} (y\delta y - y'(\delta y)')\, dx$

6. $\int_{0}^{\pi} (\delta y)' \sin y + y'\delta y \cos y)$

7. $2\int_{a}^{b} (\delta y)^2\, dx$

8. $2\int_{0}^{1} [x(\delta y)^2 + 3y'(\delta y)'^2]\, dx.$